NONNEGATIVE MATRICES IN THE MATHEMATICAL SCIENCES

This is a volume in
COMPUTER SCIENCE AND APPLIED MATHEMATICS
A Series of Monographs and Textbooks

Editor: WERNER RHEINBOLDT

A complete list of titles in this series can be obtained from the publisher.

NONNEGATIVE MATRICES IN THE MATHEMATICAL SCIENCES

ABRAHAM BERMAN

Department of Mathematics
The Technion–
Israel Institute of Technology
Haifa, Israel

ROBERT J. PLEMMONS

Departments of
Computer Science and Mathematics
The University of Tennessee
Knoxville, Tennessee

ACADEMIC PRESS New York San Francisco London 1979
A Subsidiary of Harcourt Brace Jovanovich, Publishers

ACADEMIC PRESS, INC.
111 Fifth Avenue, New York, New York 10003

United Kingdom Edition published by
ACADEMIC PRESS, INC. (LONDON) LTD.
24/28 Oval Road, London NW1 7DX

Library of Congress Cataloging in Publication Data

Berman, Abraham.
Nonnegative matrices in the mathematical sciences.

(Computer science and applied mathematics)
Bibliography: p.
1. Nonnegative matrices. I. Plemmons, Robert J., joint author. II. Title.
QA188.B47 512.9'43 78-67874
ISBN 0-12-092250-9

PRINTED IN THE UNITED STATES OF AMERICA

79 80 81 82 9 8 7 6 5 4 3 2 1

To our wives

CONTENTS

Chapter 4 Symmetric Nonnegative Matrices

Chapter 5 Generalized Inverse-Positivity

Chapter 6 M-Matrices

Chapter 7 Iterative Methods for Linear Systems

Chapter 8 Finite Markov Chains

PREFACE

Aspects of the theory of nonnegative matrices, such as the Perron–Frobenius theory, have been included in various books. However, during our work on the applications of nonnegative matrices, and in particular of inverse-positivity and monotonicity, we have felt the need for a treatise that discusses several of the applications and describes the theory from different points of view. This book was written with that goal in mind. In surveying up-to-date research, we discuss the theory of nonnegative matrices, geometrically and algebraically; we study in detail generalization of inverse-positivity and M-matrices, and describe selected applications to numerical analysis, probability, economics, and operations research. In fact, the book is intended to be used as a reference book or a graduate textbook by mathematical economists, mathematical programmers, statisticians, computer scientists, as well as mathematicians. Having in mind this wide audience, we attempted to make each chapter self-contained insofar as possible and kept the prerequisites at the level of matrix algebra through the Jordan canonical form.

To introduce the key concepts of the book we now describe a typical application to economics. Consider an economic situation involving n interdependent industries, assuming, for the sake of simplicity, that each industry produces one commodity. Let t_{ij} denote the amount of input of the ith commodity needed by the economy to produce a unit output of commodity j. The *Leontief input–output matrix* $T = (t_{ij})$ is a *nonnegative matrix*. The properties of nonnegative matrices are described in detail in Chapters 1–4. If the above model describes an economically feasible situation, then the sum of the elements in each column of T does not exceed one. Let us further assume that the model contains an open sector, where labor, profit, etc. enter in the following way. Let x_i be the total output of industry i required to meet the demand of the open sector and all n industries. Then

$$x_i = \sum_{j=1}^{n} t_{ij}x_j + d_i, \qquad i = 1, 2, \ldots n,$$

where d_i denotes the demand of the open sector from the ith industry. Here $t_{ij}x_j$ represents the input requirement of the jth industry from the ith. The output levels required of the entire set of n industries in order to meet these demands are given as the vector x that solves the *system of linear equations*

$$Ax = d, \qquad A = I - T.$$

Since the column sums of the nonnegative matrix T are at most one, it follows that its spectral radius is at most one. Whenever the spectral radius of T is less than one, T is *convergent* and A is *inverse-positive*, that is, A^{-1} is a nonnegative matrix. Inverse-positive matrices are discussed in Chapter 5. In this case x is the unique nonnegative vector given by

$$x = A^{-1}d = \left(\sum_{k=0}^{\infty} T^k\right) d.$$

Moreover, $A = I - T$ has all of its off-diagonal entries nonpositive. Inverse-positive matrices with this sign pattern are called nonsingular M-matrices. If the spectral radius of T is one, then A is a singular M-matrix. M-matrices are discussed in detail in Chapter 6.

The model discussed above and some of its generalizations are described in Chapter 9. The topics and the interdependence of the ten chapters of the book are described in the accompanying diagram.

The book in its entirety may serve as a textbook for a full year course in applied mathematics. Chapters 2 and 6 may be used, with very little extra effort by the reader or instructor, as a self-contained theoretical unit which, with any of the applications chapters, may serve as a one-quarter course in the appropriate application discipline. It should be mentioned that although the application Chapters 7–9 are almost completely self-contained, a basic knowledge of linear programming is assumed in Chapter 10.

The material in the body of each chapter is supplemented in three ways: exercises in the body of the chapter as well as exercises and notes at the end. The exercises are an integral part of the book, and we recommend that they should at least be read. Hints and references are associated with some of the more difficult problems.

A word on references: "Theorem 2.3" refers to Theorem 2.3 of the same chapter, whereas "Theorem 6.2.3" refers to Theorem 2.3 in Chapter 6. Other items are numbered in the same manner. References are given in the form "Varga [1962]" which refer to a paper (or book) by Varga published in 1962. Such references as [1976a] and [1976b] indicate that the References contains more than one work by the author published in 1976. Neumann and Plemmons [a] refers to a paper by Neumann and Plemmons that had not appeared in print at the time the manuscript of the book was submitted for publication.

Finally, we mention that the following special notation has been adopted in the book in order to describe the nonnegativity of a matrix (or vector). Let $A = (a_{ij})$ be an $m \times n$ real matrix. Then $A \geqslant 0$ means $a_{ij} \geqslant 0$ for each i and j, $A > 0$ means $A \geqslant 0$ and $A \neq 0$, $A >> 0$ means $a_{ij} > 0$ for each i and j.

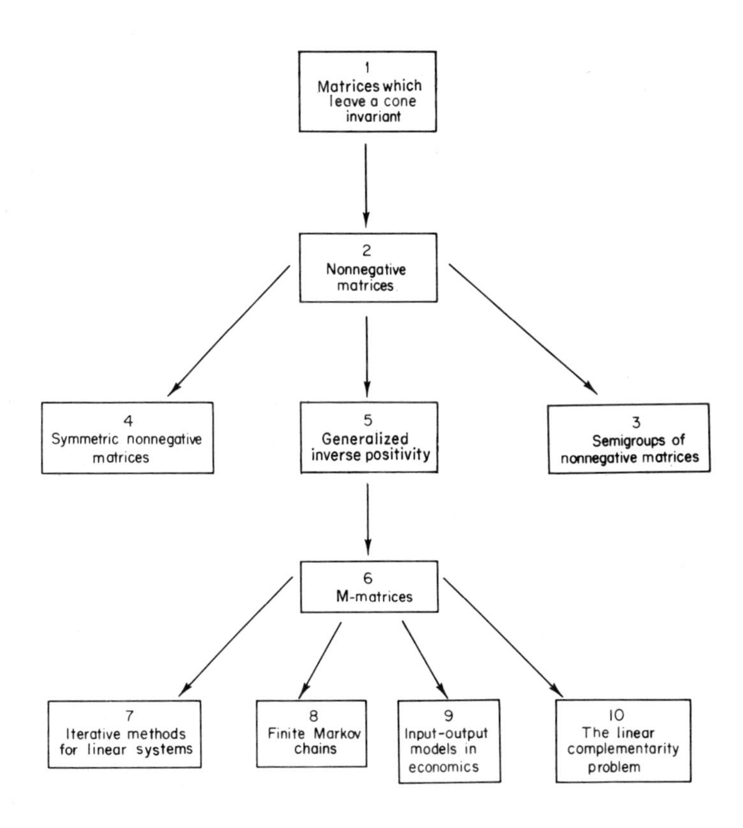

ACKNOWLEDGMENTS

The book was written during our work in the following institutes: Centre de Recherches Mathématiques, Université de Montréal; Mathematics Research Center, University of Wisconsin; Department of Mathematics and Statistics Research, Union Carbide Corporation, Oak Ridge; Department of Mathematical Sciences, Clemson University; Department of Mathematics, The Technion—Israel Institute of Technology; and the Departments of Computer Science and Mathematics, University of Tennessee. At these places we used the material of the book in lectures, seminars, and courses. We wish to thank our colleagues at these and other institutions for their interest, encouragement, and comments.

Our special thanks go to Professors D. J. Hartfiel, M. Neumann, H. Schneider, and R. S. Varga for their detailed comments on portions of the manuscript. We are especially indebted to Professor Schneider, who suggested that we write the book and who gave us moral support during its preparation.

The work of the first author was partially supported by Technion grants 100-314 and 100-372. The second author would also like to thank Dr. Paul Boggs and Dr. Jagdish Chandra of the U.S. Army Research Office (ARO) for partial financial support under contract DAHC-04-74-C-0019 and grant DAAG29-77-G-166. Most of the new material given in Chapters 6–9 was developed under support of the ARO during the period 1974–1979.

SYMBOLS

Below is a list of frequently used symbols and their definitions with the page numbers where they are first used in parentheses.

$A \geqslant 0$	each element of vector or matrix A is nonnegative (26)
$A > 0$	$A \geqslant 0$ and at least one element is positive (26)
$A >> 0$	each element is positive (26)
A^{t}	transpose (4)
A^{*}	conjugate transpose (189)
A^{-1}	inverse (10)
A^{+}	Moore–Penrose inverse (83)
A^{D}	Drazin inverse (118)
$A^{\#}$	group inverse (118)
$a_{ij}{}^{q}$	$(i\text{-}j)$ entry of A^{q} (29)
$\mathrm{tr}(A)$	trace (4)
$\mathrm{adj}(A)$	adjoint (31)
$G(A)$	directed graph of A (29)
$A[\alpha]$	principal submatrix with entries from index sequence α (39)
$A[\alpha/\beta]$	submatrix with row indices from index sequence α and column indices from index sequence β (49)
$\sigma(A)$	spectrum (54)
$\rho(A)$	spectral radius (6)
$\nu(A)$	degree (42)
$\gamma(A)$	index of primitivity (46)
$\delta(A)$	$\max \{\lvert\lambda\rvert : \lambda \in \sigma(A), \lambda \neq 1\}$ (198)
$N(A)$	null space (42)
$R(A)$	range or column space (118)
$M(A)$	comparison matrix for A (142)
$\Omega(A)$	equimodular set for A (142)
I_r	identity matrix of order r (68)
R^{n}	n-dimensional real space (1)
R_{+}^{n}	nonnegative orthant (1)
$R^{n \times n}$	all $n \times n$ real matrices (1)
$C^{n \times n}$	all $n \times n$ complex matrices (142)
$Z^{n \times n}$	all $A \in R^{n \times n}$ with $a_{ij} \leqslant 0$ for all $i \neq j$ (132)
Ω_n	all $n \times n$ doubly stochastic matrices (49)
$\mathscr{D}_n$	semigroup of all $n \times n$ doubly stochastic matrices (63)

$\pi(R^n_+)$	all $A \in R^{n\times n}$ with $A \geqslant 0$ (1)
$\pi(K)$	all $A \in R^{n\times n}$ with $AK \subseteq K$; K is a proper cone (1)
$Int(K)$	interior of cone K (3)
$\mathrm{bd}(K)$	boundary of cone K (3)
$x \overset{K}{\leqslant} y$	means $(y - x) \in K$, K is a proper cone (3)
$S_0(K_1,K_2)$	all $A \in R^{n\times n}$ such that $Ax \in K_2$ for some $0 \neq x \in K_1$ (9)
$\mathrm{Re}\ \lambda$	real part of complex λ (21)
$\| \ \|$	Euclidean norm unless specified otherwise (18)

NONNEGATIVE MATRICES IN THE MATHEMATICAL SCIENCES

CHAPTER 1

MATRICES WHICH LEAVE A CONE INVARIANT

1 INTRODUCTION

Let R^n_+ denote the *nonnegative orthant*, the set of all nonnegative vectors in n-dimensional Euclidean space R^n. Let $R^{n \times n}$ denote the set of $n \times n$ real matrices and let $\pi(R^n_+)$ denote the set of $n \times n$ matrices with nonnegative entries. The set R^n_+ is a proper cone (see Section 2) in R^n. Every matrix in $\pi(R^n_+)$ maps R^n_+ into itself. The set $\pi(R^n_+)$ is a proper cone in $R^{n \times n}$ and is closed under matrix multiplication (see Chapter 3). In general, if K is a proper cone in R^n and $\pi(K)$ denotes the set of $n \times n$ matrices which leave K invariant, then $\pi(K)$ is closed under multiplication and is a proper cone in $R^{n \times n}$.

The Perron-Frobenius theorems on nonnegative matrices (see Chapter 2) have been extended to operators which leave a cone invariant in infinite-dimensional spaces. Our interest in this chapter will focus upon finite-dimensional extensions of this sort. Using matrix theory methods we study the spectral properties of matrices in $\pi(K)$, where K is a proper cone in R^n. We combine the use of the Jordan form of a matrix (Birkhoff [1967b], Vandergraft [1968]) and of matrix norms (Rheinboldt and Vandergraft [1973]). The needed background on cones is described, without proofs, in Section 2. In Section 3 we study matrices in $\pi(K)$, in particular K-irreducible matrices. Cone-primitive matrices are discussed in Section 4.

2 CONES

(2.1) Definitions With $S \subseteq R^n$ we associate two sets: S^G, *the set generated* by S, which consists of all finite nonnegative linear combinations of elements of S, and S^*, *the dual of* S, defined by

$$S^* = \{y \in R^n; x \in S \rightarrow (x, y) \geq 0\},$$

where (,) denotes inner product. A set K is defined to be a *cone* if $K = K^{\mathrm{G}}$. A set is *convex* if it contains, with any two of its points, the line segment between the points. Dual sets and convex cones are examples of convex sets.

(2.2) Examples of Convex Cones (a) R^n, (b) $\{0\}$, (c) $R^n_+ = \{x \in R^n; x_i \geq 0\}$, (d) $\{0\} \cup \{x \in R^n; x_i > 0\}$ (e) $K_n = \{x \in R^n; (x_2^2 + \cdots + x_n^2)^{1/2} \leq x_1\}$, the *ice cream cone.*

All but example (d) are closed. The dual of a subspace L is its orthogonal complement $L^{\perp}$. Thus the dual of R^n is $\{0\}$ and the dual of $\{0\}$ is R^n. Notice that R^n_+ and K_n are *self-dual.* The dual of (d) is R^n_+. For every S, S^* is a closed convex cone. By the definition of the operator G, $S^{\mathrm{GG}} = S^{\mathrm{G}}$.

For $*$ we have the following result of Farkas.

(2.3) Exercise

$$S^{**} = \mathrm{cl}\, S^{\mathrm{G}},$$

where cl denotes closure, or equivalently, K is a closed convex cone if and only if $K = K^{**}$ (e.g., Berman [1973]).

(2.4) Definition The cone S^{G} is called a *polyhedral cone* if S is finite.

Thus K is a polyhedral cone if $K = BR^k_+$ for some natural number k and an $n \times k$ matrix B.

The first three examples in (2.2) are of polyhedral cones. We state, without proof, some of the basic properties of such cones.

(2.5) Theorem (a) A nonempty subset K of R^n is a polyhedral cone if and only if it is the intersection of finitely many closed half spaces, each containing the origin on its boundary.

(b) A polyhedral cone is a closed convex cone.

(c) A nonempty subset K of R^n is a polyhedral cone if and only if K^* is a polyhedral cone.

(2.6) Definitions A convex cone is

(a) *pointed* if $K \cap (-K) = \{0\}$,

(b) *solid* if int K, the interior of K, is not empty, and

(c) *reproducing* if $K - K = R^n$.

The proofs of the following statements are left as exercises.

(2.7) Exercise A closed convex cone in R^n is *solid* if and only if it is reproducing (e.g., Krasnoselskii [1964]).

(2.8) Exercise A closed convex cone K is pointed if and only if K^* is solid, (e.g., Krein and Rutman [1948]).

Let K be a pointed closed convex cone. Then the interior of K^* is given by

$$\text{int}\, K^* = \{y \in K^*;\, 0 \neq x \in K \to (x, y) > 0\}. \tag{2.9}$$

(2.10) Definition A closed, pointed, solid convex cone is called a *proper cone.*

A proper cone induces a partial order in R^n via $y \overset{K}{\leq} x$ if and only if $x - y \in K$. In addition we shall use the notation $y \overset{K}{<} x$ if $x - y \in K$ and $x \neq y$ and $y \overset{K}{\ll} x$ if $x - y \in \text{int}\, K$.

Of the five cones in (2.2) only R^n_+ and K_n are proper. The entire space R^n is not pointed, $\{0\}$ is not solid, and $\{0\} \cup \text{int}\, R^n_+$ is not closed.

(2.11) Definition Let K be a closed convex cone. A vector x is an *extremal* of K if $0 \overset{K}{\leq} y \overset{K}{\leq} x$ implies that y is a nonnegative multiple of x.

If K has an extremal vector x then clearly K is pointed and $x \in \text{bd}\, K$, the boundary of K.

(2.12) Exercise A proper cone is generated by its extremals (e.g., Vandergraft [1968], Loewy and Schneider [1975a]). This is a special case of the Krein–Milman theorem.

(2.13) Definitions If x is an extremal vector of K, then $\{x\}^G$ is called an *extremal ray* of K. A proper cone in R^n which has exactly n extermal rays is called *simplicial.* In other words, $K \subseteq R^n$ is a simplicial cone if $K = BR^n_+$, where B is a nonsingular matrix of order n.

Clearly $(BR^n_+)^* = (B^{-1})^t R^n_+$. In R^2 a polyhedral cone is proper if and only if it is simplicial. In R^3, however, one can construct a polyhedral cone with k extremal rays for every natural number k.

(2.14) Definition Let K and $F \subseteq K$ be pointed closed cones. Then F is called a *face* of K if

$$x \in F. \quad 0 \overset{K}{\leq} y \overset{K}{\leq} x \to y \in F.$$

The face F is *nontrivial* if $F \neq \{0\}$ and $F \neq K$.

The faces of R^n_+ are of the form

$$F_I = \{x \in R^n_+ : x_i = 0 \text{ if } i \notin I\} \qquad \text{where} \quad I \subseteq \{1, \ldots, n\}.$$

The nontrivial faces of the ice cream cone K_n are of the form $\{x\}^G$, where $x \in \text{bd}\, K_n$. The *dimension of a face* F is defined to be the dimension of the subspace $F - F$, the *span* of F. Thus the extremal rays of K are its one-dimensional faces.

(2.15) Exercise If F is a face of K, then $F = K \cap (F - F)$. If F is nontrivial then $F \subseteq \text{bd}\, K$ (e.g., Barker [1973]).

Denote the interior of F, relative to its span $F - F$, by $\text{int}\, F$.

(2.16) Exercise For $x \in K$ let

$$F_x = \{y \in K; \text{there exists a positive } \alpha \text{ such that } \alpha y \overset{K}{\leq} x\}.$$

Then

(a) F_x is the smallest face which contains x and it is nontrivial if and only if $0 \neq x \in \text{bd}\, K$.

(b) F is a face of K and $x \in \text{int}\, F$ if and only if $F = F_x$, (e.g., Vandergraft [1968]).

The set F_x is called *the face generated by x*.

Let K_1 and K_2 be proper cones, in R^n and R^m, respectively. Denote by $\pi(K_1,K_2)$ the set of matrices $A \in R^{m \times n}$ for which $AK_1 \subseteq K_2$.

(2.17) Exercise The set $\pi(K_1,K_2)$ is a proper cone in $R^{m \times n}$. If K_1 and K_2 are polyhedral then so is $\pi(K_1,K_2)$, (Schneider and Vidyasagar [1970]).

For $K_1 = R^n_+$ and $K_2 = R^m_+$, $\pi(K_1,K_2)$ is the class of $m \times n$ nonnegative matrices. If $A \in \pi(K_1,K_2)$ then $A^t \in \pi(K_2^*,K_1^*)$ by the definitions of the transpose operator t and the dual cone. The interior of $\pi(K_1,K_2)$ can be shown to be the following.

(2.18) Exercise $\text{int}\, \pi(K_1,K_2) = \{A \in R^{m \times n}; A(K_1 - \{0\}) \subseteq \text{int}\, K_2\}$ (e.g., Barker [1972]).

Let

$$(A,B) = \text{tr}\, AB^t. \tag{2.19}$$

Then (2.19) is an inner product in $R^{m \times n}$.

Let

$$Q = \{uv^t; u \in K_2^*, v \in K_1\}$$

Then we have the following.

(2.20) Exercise (a) $\pi(K_1,K_2) = Q^*$.

(b) $(\pi(K_1,K_2))^* = Q^G$.

(c) $(\pi(K_1,K_2))^* \subseteq \pi(K_1^*,K_2^*)$ (e.g., Berman and Gaiha [1972], Tam [1977]).

For $m = n$ and $K_1 = K_2 = K$ we use $\pi(K)$ as a short notation for $\pi(K,K)$. Thus $\pi(R^n_+)$ is the class of nonnegative matrices of order n. This set is denoted by $\mathcal{N}_n$ in the literature on semigroups and this notation is used in Chapter 3. The cone $\pi(K_n)$ *contains* the diagonal matrices

$$D = \operatorname{diag}\{d_1, \ldots, d_n\},$$

where $d_1 \geq |d_i|$, $i = 2, \ldots, n$.

(2.21) Definitions The matrices in $\pi(K)$ are called K-*nonnegative* and are said to *leave K invariant*. A matrix A is K-*positive* if

$$A(K - \{0\}) \subseteq \operatorname{int} K.$$

It is easy to check that the following is true.

(2.22) A is K-nonnegative if and only if A^t is K^*-nonnegative

and

(2.23) A is K-positive if and only if A^t is K^* positive.

In the next section we shall use norms induced by partial orders.

(2.24) Exercise Let K be a proper cone and let $u \in \operatorname{int} K$.

(a) The order interval

$$B_u = \{x \in R^n;\ -u \overset{K}{\leq} x \overset{K}{\leq} u\}$$

is a *convex body* in R^n, i.e., B_u is closed and convex and for any $x \in R^n$, there exists a positive t such that $x \in tB_u$, and

$$x \in B_u,\quad |\alpha| \leq 1 \rightarrow \alpha x \in B_u.$$

(b) B_u being a convex body defines a norm on R^n,

$$\|x\|_u = \inf\{t \geq 0;\ x \in tB_u\}$$

and thus

$$\|u\|_u = 1.$$

(For $u^t = (1, \ldots, 1)$ and $K = R^n_+$ one gets the l_∞ norm.) (See Householder [1964], Rheinboldt and Vandergraft [1973].)

In the next section we shall need the following observations:

$$tu \overset{K}{\geq} y \rightarrow tu - x = tu - y + y - x \in K \qquad \text{and} \qquad tu + x = tu - x + x + x \in K.$$

Thus the norm is monotonic with respect to K; namely,

$$0 \overset{K}{\leq} x \overset{K}{\leq} y \rightarrow \|x\|_u \leq \|y\|_u. \tag{2.25}$$

Similarly, the induced operator norm

$$\|A\|_u = \sup_{\|x\|_u = 1} \|Ax\|_u$$

satisfies

$$\|A\|_u = \|Au\|_u \qquad \text{if} \quad A \in \pi(K), \tag{2.26}$$

since $tu - Ax = tu - Au + A(u - x)$ and $tu + Ax = tu - Au + A(u + x)$, if $tu \overset{K}{\geq} Au$, $x \in B_u$ and $A \in \pi(K)$.

3 SPECTRAL PROPERTIES OF MATRICES IN $\pi(K)$

In this section and the next, K denotes a proper cone in R^n, $n > 1$. Let $A \in \pi(K)$. By the (finite-dimensional case of the) Krein–Rutman theorem, A has an eigenvector in K which corresponds to $\rho(A)$, the spectral radius of A. This suggests the following.

(3.1) Question Let A be a matrix such that $\rho(A)$ is an eigenvalue of A. Is there a proper cone which A leaves invariant?

Here we offer a proof, due to Birkhoff, of the finite-dimensional version of the Krein–Rutman theorem. This proof specifies another property of $\rho(A)$, which allows an answer to Question 3.1.

Let λ be an eigenvalue of A. The *degree of* λ, $\deg \lambda$, is the size of the largest diagonal block in the Jordan canonical form of A, which contains λ (the multiplicity of λ in the minimal polynomial of A). With this definition we restate the following theorem.

(3.2) Theorem If $A \in \pi(K)$, then

(a) $\rho(A)$ is an eigenvalue,
(b) if λ is an eigenvalue of A such that $|\lambda| = \rho(A)$, then $\deg \lambda \leq \deg \rho(A)$,
(c) K contains an eigenvector of A corresponding to $\rho(A)$, and
(d) K^* contains an eigenvector of A^t which corresponds to $\rho(A)$.

Proof If $\rho(A) = 0$, A is nilpotent so $A^r = 0$ for some minimal r, and there is $0 \neq x \in K$ such that $w = A^{r-1}x \neq 0$. Clearly $w \in K$ and $Aw = 0$, so that w is the eigenvector in (c).

If $\rho(A) > 0$, let $\{x_{ij}\}$, $i = 1, \ldots, k$, $j = 1, \ldots, m_i$; $\sum_{i=1}^{k} m_i = n$, be a Jordan canonical basis (of C^n); i.e.,

$$Ax_{ij} = \lambda_i x_{ij} + x_{ij-1}, \qquad x_{i0} = 0,$$

where the eigenvalues λ_i are ordered by the following rules:

$$\rho = \rho(A) = |\lambda_1| = \cdots = |\lambda_\nu| > |\lambda_{\nu+1}| \geq \cdots \geq |\lambda_k|,$$

$$m = m_1 = m_2 = \cdots = m_h > m_{h+1} \geq \cdots \geq m_\nu,$$

$$\lambda_l = \rho e^{i\theta_l}, \qquad 0 \leq \theta_l < 2\pi, \qquad l = 1, \ldots, h,$$

$$0 \leq \theta_1 \leq \cdots \leq \theta_h.$$

The principal eigenvectors $\{x_{ij}\}$ are either real or occur in conjugate pairs since A is real and every vector $y \in R^n$ can be written as

$$y = \sum_{i=1}^{k} \sum_{j=1}^{m_i} \alpha_{ij} x_{ij}, \qquad \alpha_{ij} = \bar{\alpha}_{pq} \qquad \text{if} \quad x_{ij} = \bar{x}_{pq}.$$

Since K is solid we can choose $y \in \operatorname{int} K$ and a small enough δ such that for all i and j, $c_{ij} = \alpha_{ij} + \delta \neq 0$, and

$$z = \sum_{i=1}^{k} \sum_{j=1}^{m_i} c_{ij} x_{ij} = y + \sum_{i=1}^{k} \sum_{j=1}^{m_i} x_{ij} \in \operatorname{int} K.$$

Our aim now is to show that K contains a nonzero vector which is a linear combination of the eigenvectors $x_{11}, \ldots, x_{h1}$. To do this we observe that

$$A^r x_{ij} = \sum_{k=0}^{j-1} \binom{r}{k} \lambda_i^{r-k} x_{i,\, j-k} \qquad \text{(induction on } r\text{)}$$

and thus

$$A^r z = \sum_{i=1}^{k} \sum_{j=1}^{m_i} c_{ij} \sum_{s=0}^{j-1} \binom{r}{s} \lambda_i^{r-s} x_{i,\, j-s}.$$

For large values of r the dominant summands will be $c_{im}\binom{r}{m-1}\lambda_i^{r-m+1} x_{i1}$, $i = 1, \ldots, h$, and thus a good approximation to $A^r z$ is

$$A^r z \sim \binom{r}{m-1} \rho^{r-m-1} \sum_{l=1}^{h} c_{lm} e^{i\theta_l} x_{l1} \tag{3.3}$$

The right-hand side of (3.3) is clearly different from zero since the eigenvectors are linearly independent and all the coefficients are nonzero. Thus for every r, $A^r z \neq 0$ and $A^r z \in K$ since $A \in \pi(K)$. The set of rays in K is compact since K is closed, thus the sequence of rays $\{(A^r z)^{\mathrm{G}}\}$ has a convergent subsequence, converging, say, to $\{x_h\}^{\mathrm{G}}$. By (3.3),

$$x_h = \sum_{i=1}^{h} \beta_{ih} x_{i1},$$

and this is a nonzero vector in K.

We now make use of the following lemma whose proof is left as an exercise.

(3.4) Lemma For every complex number α off the nonnegative real axis there exist positive numbers $w_0, \ldots, w_q$ such that $\sum_{p=0}^{q} w_p \alpha^p = 0$.

If $\lambda_h \neq \rho$ then by the lemma there exist positive numbers $w_0, \ldots, w_q$ such that $\sum_{p=0}^{q} w_p \lambda_h^p = 0$. In this case we let

$$x_{h-1} = \sum_{p=0}^{q} w_p A^p x_h = \sum_{p=0}^{q} w_p \sum_{i=1}^{h} \beta_{ih} \lambda_i^p x_{i1}$$

$$= \sum_{i=1}^{h} \beta_{ih} \sum_{p=0}^{q} w_p \lambda_i^p x_{i1} = \sum_{i=1}^{h-1} \beta_{ih-1} x_{i1},$$

where $\beta_{ih-1} = \beta_{ih} \sum_{p=0}^{q} w_p \lambda_i^p$.

The vector x_{h-1} is a nonzero vector in K. This follows from $w_p A^p x_h \in K$ and $w_0 x_h \neq 0$. This proves that $\lambda_1 = \rho$, since otherwise we could use the same process to generate a sequence of nonzero vectors $x_{h-2}, \ldots, x_1, x_0$ but $x_0 = 0$ by Lemma 3.4. If $\lambda_f = \rho$ but $\lambda_{f+1} \neq \rho$, then $x_f = \sum_{i=1}^{f} \beta_{if} x_{i1}$ is a nonzero vector in K and clearly

$$Ax_f = \rho(A) x_f$$

which proves (a), (b), and (c).

Statement (d) follows from (2.29). ∎

We now answer Question 3.1.

(3.5) Theorem If $\rho(A)$ is an eigenvalue of A, and if $\deg \rho(A) \geq \deg \lambda$ for every eigenvalue λ such that $|\lambda| = \rho(A)$, then A leaves a proper cone invariant.

Proof In the notation of the proof of Theorem 3.2, $\lambda_i = \rho(A)$ and $m_1 \geq m_i$, $i = 1, \ldots, v$. Let the vectors $\{x_{ij}\}$, $j \geq 1$, be normalized so that

$$Ax_{ij} = \lambda_i x_{ij} + \delta x_{ij-1}, \qquad i = 1, \ldots, k, \quad j = 1, \ldots, m_i,$$

where $x_{i0} = 0$ and

$$\delta = \begin{cases} 1 & \text{if } v = k, \\ \lambda_1 - |\lambda_{v+1}| & \text{if } v < k. \end{cases}$$

Let

$$K = \Big\{ x \in R^n;\ x = \sum_{i=1}^{k} \sum_{j=1}^{m_i} \alpha_{ij} x_{ij},\ |\alpha_{ij}| \leq \alpha_{1j} \text{ if } j \leq m_1,$$

$$|\alpha_{ij}| \leq \alpha_{1m_1} \text{ if } j \geq m_1,$$

$$\alpha_{ij} = \bar{\alpha}_{pq} \text{ if } x_{ij} = \bar{x}_{pq} \Big\}.$$

We leave it to the reader to complete the proof by checking the following.

(3.6) Exercise K is a proper cone in R^n and $A \in \pi(K)$ (Vandergraft [1968]). ∎

Two simple corollaries of Theorem 3.5 are that every strictly triangular matrix has an invariant proper cone and that if all the eigenvalues of the matrix A are real, as in the case when A is real and symmetric, then A or $-A$ is K-nonnegative for some proper cone K.

We now collect some results on order inequalities of the form

$$Bx \overset{K}{\leq} \alpha x, \qquad 0 \neq x \in K,$$

where B is K-nonnegative. As a preparation for these results we start with the following.

(3.7) Exercise Let K_1 and K_2 be proper cones in R^n and R^m, respectively, and let $A \in R^{m \times n}$. Consider the following systems:

(i) $Ax \in \operatorname{int} K_2,\ x \in \operatorname{int} K_1$,
(ii) $A^t y \in K_1^*,\ 0 \neq y \in -K_2^*$.
(i_0) $Ax \in K_2,\ 0 \neq x \in K_1$,
(ii_0) $A^t y \in \operatorname{int} K_1^*,\ y \in -\operatorname{int} K_2^*$.

Then, exactly one of the systems (i) and (ii) is consistent and exactly one of the systems (i_0) and (ii_0) is consistent (Berman and Ben-Israel [1971]).

(3.8) Notation and Definition The set of matrices for which (i) is consistent is denoted by $S(K_1,K_2)$. The set of matrices for which (i_0) is consistent is denoted by $S_0(K_1,K_2)$. A square matrix A is said to be K-*semipositive* if $A \in S(K,K)$.

A relation between these definitions and positive definiteness is given by the following.

(3.9) Theorem Let A be a square matrix of order n. Then for every proper cone K in R^n:

(a) If $A + A^t$ is positive definite, then $A \in S(K,K^*)$.
(b) If $A + A^t$ is positive semidefinite, then $A \in S_0(K,K^*)$.

Proof (a) Suppose $A \notin S(K,K^*)$. By Exercise 3.7, there exists $0 \neq y \in K$ such that $-A^t y \in K^*$. But then $((A + A^t)y, y) = (A^t y, y) + (y, A^t y) \leq 0$. The proof of (b) is similar. ∎

(3.10) Definition A matrix B is said to be *convergent* if $\lim_{k\to\infty} B^k$ exists and is the zero matrix.

(3.11) Exercise (a) Show that B is convergent if and only if $\rho(B) < 1$.

(b) Show that B is convergent if and only if $I - B$ is nonsingular and $(I - B)^{-1} = \sum_{k=0}^{\infty} B^k$ (e.g., Varga [1962], Oldenburger [1940]).

The relation between convergence, semipositivity, and similar properties is now described.

(3.12) Theorem Let $A = \alpha I - B$, where $B \in \pi(K)$. Then

(a) If $Ax \in K$ for some $x \in \operatorname{int} K$, then $\rho(B) \leq \alpha$. If, in addition $\alpha > 0$, then $\lim_{k\to\infty}(\alpha^{-1}B)^k x = x^*$ exists and $x^* \overset{K}{\leq} x$. Moreover $x^* = 0$ if and only if $\alpha^{-1}B$ is convergent, i.e., $\rho(B) < \alpha$.

(b) The matrix A is K-semipositive if and only if $\alpha^{-1}B$ is convergent.

(c) If $\rho(B) \leq \alpha$, then $A \in S_0(K,K)$.

Proof (a) The spectral radius is bounded by all norms. In particular,

$$\rho(B) \leq \|B\|_x = \|Bx\|_x = \|\alpha x\|_x = \alpha$$

by (2.26). If α is positive, then the sequence $\{(\alpha^{-1}B)^k x\}$ decreases in the partial order induced by K, and is bounded by $\{0\}$, which assures the existence of x^*, and by x which implies $x^* \overset{K}{\leq} x$. If $\rho(\alpha^{-1}B) < 1$, then $\lim_{k\to\infty}(\alpha^{-1}B)^k = 0$. Conversely, by (2.26), $\|(\alpha^{-1}B)^k\|_x = \|(\alpha^{-1}B)^k x\|_x$ and thus if $x^* = 0$, $\|(\alpha^{-1}B)^k\| \to 0$ so $\rho(B) < 1$.

(b) If: Let $y \in \operatorname{int} K$. Then

$$x = \alpha A^{-1} y = (I - \alpha^{-1}B)^{-1} y = y + \textstyle\sum_{k=1}^{\infty} (\alpha^{-1}B) y \in \operatorname{int} K \quad \text{and} \quad Ax \in \operatorname{int} K.$$

Only if: If A is K-semipositive, then the proof follows by Exercise 3.7.

$$(B^{\mathrm{t}} - \alpha I)y \in K^*, \qquad y \in K^* \to y = 0. \tag{3.13}$$

Let z be an eigenvector of B^{t} which lies in K^* and corresponds to $\rho(B)$ (Theorem 3.2(d)). Then if $\rho(B) \geq \alpha$, z is a counterexample to (3.13).

(c) For every natural number k, $\rho(B) < \alpha + (1/k)$. By (b) there exists $x^{(k)} \in K$ such that $Bx^{(k)} \overset{K}{<} (\alpha + (1/k))x^{(k)}$. Since we can normalize the vectors $x^{(k)}$ so that $\|x^{(k)}\| = 1$ for all k, there exists a limit x^* of a converging subsequence, satisfying

$$x^* \in K, \quad \|x^*\| = 1 \qquad \text{and thus} \qquad x^* \neq 0 \quad \text{and} \quad Ax^* \overset{K}{\leq} \alpha x^*. \quad \blacksquare$$

The assumption $x \in \operatorname{int} K$ cannot be replaced by $x \overset{K}{>} 0$. This can be demonstrated by taking

$$K = R^n_+, \qquad B = \begin{bmatrix} 1 & 1 \\ 0 & 2 \end{bmatrix}, \qquad \alpha = 1, \qquad \text{and} \qquad x = \begin{bmatrix} 1 \\ 0 \end{bmatrix}.$$

A strengthening of the previous results is possible for a subclass of $\pi(K)$ which we now study.

(3.14) Definitions A matrix in $\pi(K)$ is K-irreducible if the only faces of K that it leaves invariant are $\{0\}$ or K itself. A matrix in $\pi(K)$ is K-reducible if it leaves invariant a nontrivial face of K.

Before we state and prove analogues of the previous theorems, we give some characterizations of K-irreduciblity.

(3.15) Theorem A matrix $A \in \pi(K)$ is K-irreducible if and only if no eigenvector of A lies on the boundary of K.

Proof If: Suppose F is a nontrivial face of K. F is a proper cone in $F - F$. Applying Theorem 3.2, part (c), to A_F, the restriction of A to $F - F$, we see that it has an eigenvector $x \in F$, but x is also an eigenvector of A and $x \in \operatorname{bd} K$.

Only if: Let F_x be the face of K generated by x, defined in Lemma 2.16. If x is an eigenvector of A, then $AF_x \subseteq F_x$. If $0 \neq x \in \operatorname{bd} K$, then F_x is nontrivial. ■

(3.16) Theorem A matrix $A \in \pi(K)$ is K-irreducible if and only if A has exactly one (up to scalar multiples) eigenvector in K, and this vector is in $\operatorname{int} K$.

Proof If: The proof follows from Theorem 3.15.

Only if: By Theorem 3.15, A has no eigenvector on $\operatorname{bd} K$. Being K-nonnegative it has an eigenvector in K which has to be in $\operatorname{int} K$. The uniqueness of this eigenvector follows from the first part of the following lemma.

(3.17) Lemma If $A \in \pi(K)$ has two eigenvectors in $\operatorname{int} K$, then A has an eigenvector on the boundary. Furthermore, the corresponding eigenvalues are all equal.

Proof Let

$$Ax_1 = \lambda_1 x_1, \qquad x_1 \in \operatorname{int} K,$$

$$Ax_2 = \lambda_2 x_2, \qquad x_2 \in \operatorname{int} K.$$

The eigenvalues λ_1 and λ_2 are nonnegative since A is K-nonnegative. Assume $\lambda_1 \geq \lambda_2$ and let

$$t_0 = \min\{t > 0 : tx_2 - x_1 \in K\}.$$

This minimum exists since $x_2 \in \operatorname{int} K$. Let $x_3 = t_0 x_2 - x_1$. Clearly $x_3 \in \operatorname{bd} K$. If $\lambda_1 = 0$ then so is λ_2 and $Ax_3 = 0$. If $\lambda_1 > 0$ then

$$Ax_3 = t_0 \lambda_2 - \lambda_1 x_1 = \lambda_1 \{t_0(\lambda_2/\lambda_1)x_2 - x_1\}.$$

The vector $Ax_3 \in K$. Thus by the definition of t_0, $\lambda_2 \geq \lambda_1$. Thus $\lambda_1 = \lambda_2$ and $Ax_3 = \lambda_1 x_3$, which completes the proof of the lemma and the theorem. ■

The following characterization is given in terms of order inequalities.

(3.18) Theorem A matrix $A \in \pi(K)$ is K-irreducible if and only if

$$Ax \overset{K}{\leq} \alpha x \qquad \text{for some} \quad 0 \neq x \in K \tag{3.19}$$

implies that $x \in \operatorname{int} K$.

Proof If: Suppose A is K-reducible. Then

$$Ax = \lambda x \qquad \text{for some} \quad 0 \neq x \in \operatorname{bd} K.$$

Thus x satisfies (3.19) but is not in $\operatorname{int} K$.

Only if: $AF_x \subseteq F_x$ for every x which satisfies (3.19). ■

Every K-positive matrix is K-irreducible. Conversely we have the following.

(3.20) Theorem An $n \times n$ matrix $A \in \pi(K)$ is K-irreducible if and only if $(I + A)^{n-1}$ is K-positive.

Proof If: Suppose A is K-irreducible, so $x \in \operatorname{bd} K$ is an eigenvector of A. Then $(I + A)^{n-1}x \in \operatorname{bd} K$.

Only if: Let y be an arbitrary nonzero element on $\operatorname{bd} K$. By K-irreducibility $y_1 = (I + A)y$ does not lie in $Fy - Fy$ and the dimension of Fy_1 is greater than the dimension of Fy. Repeating this argument shows that $(I + A)^k y \in \operatorname{int} K$ for some $k \leq n - 1$, hence $(I + A)^{n-1}$ is K-positive. ■

As corollaries of Theorem 3.20 and of statements (2.22) and (2.23) we have the following.

(3.21) Corollary If A and B are in $\pi(K)$ and A is K-irreducible, then so is $A + B$.

(3.22) Corollary A K-nonnegative matrix A is K-irreducible if and only if A^t is K^*-irreducible.

We now state the analogs of Theorems 3.2 and 3.5.

(3.23) Theorem If $A \in \pi(K)$ is K-irreducible, then

(a) $\rho(A)$ is a simple eigenvalue and any other eigenvalue with the same modulus is also simple, and

(b) there is an eigenvector corresponding to $\rho(A)$ in int K, and no other eigenvector (up to scalar multiples) lies in K.

Furthermore, (a) is sufficient for the existence of a proper cone K, for which A is K-nonnegative and K-irreducible.

Proof Part (b) is a restatement of Theorem 3.12. Part (a) follows from Theorem 3.2, provided $\rho(A)$ is simple. Suppose $\rho(A)$ is not simple. Then there exist linearly independent vectors x_1 and x_2, with $x_1 \in \operatorname{int} K$

$$Ax_1 = \rho(A)x_1$$

and either,

$$Ax_2 = \rho(A)x_2 \tag{3.24}$$

or

$$Ax_2 = \rho(A)x_2 + x_1. \tag{3.25}$$

If (3.24) were true, then, for large enough $t > 0$, $x_3 = tx_1 + x_2 \in K$, and x_3 is another eigenvector; this contradicts the uniqueness of the eigenvector. If (3.25) holds, then $-x_2 \notin K$, and we can define

$$t_0 = \min\{t > 0;\ tx_1 - x_2 \in K\}.$$

Then, $\rho(A) \neq 0$ because A is K-irreducible; but $\rho(A) \neq 0$ implies

$$\begin{aligned} A(t_0x_1 - x_2) &= t_0\rho(A)x_1 - \rho(A)x_2 - x_1 \\ &= \rho(A)\left\{\left(t_0 - \frac{1}{\rho(A)}\right)x_1 - x_2\right\} \in K, \end{aligned}$$

which contradicts the definition of t_0. Hence $\rho(A)$ must be simple. To prove the "furthermore" part, we use the proof of Theorem 3.5. The cone K defined there contains only elements of the form $\alpha x_1 + y$, where x_1 is the eigenvector corresponding to $\rho(A)$ and $\alpha = 0$ only if $y = 0$. Hence no other eigenvector can lie in K, so by Theorem 3.16, A is K-irreducible. ■

Part (a) of Theorem 3.23 can be strengthened if A is K-positive.

(3.26) Theorem If A is K-positive, then

(a) $\rho(A)$ is a simple eigenvalue, greater than the magnitude of any other eigenvalue and
(b) an eigenvector corresponding to $\rho(A)$ lies in int K.

Furthermore, condition (a) assures that A is K-positive for some proper cone K.

Proof Part (b) and the simplicity of $\rho(A)$ follow from the previous theorem. Let λ_2 be an eigenvalue of A with eigenvector x_2, and assume $\rho(A) = \rho$ and $\lambda_2 = \rho e^{i\theta}$, $2\pi > \theta > 0$. For any ϕ, either $\operatorname{Re} e^{i\phi}x_2 \in K$ or else one can define a positive number t_ϕ by

$$t_\phi = \min\{t > 0; tx_1 + \operatorname{Re} e^{i\phi}x_2 \in K\},$$

where x_1 is the eigenvector in $\operatorname{int} K$ corresponding to $\rho(A)$. The nonzero vector $y = t_\phi x_1 + \operatorname{Re} e^{i\phi}x_2$ lies on $\operatorname{bd} K$, and

$$Ay = \rho(A)(t_\phi x_1 + \operatorname{Re} e^{i(\phi+\theta)}x_2) \in \operatorname{int} K.$$

Hence $\operatorname{Re} e^{i(\phi+\theta)}x_2 \in K$ or $t_\phi > t_{\phi+\theta}$. By repeating this argument it follows that for some ϕ_0,

$$y_0 = \operatorname{Re} e^{i\phi_0}x_2 \in K.$$

By Exercise 3.4, $\theta \neq 0$ implies the existence of a finite set of positive numbers $\{\xi_k\}$ such that

$$\sum_{k=0} \xi_k \rho^k e^{ik\theta} = 0.$$

Hence,

$$\sum_{k=0} \xi_k A^k y_0 = \sum_k \xi_k \rho^k \operatorname{Re}(e^{ik\phi}e^{i\phi_0}x_2) = \operatorname{Re}\left(\sum_k \xi_k \rho^k e^{ik\theta}\right) e^{i\phi_0}x_2 = 0.$$

Thus $y_0 = 0$; i.e., $e^{i(\phi_0+\pi/2)}x_2 = y_2$ is real. Since $Ay_2 = \lambda_2 y_2$, λ_2 is real. Since $\lambda_2 \neq \rho$, $\lambda_2 = -\rho$. Thus $y_2 \notin K \cup (-K)$.

Let

$$t_0 = \min\{t > 0; tx_1 + y_2 \in K\}.$$

Then $0 \neq t_0 x_1 + y_2 \in \operatorname{bd} K$, but

$$A^2(t_0 x_1 + y_2) = \rho^2(t_0 x_1 + y_2) \in \operatorname{int} K,$$

which contradicts the definition of t_0. Hence $|\lambda_2| < \rho(A)$.

To prove the last statement of the theorem, we again use the notation of Theorem 3.5. The cone K becomes

$$K = \left\{x; x = \alpha_1 x_1 + \sum_{i=2}^{k} \sum_{j=1}^{m_i} \alpha_{ij}x_{ij}, |\alpha_{ij}| \leq \alpha_1, \alpha_{ij} - \bar{\alpha}_{pq} \text{ if } x_{ij} = \bar{x}_{pq}\right\}.$$

It is easy to check that

(3.27) Exercise $A(K - \{0\}) \subseteq \operatorname{int} K$ (Vandergraft [1968]). ■

K-irreducibility allows a strengthening of the first part of Theorem 3.12.

(3.28) Theorem Let $A \in \pi(K)$ be K-irreducible. Then the existence of a real number α and of $x \in \operatorname{int} K$ such that $0 \neq \alpha x - Ax \in K$ implies that $\rho(A) < \alpha$.

Proof By Theorem 3.12, $\rho(A) \leq \alpha$. If $\rho(A) = 0$ then $\rho(A) < \alpha$. Suppose that $\rho(A) = \alpha \neq 0$. Let x_1 be the eigenvector of A in $\operatorname{int} K$ and

$$z = \|x_1\|_x x - x_1.$$

Then $z \in \operatorname{bd} K$ and

$$0 \overset{K}{\leq} Az = \|x_1\|_x Ax - Ax_1 \overset{K}{\leq} \|x_1\|_x \alpha x - \alpha x_1 = \alpha z,$$

which contradicts the K-irreducibility of A. ■

(3.29) Corollary Let $0 \overset{\pi(K)}{\leq} A \overset{\pi(K)}{\leq} B$, where A is K-irreducible and $A \neq B$. Then $\rho(A) < \rho(B)$.

Proof By (3.21) B too is K-irreducible. Let x be the eigenvector of B in $\operatorname{int} K$. Then

$$Ax \overset{K}{\leq} Bx = \rho(B)x.$$

Since $B \neq A$, $\|B - A\|$ is positive for any norm. In particular $\|B - A\|_x > 0$. Thus, using (2.26)

$$0 < \|B - A\|_x = \|(B - A)x\|_x = \|\rho(B)x - Ax\|_x,$$

so that $\rho(B)x \neq Ax$. Applying (3.28) proves that $\rho(B) > \rho(A)$. ■

(3.30) Corollary If $0 \overset{K}{\leq} A \overset{K}{\leq} B$, then $\rho(A) \leq \rho(B)$.

Proof Let C be a K-positive matrix and define $A_t = A + tC$, $B_t = B + tC$, $t > 0$. Being K-positive, A_t is K-irreducible and by the previous corollary $\rho(A_t) < \rho(B_t)$. Letting $t \to 0$ yields $\rho(A) \leq \rho(B)$. ■

Theorems 3.12 and 3.28 give upper bounds for the spectral radius of K-nonnegative matrices. We now complement them with lower bounds. Here we start in the K-irreducible case.

(3.31) Theorem Let $A \in \pi(K)$ be K-irreducible. If

(3.32) $$Ax \overset{K}{>} \alpha x, \qquad \text{for some} \quad x \in K, \quad \alpha > 0$$

then $\rho(A) > \alpha$. Conversely, if $\rho(A) > \alpha$, then $Ax \overset{K}{\gg} \alpha x$ for some $x \in \operatorname{int} K$.

Proof Let $\hat{A} = \alpha^{-1}A$. By (3.32) $\hat{A}x \overset{K}{>} x$. Let x_1 be the eigenvector of $\hat{A}$ in $\operatorname{int} K$. Now $\|x\|_{x_1}x_1 - x \in K$, and hence

$$(3.33) \qquad 0 \overset{K}{\leq} \hat{A}(\|x\|_{x_1}x_1 - x) \overset{K}{<} \|x\|_{x_1}\rho(\hat{A})x_1 - x$$

or

$$x \overset{K}{<} \rho(\hat{A})\|x\|_{x_1}x_1,$$

which, by definition of $\|x\|_{x_1}$, implies that $\rho(\hat{A}) \leq 1$. Equality is impossible because of the K-irreducibility of A and (3.33), thus $\rho(\hat{A}) < 1$; that is, $\rho(A) < \alpha$. Conversely, let x_1 be the eigenvector of A in $\operatorname{int} K$. Then $\rho(A) > \alpha$ implies that $Ax_1 = \rho(A)x_1 \gg \alpha x_1$. ∎

Using the continuity argument of (3.30) we can drop the K-irreducibility.

(3.34) Corollary Let $A \in \pi(K)$. Then $Ax \overset{K}{\geq} \alpha x$ for some $0 \neq x \in K$ with $\alpha > 0$, if and only if $\rho(A) \geq \alpha$.

Combining the lower and upper bounds yields the following.

(3.35) Theorem Let $A \in \pi(K)$. Then

$$(3.36) \qquad \alpha x \overset{K}{\leq} Ax \overset{K}{\leq} Bx \qquad \text{for some} \quad x \in \operatorname{int} K$$

implies that $\alpha \leq \rho(A) \leq \beta$.

If in addition A is K-irreducible, $\alpha x \neq Ax$, and $Ax \neq \beta x$, then $\alpha < \rho(A) < \beta$.

Notice that if A is K-irreducible, then by Theorem 3.19, $x \gg 0$ may be replaced by $x > 0$ in (3.36).

4 CONE PRIMITIVITY

In this section we study a subclass of the K-irreducible matrices which contain the K-positive ones.

(4.1) Definition A matrix A in $\pi(K)$ is K-*primitive* if the only nonempty subset of $\operatorname{bd} K$ which is left invariant by A is $\{0\}$.

The spectral structure of K-primitive matrices is due to the following result.

(4.2) Theorem $A \in \pi(K)$ is K-primitive if and only if there exists a natural number m such that A^m is K-positive.

Proof Let

$$A^m(K - \{0\}) \subseteq \operatorname{int} K \tag{4.3}$$

and let S be a nonempty subset of $\operatorname{bd} K$ which is invariant under A. Then

$$A^m S \subseteq S \subseteq \operatorname{bd} K$$

so by (4.3), $S = \{0\}$.

Conversely, if $x \in \operatorname{int} K$ then $A^m x \in \operatorname{int} K$ for all m. For $x \in \operatorname{bd} K - \{0\}$, consider the sequence $S = \{A^i x\}$, $i = 0,1,2, \ldots$. If A is K-primitive, then there is $m(x)$ such that $A^{m(x)}x \in \operatorname{int} K$, otherwise the nonzero set S, which is contained in $\operatorname{bd} K$, is invariant under A.

Let Q be the compact set $\{x \in K, x^t x = 1\}$. For each $x \in Q$, there is an integer $m(x)$ and a set $U(x)$ open in the relative topology of Q such that

$$A^{m(x)}U(x) \subseteq \operatorname{int} K.$$

The collection $\{U(x); x \in Q\}$ is an open cover of Q from which we may extract a finite subcover, say $U(x_1), \ldots, U(x_n)$, with corresponding exponents $m(x_1), \ldots, m(x_n)$. Let $m = \max\{m(x_1), \ldots, m(x_n)\}$. Let $x \in \operatorname{bd} K - \{0\}$. Then $y = (x^t x)^{-1/2} x \in Q$, and there exists x_i such that $y \in U(x_i)$. Thus

$$(x^t x)^{-1/2} A^m x = A^m y = A^{m - m(x_i)}(A^{m(x_i)} y) \in \operatorname{int} K$$

implying that $A^m x \in \operatorname{int} K$ and thus that A^m is K-positive. ■

(4.4) Corollary A is K-primitive if and only if A^t is K^*-primitive.

Proof The proof follows from statement (2.15). ■

(4.5) Corollary The sum of a K-primitive matrix and a K-nonnegative matrix is K-primitive.

(4.6) Corollary If A is K-primitive and l is a natural number then A^l is K-primitive.

(4.7) Corollary If A is K-primitive then A^l is K-irreducible for every natural number l.

(4.8) Remark The converse of Corollary 4.7 is not true, for let A be a rotation of the ice cream cone K_3 through an irrational multiple of 2π, then A^k is K_3-irreducible for all k but $A(\operatorname{bd} K) = \operatorname{bd} K$, so A is not primitive.

A converse of Corollary (4.7) does exist for polyhedral cones.

(4.9) Exercise Let K be a polyhedral cone having p generators. Then A is K-primitive if and only if the matrices $A, A^2, \ldots, A^{2^p-1}$ are K-irreducible (Barker, [1972]).

We now state a spectral characterization of K-primitive matrices.

(4.10) Theorem A K-irreducible matrix in $\pi(K)$ is K-primitive if and only if $\rho = \rho(A)$ is greater in magnitude than any other eigenvalue.

Proof If: Let v, the eigenvector of A in int K, and ψ, the eigenvector of A^t in int K^*, be normalized so that $\psi^t v = 1$. Define A_1 by

$$A_1 x = Ax - \rho\psi^t xv.$$

It can be shown that the following is true.

(4.11) Exercise λ is an eigenvalue of A_1 if and only if $\lambda \neq \rho$ and λ is an eigenvalue of A.

Let ρ_1 be the spectral radius of A_1. Then

$$\lim_{n\to\infty} (\|A_1\|^n)^{1/n} = \rho_1 < \rho. \tag{4.12}$$

Now, $\psi^t A_1 x = \psi^t Ax - \psi^t \rho\psi^t xv = \rho\psi^t x - \rho\psi^t x\psi^t v = 0$, since $\psi^t v = 1$. Thus

$$A^n x = A_1^n x + \rho^n \psi^t xv,$$

so that

$$\|\rho^{-n} A^n x - \psi^t xv\| \leq \rho^{-n} \|A_1^n\| \, \|x\| \to 0.$$

If $v \neq x \in K$, then $\psi^t xv \in \text{int } K$. Thus for every such x there is an n such that $\rho^{-n} A^n x$, and therefore $A^n x$ is in int K. Using the argument in the proof of Theorem 4.2, this implies that A is K-primitive.

The "only if" part follows from Theorems 3.22 and 4.2. ■

By the last part of Theorem 3.22, Theorem 4.10 can be restated as follows.

(4.13) Corollary A is K-primitive if and only if it is $\tilde{K}$-positive for some proper cone $\tilde{K}$.

If K is simplicial and $A \in \pi(K)$ is K-irreducible and has m eigenvalues with modulus $\rho = \rho(A)$, then these eigenvalues are ρ times the unit roots of order m and the spectrum of A is invariant under rotation of $2\pi/m$. Since simplicial cones are essentially nonnegative orthants, we shall defer this elegant study to Chapter 2.

5 EXERCISES

(5.1) Prove or give a counterexample:

(a) The sum of proper cones is a proper cone.
(b) The sum of closed convex cones is a closed convex cone.
(c) The sum of polyhedral cones is a polyhedral cone.
(d) The sum of simplicial cones is a simplicial cone.

(5.2) Let K be a closed convex cone. Show that $K \cap K^* = \{0\}$ if and only if K is a subspace (Gaddum, [1952]).

(5.3) Let K be a closed convex cone in R^n.

(a) Show that every point $x \in R^n$ can be represented uniquely as $x = y + z$, where $y \in K$, $z \in -K^*$, and $(y,z) = 0$ (Moreau [1962]).

(b) Show that K contains its dual, K^*, if and only if, for each vector $x \in R^n$ there exist vectors y and t in K such that $x = y - t$, $(y,t) = 0$ (Haynsworth and Hoffman [1969]).

(5.4) Let K be the cone generated by the vectors (1,1,1), (0,1,1), $(-1,0,1)$, $(0,-1,1)$, and $(1,-1,1)$. Show that K is self-dual.

(5.5) Show that every self-dual polyhedral cone in R^3 has an odd number of extremals (Barker and Foran [1976]).

(5.6) Show that $\pi(K)$ is self-dual if and only if K is the image of the nonnegative orthant under an orthogonal transformation (Barker and Loewy [1975]).

(5.7) Let K be the cone generated by the five vectors $(\pm 1,0,1,0)$, $(0,\pm 1,1,0)$, and (0,0,0,1). Let F be the cone generated by $(\pm 1,0,1,0)$. Show that F *is not* a face of K.

(5.8) Let F and G be faces of a proper cone K. Define $F \wedge G = F \cap G$ and let $F \vee G$ be the smallest face of K which contains $F \cup G$. Show that with these definitions $F(K)$, the set of all faces of K, is a complete lattice and that $F(K)$ is distributive if and only if K is simplical (Barker [1973], Birkhoff [1967a]).

(5.9) A cone K is a *direct sum* of K_1 and K_2, $K = K_1 \oplus K_2$, if span $K_1 \cap$ span $K_2 = \{0\}$ and $K = K_1 + K_2$. Show that in this case, K_1 and K_2 are faces of K (Loewy and Schneider [1975b]).

(5.10) Let ext K denote the set of extreme vectors of K and let ΔK be the closure of the convex hull of $\{xy^t;\ x \in K,\ y \in K^*\}$. A cone K is *indecomposable* if $K = K_1 + K_2 \rightarrow K_1 = 0$ or $K_2 = 0$. Let K be a proper cone in R^n. Show that the following are equivalent:

(i) K is indecomposable,
(ii) K^* is indecomposable,
(iii) $\pi(K)$ is indecomposable,
(iv) $\Delta(K)$ is indecomposable,
(v) $I \in \operatorname{ext} \pi(K)$,
(vi) A nonsingular, $A\{\operatorname{ext} K\} \subseteq \operatorname{ext} K \rightarrow A \in \operatorname{ext} \pi(K)$, and
(vii) A nonsingular, $AK = K \rightarrow A \in \operatorname{ext} \pi(K)$ (Barker and Loewy [1975], Loewy and Schneider [1975a]).

(5.11) Show that a proper cone is simplicial if and only if $I \in \Delta(K)$ (Barker and Loewy [1975]).

(5.12) Let H be the space of $n \times n$ hermitian matrices with the inner product $(A,B) = \operatorname{tr} AB$. Show that PSD, the set of positive semidefinite matrices in H, is a self-dual proper cone and that the interior of PSD consists of the positive definite matrices (e.g., Berman and Ben-Israel [1971], Hall [1967]).

(5.13) An $n \times n$ symmetric matrix A is

(a) *copositive if* $x \geq 0 \rightarrow (Ax,x) \geq 0$,
(b) *completely positive* if there are, say, k nonnegative vectors, a_i $(i = 1, \ldots, k)$, such that

$$(Ax,x) = \sum_{i=1}^{k} (a_i,x)^2 \qquad \text{for all} \quad x \in R^n.$$

Let S, CP, and C denote the sets of symmetric, completely positive, and copositive matrices of order n, respectively. Show that with the inner product in S: $(A,B) = \operatorname{tr} AB$, that C and CP are dual cones (Hall [1967]).

(5.14) Let $C^{n\times n}(R)$ be the set of $n \times n$ complex matrices, considered as a real vector space. Which of the following sets is a proper cone in $C^{n\times n}(R)$?

(a) $CDD = \{A \in C^{n\times n};\ |a_{jj}| \geq \sum_{k\neq j}|a_{jk}|,\ j = 1, \ldots, n\}$,
(b) $D_1 = \{A \in C^{n\times n};\ a_{jj} \geq \sum_{k\neq j}|a_{jk}|,\ j = 1, \ldots, n\}$,
(c) $D_2 = \{A \in C^{n\times n};\ \operatorname{Re} a_{jj} \geq \sum_{k\neq j}|a_{jk}|;\ \operatorname{Im} a_{jj} \geq 0,\ j = 1, \ldots, n\}$,
(d) $D_3 = \{A \in C^{n\times n};\ \operatorname{Re} a_{jj} \geq \sum_{k\neq j}|a_{jk}|, j = 1, \ldots, n\}$,
(e) $D_H = \{A \in C^{n\times n};\ A = A^H,\ a_{jj} \geq \sum_{k\neq j}|a_{jk}|,\ j = 1, \ldots, n\}$ (Barker and Carlson [1975]).

(5.15) Let K be a proper cone, $u \in \operatorname{int} K$, $v \in \operatorname{int} K^*$. Show that uv^t is K-positive.

(5.16) Let $A \in \pi(K)$ where K is a proper cone. Let $\operatorname{core} A = \bigcap_{m \geq 0} A^m K$. Show that $\operatorname{core} A$ is a pointed closed convex cone. When is $\operatorname{core} A$ a proper cone? (See Pullman [1971].)

(5.17) Let K be a proper polyhedral cone in R^n, and let $A \in \pi(K)$. Show that there exist cones $K_1, \ldots, K_r$ such that $\dim(K_i - K_i) = n_i$, $\sum_{i=1}^r n_i = n$, and A is similar to a block triangular matrix

$$\begin{bmatrix} A_r & & & 0 \\ & A_{r-1} & & \\ & & \ddots & \\ * & & & A_1 \end{bmatrix},$$

where A_j is $n_j \times n_j$ and $A_j = 0$ or A_j is A restricted to $\operatorname{span} K_j$ and is K_j-irreducible (Barker [1974]).

(5.18) Let K be a proper cone in R^n. An $n \times n$ matrix A is called *cross-positive on* K if

$$y \in K, \qquad z \in K^*, \qquad (y,z) = 0 \rightarrow (z,Ay) \geq 0.$$

A is *strongly cross-positive* on K if it is cross-positive on K and for each $0 \neq y \in \operatorname{bd} K$, there exists $z \in K^*$ such that $(y,z) = 0$ and $(z,Ay) > 0$. A is strictly cross-positive on K if

$$0 \neq y \in K, \qquad 0 \neq z \in K^*, \qquad (y,z) = 0 \rightarrow (z,Ay) > 0.$$

Let $\lambda = \max\{\operatorname{Re} \mu;\ \mu \in \operatorname{spectrum}(A)\}$. Prove the following.

(a) A is cross-positive on K if and only if $\alpha x \overset{K}{\gg} Ax$, for some $x \in K$ some α, implies $x \in \operatorname{int} K$.

(b) If A is cross-positive on K, then λ is an eigenvalue of A and a corresponding eigenvector lies in K.

(c) If A is strongly cross-positive on K, then λ is a simple eigenvalue of A, the unique eigenvector of A corresponding to λ lies in $\operatorname{int} K$, and A has no other eigenvector in K.

(d) If A is strictly cross-positive on K, then λ is a simple eigenvalue of A, the unique eigenvector of A corresponding to λ lies in $\operatorname{int} K$, A has no other eigenvector in K and $\lambda > \operatorname{Re} \mu$ for any other eigenvector μ of A (Schneider and Vidyasagar [1970], Tam[1977]).

(5.19) Show that

$$A = \begin{bmatrix} 2 & 0 \\ 0 & 1 \end{bmatrix}$$

is (R_+^2)-nonnegative and (R_+^2)-reducible. Show that A is K_2-positive (and thus K_2-irreducible), where K_2 is the ice cream cone in R_2.

(5.20) Prove or give a counterexample of the following.

(a) The product of two K-irreducible matrices is K-irreducible.
(b) The product of two K-primitive matrices is K-primitive.

(5.21) Show that

$$A = \begin{bmatrix} 1 & 0 & 0 \\ 0 & 0 & 0 \\ 0 & 0 & -1 \end{bmatrix}$$

is K_3-irreducible but not K_3-primitive where K_3 is the ice cream cone in R^3.

(5.22) Let $A \in \pi(K)$. Show that the following are equivalent.

(i) A is K-irreducible;
(ii) for some $\lambda > \rho(A)$, $A(\lambda I - A)^{-1}$ is K-positive;
(iii) for every $0 \neq x \in K$ and $0 \neq y \in K^*$ there exists a natural number p such that $y^t A^p x > 0$;
(iv) $\rho(A)$ is simple and A and A^t have corresponding eigenvectors in $\operatorname{int} K$ and $\operatorname{int} K^*$, respectively (Vandergraft [1968], Barker [1972]).

(5.23) $A \in \pi(K)$ is *u-positive* if there exists a vector $0 \neq u \in K$ such that for every $0 \neq x \in K$ there exist positive α, β, k where k is an integer such that

$$\alpha u \overset{K}{\leq} A^k x \overset{K}{\leq} \beta u.$$

Show that

(a) if $A \in \pi(K)$ is u-positive and $u \in \operatorname{int} K$, then A is K-primitive;
(b) if $A \in \pi(K)$ is K-irreducible then u-positivity is equivalent to K-primitivity.

Let K be the nonnegative orthant. Check that

$$A = \begin{bmatrix} 0 & 1 \\ 1 & 0 \end{bmatrix}$$

is K-reducible but not K-primitive, and thus not u-positive, and that the K reducible

$$A = \begin{bmatrix} 1 & 1 \\ 0 & 0 \end{bmatrix}$$

is u-positive for

$$u = \begin{bmatrix} 1 \\ 0 \end{bmatrix}$$

(Barker [1972]).

(5.24) Let $A \in \pi(K)$ have a complex eigenvalue $\lambda = \mu + i\nu$, $\nu \neq 0$. If $z = x + iy$ is a corresponding eigenvector, show that x and y are linearly independent and that $\operatorname{span}\{x, y\} \cap K = \{0\}$ (Barker and Turner [1973], Barker and Schneider [1975]).

(5.25) Prove that if K is polyhedral, $A \in \pi(K)$ is K-irreducible and $\rho(A) = 1$, then every eigenvalue of A of modulus 1 is a root of unity (Barker and Turner [1973], Barker [1974]).

(5.26) Let A and B be K-irreducible matrices in $\pi(K)$. Let $\alpha > \rho(A)$. Show that there exist a unique $\lambda > 0$ such that

$$\rho(A + (B/\lambda)) = \alpha.$$

(5.27) A symmetric matrix A is *copositive with respect to a proper cone* K if

$$x \in K \rightarrow x^t A x \geq 0$$

Let A be a symmetric matrix. Prove that $\rho(A)$ is an eigenvalue if and only if A is copositive with respect to a self-dual cone (Haynsworth and Hoffman [1969]).

(5.28) Let $e \in \operatorname{int} K$. $A \in \pi(K)$ is called *K-stochastic* if $y \in K^*$, $y^t e = 1 \rightarrow y^t A e = 1$. Show that if A is K-stochastic then $\rho(A) = 1$ and is an eigenvalue with linear elementary divisors (Marek [1971], Barker [1972]).

6 NOTES

(6.1) Many of the cone-theoretic concepts introduced in this chapter are known by other names. Our convex cone is called a linear semigroup in Krein and Rutman [1948] and a wedge in Varga [a]. Our proper cone is also called cone (Varga [a]), full cone (Berman [1973]), good cone, and positive cone. Dual is polar in Ben-Israel [1969], Haynsworth and Hoffman [1969], and Schneider and Vidyasagar [1970] and conjugate semigroup in Krein and Rutman [1948]. Equivalent terms for polyhedral cone are finite cone (Gale [1960]) and coordinate cone (Smith [1974]). An equivalent term for simplicial cone is minihedral cone (Varga [a]).

(6.2) The ice cream cone K_n is called the circular Minkowski cone in Krein and Rutman [1948]. Usually, K_n is defined as

$$\{x \in R^n : (x_1^2 + \cdots + x_{n-1}^2)^{1/2} \leq x_n\}.$$

Our slight change of definition makes K_n top heavy. (See Fiedler and Haynsworth [1973].)

(6.3) An n-dimensional space which is partially ordered by a proper cone is called a Kantorovich space of order n in the Russian literature, e.g., Glazman and Ljubic [1974].

(6.4) Theorem 2.5 is borrowed from Klee [1959] and Rockafellar [1970]. Of the many other books on convexity and cones we mention Berman [1973], Fan [1969], Gale [1960], Glazman and Ljubic [1974], Grunbaum [1967], Schaefer [1971, 1974], and Stoer and Witzgal [1970].

(6.5) Many questions are still open, at the writing of this book, concerning the structure of $\pi(K)$ where K is a proper cone. A conjecture of Loewy and Schneider [1975a] states that if $A \in \text{ext}\ \pi(K)$, the set of extremals of $\pi(K)$, then $A(\text{ext}\ K) \subseteq \text{ext}\ K$. The converse is true for a nonsingular A and indecomposable K. (See Exercise 5.9.)

(6.6) The first extension of the Perron [1907] and Frobenius [1908, 1909, and 1912] theorems to operators in partially ordered Banach space is due to Krein and Rutman [1948]. There is an extensive literature on operators that leave a cone invariant in infinite-dimensional spaces. The interested reader is referred to the excellent bibliographies in Barker and Schneider [1975] and in Marek [1970].

(6.7) Theorem 3.2 is due to Birkhoff [1967b]. Theorems 3.5, 3.15, 3.16, 3.20, 3.23, and 3.26 are due to Vandergraft [1968].

(6.8) The concept of irreducibility of nonnegative matrices ($K = R_+^n$) was introduced independently by Frobenius [1912] and Markov [1908]. (See the interesting comparison in Schneider [1977] and Chapter 2.)

(6.9) The definition of a face, given in this chapter, is the one used by Schneider. Vandergraft [1968] defined a face of a cone K to be a subset of $\mathrm{bd}\,K$ which is a pointed closed convex cone generated by extremals of K. Thus K itself is not a face by Vandergraft's definition. Except for K, every face, by Schneider's definition, is a face by Vandergraft's. That the converse is not true is shown in Exercise 5.6. The concepts of K-irreducibility which the definitions of a face yield are, however, the same.

(6.10) Several concepts related or equivalent to K-irreducibility are surveyed in Vandergraft [1968] and Barker [1972]. (See Exercise 5.21.)

(6.11) The sets S and S_0 for nonnegative orthants are discussed in Fiedler and Ptak [1966] and for general cones in Berman and Gaiha [1972]. The concept of semipositivity is studied by Vandergraft [1972].

(6.12) The results on order inequalities and the consequent corollaries are borrowed from Rheinboldt and Vandergraft [1973].

(6.13) Most of Section 4 is based on Barker [1972]. Theorem 4.10 is taken from Krein and Rutman [1948]. Concepts which are equivalent to K-positivity are described in Barker [1972].

CHAPTER 2

NONNEGATIVE MATRICES

1 INTRODUCTION

In this chapter we consider square *nonnegative matrices*, i.e., square matrices all of whose elements are nonnegative. The material developed here will be used extensively in Chapter 6 and in the application chapters.

In conformity with the notation introduced after Definition 1.2.10 we write

$$A \geq B \quad \text{if} \quad a_{ij} \geq b_{ij} \quad \text{for all} \quad i \text{ and } j,$$

$$A > B \quad \text{if} \quad A \geq B \quad \text{and} \quad A \neq B,$$

and

$$A \gg B \quad \text{if} \quad a_{ij} > b_{ij} \quad \text{for all} \quad i \text{ and } j.$$

The matrices A which satisfy $A \gg 0$ are called *positive matrices.*

The two basic approaches to the study of nonnegative matrices are geometrical and combinatorial. The first approach was taken in Chapter 1 in the study of operators mapping a cone into itself. Observing that the nonnegative matrices of order n are those that map R^n_+, the nonnegative orthant in R^n, onto itself, one may use the results of Chapter 1. For the convenience of the reader who is primarily concerned with the applications and may not wish to read Chapter 1 in detail, we now summarize the appropriate definitions and results in the context of nonnegative matrices.

(1.1) Theorem (See Theorem 1.3.2.) If A is a nonnegative square matrix, then

(a) $\rho(A)$, the spectral radius of A, is an eigenvalue,
(b) A has a nonnegative eigenvector corresponding to $\rho(A)$,
(c) A^t has a nonnegative eigenvector corresponding to $\rho(A)$.

(1.2) Definition (See Definition 1.3.14.) An $n \times n$ matrix A is *cogredient* to a matrix E if for some permutation matrix P, $PAP^t = E$. A is *reducible* if it is cogredient to

$$E = \begin{bmatrix} B & 0 \\ C & D \end{bmatrix},$$

where B and C are square matrices, or if $n = 1$ and $A = 0$. Otherwise, A is *irreducible.*

(1.3) Theorem (See Theorems 1.3.15, 1.3.16, 1.3.18, and 1.3.20, and Corollary 1.3.22.) Each of the following conditions characterize the irreducibility of a nonnegative matrix A of order n ($n > 1$).

(a) No eigenvector of A has a zero coordinate.
(b) A has exactly one (up to scalar multiplication) nonnegative eigenvector, and this eigenvector is positive.
(c) $\alpha x \geq Ax, x > 0 \rightarrow x \gg 0$.
(d) $(I + A)^{n-1} \gg 0$.
(e) A^t is irreducible.
The nonnegative irreducible matrices include the positive ones.

(1.4) Theorem (See Theorems 1.3.23 and 1.3.26.)

(a) If A is positive, then $\rho(A)$ is a simple eigenvalue, greater than the magnitude of any other eigenvalue.
(b) If $A \geq 0$ is irreducible then $\rho(A)$ is a simple eigenvalue, any eigenvalue of A of the same modulus is also simple, A has a positive eigenvector x corresponding to $\rho(A)$, and any nonnegative eigenvector of A is a multiple of x.

Theorem 1.4 is part of the classical Perron–Frobenius theorem. Perron [1907] proved it for positive matrices and Frobenius [1912] gave the extension to irreducible matrices. The second part of the Perron–Frobenius theorem will be given in the next section.

(1.5) Corollary (See Corollaries 1.3.29 and 1.3.30.)

(a) If $0 \leq A \leq B$, then $\rho(A) \leq \rho(B)$.
(b) If $0 \leq A < B$ and $A + B$ is irreducible (see Corollary 1.10) then $\rho(A) < \rho(B)$.

(1.6) Corollary (a) If B is a principal submatrix of $A \geq 0$, then $\rho(B) \leq \rho(A)$.

(b) $\rho(A)$ is an eigenvalue of some proper principal submatrix of a nonnegative matrix A if and only if A is reducible.

(1.7) Theorem (See Definition 1.4.1 and Theorems 1.4.2 and 1.4.10.) The following conditions, on a nonnegative matrix A, are equivalent:

(a) A is irreducible and $\rho(A)$ is greater in magnitude than any other eigenvalue.

(b) The only nonempty subset of bd R^n_+ which is left invariant by A is $\{0\}$.

(c) There exists a natural number m such that A^m is positive.

(1.8) Definition (See Theorem 1.4.2.) Matrices that satisfy the conditions in Theorem 1.7 are called *primitive*.

(1.9) Corollary (See Corollaries 1.4.4 and 1.4.6.) If A is primitive and l is a natural number, then A^t and A^l are (irreducible and) primitive.

(1.10) Corollary (See Corollaries 1.3.21 and 1.4.5.)

(a) If $A \geq 0$ is irreducible and $B \geq 0$, then $A + B$ is irreducible.
(b) If $A \geq 0$ is primitive and $B \geq 0$, then $A + B$ is primitive.

(1.11) Theorem (See Theorems 1.3.34 and 1.3.35.) For $A \geq 0$,

$$\alpha x \leq Ax, \; x > 0 \quad \text{implies} \quad \alpha \leq \rho(A)$$

and

$$Ax \leq \beta x, \; x \gg 0 \quad \text{implies} \quad \rho(A) \leq \beta.$$

If, in addition, A is irreducible, then

$$\alpha x < Ax < \beta x, \; x > 0 \quad \text{implies} \quad \alpha < \rho(A) < \beta \quad (\text{and} \quad x \gg 0).$$

(1.12) Corollary If x is a positive eigenvector of a nonnegative matrix A then x corresponds to $\rho(A)$.

The approach in the main body of this chapter is combinatorial, using the elementwise structure in which the zero–nonzero pattern plays an important role. Irreducible matrices are studied in Section 2 while Section 3 is concerned with the reducible case. We return to the irreducible case and study primitivity in Section 4. Stochastic and doubly stochastic matrices are introduced in Section 5. The text is supplemented with many exercises and the chapter is concluded with notes.

2 IRREDUCIBLE MATRICES

We start with combinatorial characterizations of irreducibility. Let $a_{ij}^{(q)}$ denote the (i,j) element of A^q.

(2.1) Theorem A nonnegative matrix A is irreducible if and only if for every (i,j) there exists a natural number q such that

$$a_{ij}^{(q)} > 0 \tag{2.2}$$

Proof The "only if" part follows from Definition 1.2. Conversely, by Theorem 1.3, $(I + A)^{n-1} \gg 0$. Let $B = (I + A)^{n-1}A$. As a product of a positive and an irreducible matrix, B itself is positive. Let $B = A^n + C_{n-1}A^{n-1} + \cdots + C_1 A$. Then $b_{ij} = a_{ij}^{(n)} + c_{n-1}a_{ij}^{(n-1)} + \cdots + c_1 a_{ij} > 0$, for all (i,j), so for each (i,j) there must exist a positive integer q such that $a_{ij}^{(q)} > 0$. ∎

(2.3) Exercise Let m be the degree of the minimal polynomial of A. Show that q in (2.2) can be chosen so that $q \leq m$ if $i = j$ and $q < m$ if $i \neq j$ and that in Theorem 1.3(d), n can be replaced by m (Gantmacher [1959]).

The (i,j,q) characterization of irreducibility has a beautiful graph theoretic interpretation.

(2.4) Definition The associated *directed graph*, $G(A)$, of an $n \times n$ matrix A, consists of n vertices $P_1, P_2, \ldots, P_n$ where an edge leads from P_i to P_j if and only if $a_{ij} \neq 0$.

(2.5) Example Let

$$A = \begin{bmatrix} 0 & 1 & 1 \\ 1 & 0 & 0 \\ 1 & 0 & 0 \end{bmatrix},$$

$$B = \begin{bmatrix} 1 & 0 & 1 & 0 \\ 0 & 1 & 1 & 1 \\ 1 & 0 & 1 & 0 \\ 1 & 1 & 0 & 1 \end{bmatrix}, \qquad C = \begin{bmatrix} 1 & 0 & 1 & 0 \\ 0 & 0 & 0 & 1 \\ 0 & 1 & 0 & 0 \\ 1 & 0 & 0 & 1 \end{bmatrix}.$$

Then $G(A)$ is

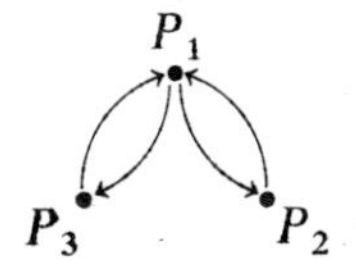

$G(B)$ is

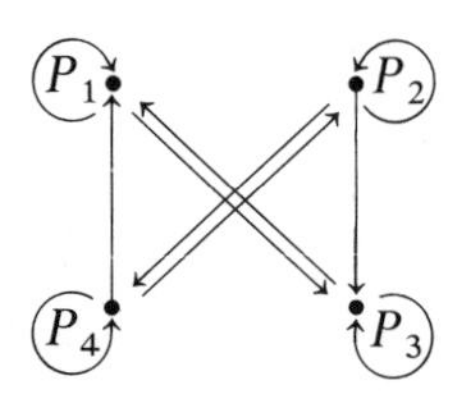

and $G(C)$ is

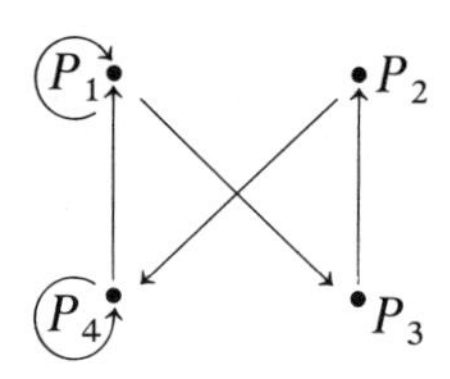

(2.6) Definition A directed graph G is *strongly connected* if for any ordered pair (P_i, P_j) of vertices of G, there exists a sequence of edges (a *path*) which leads from P_i to P_j.

Since $a_{ij}^{(q)} > 0$ if and only if there exists a sequence of q edges from P_i to P_j, Theorem 2.1 translates into the following theorem.

(2.7) Theorem A matrix A is irreducible if and only if $G(A)$ is strongly connected.

(2.8) Example In Example 2.5, A and C are irreducible but B is reducible.

We now deepen the study of the spectral radii of irreducible matrices. Let $A \geq 0$ be irreducible. For every vector $x > 0$ define

$$r_x = \min_{x_i \neq 0} \frac{(Ax)_i}{x_i}.$$

Obviously $r_x \geq 0$ and r_x is the largest real number ρ such that $\rho x \leq Ax$.

(2.9) Exercise The function r_x assumes a maximum value r for some vector $z \gg 0$. [Hint: The function r_y is continuous on the compact set of vectors y of the form $y = (I + A)^{n-1}x$, $x^t x = 1$.]

(2.10) Theorem Let $r = r_z = \max_{x > 0} r_x$. Then

(a) $r > 0$,
(b) $Az = rz$,
(c) $z \gg 0$.

Proof (a) Let $u = (1, \ldots, 1)^t$. Then $r_u = \min_i \sum_{j=1}^n a_{ij}$ is positive since an irreducible matrix cannot have a zero row. Therefore $r \geq r_u > 0$.

(b) Suppose $Az > rz$. Let $x = (I + A)^{n-1}z$. Then $x \gg 0$, $(I + A)^{n-1}(Az - rz) \gg 0$, and $Ax \gg rx$, but the last inequality contradicts the definition of r.

(c) Since $Az = rz$, $0 \ll x = (I + A)^{n-1}z = (1 + r)^{n-1}z$, so z too is positive. ■

Since, by Theorem 1.4, the only positive eigenvector of an irreducible matrix A corresponds to $\rho(A)$, it follows that $r = \rho(A)$ so one obtains the following max min characterization for $\rho(A)$:

$$\rho(A) = \max_{x > 0} \left\{ \min_{x_i > 0} \frac{(Ax)_i}{x_i} \right\} \tag{2.11}$$

(2.12) Exercise Show that for an irreducible $A \geq 0$

$$\rho(A) = \min_{x > 0} \left\{ \max_{x_i > 0} \frac{(Ax)_i}{x_i} \right\}$$

and that if $\rho(A) = \max((Az)_i/z_i)$, then z is an eigenvector that corresponds to $\rho(A)$.

In the remainder of the section let $\rho(A) = r$. Let adj A denote the adjoint (the adjugate) of A. Another corollary on the positivity of the eigenvectors of A and A^t that correspond to r follows.

(2.13) Corollary Let $B(x) = \operatorname{adj}(xI - A)$. Then $B(r) \gg 0$.

Proof Let $\Delta(x)$ be the characteristic polynomial of A. Then $(xI - A)B(x) = \Delta(x)I$. For $x = r$ it follows that the columns of $B(r)$ are eigenvectors of A corresponding to r, so in each column all elements are different from zero and have the same sign. The same argument applied to A^t shows that all elements of $B(r)$ have the same sign. By the definition of $\rho(A)$, $\Delta'(r) > 0$, but $\Delta'(r) = \operatorname{tr} B(r)$ and thus $B(r) \gg 0$. ■

For any complex matrix C let $|C|$ denote the matrix with entries $|c_{ij}|$.

(2.14) Theorem (a) If A is irreducible and $A \geq |C|$, then for every eigenvalue γ of C

$$|\gamma| \leq r. \tag{2.15}$$

(b) Equality holds in (2.15) if and only if $C = e^{i\theta}DAD^{-1}$ where $e^{i\theta} = \gamma/r$ and $|D| = I$.

Proof (a) Let y be an eigenvector of C corresponding to γ

$$Cy = \gamma y, \qquad y \neq 0. \tag{2.16}$$

Taking absolute values one gets

$$|\gamma||y| \leq |C||y| \leq A|y| \tag{2.17}$$

and thus $|\gamma| \leq r|y| \leq r$.

(b) If $|\gamma| = r$, then (2.17) implies that $A|y| = r|y| = |C||y|$ and $|y| \gg 0$. Thus by Corollary 1.5

$$|C| = A. \tag{2.18}$$

Let $y_j = |y_j|e^{i\theta_j}$ and define $D = \operatorname{diag}\{e^{i\theta_1}, e^{i\theta_2}, \ldots, e^{i\theta_n}\}$. Then $y = D|y|$. Let $\gamma = re^{i\theta}$ and $F = e^{-i\theta}D^{-1}CD$. Substitution in (2.15) yields

$$F|y| = r|y|. \tag{2.19}$$

Again $|F| \leq A$ and thus $|F| = A$. Thus $F|y| = |F||y|$ but since $|y| \gg 0$, $F = |F|$; i.e., $e^{-i\theta}D^{-1}CD = A$. Hence $C = e^{i\theta}DAD^{-1}$. ■

Now we can state and prove the second part of the Perron–Frobenius theorem. (The first part was Theorem 1.4.)

(2.20) Theorem (a) If an irreducible $A \geq 0$ has h eigenvalues

$$\lambda_0 = re^{i\theta_0},\ \lambda_1 = re^{i\theta_1}, \ldots, \lambda_{h-1} = re^{i\theta_{h-1}} \quad \text{of modulus } \rho(A) = r,$$

$$0 = \theta_0 < \theta_1 < \cdots < \theta_{h-1} < 2\pi,$$

then these numbers are the distinct roots of $\lambda^h - r^h = 0$.

(b) More generally, the whole spectrum $S = \{\lambda_0, \lambda_1, \ldots, \lambda_{n-1}\}$ of A goes over into itself under a rotation of the complex plane by $2\pi/h$.

(c) If $h > 1$, then A is cogredient to

$$PAP^{\mathrm{t}} = \begin{bmatrix} 0 & A_{12} & 0 & \cdots & 0 \\ 0 & 0 & A_{23} & \cdots & 0 \\ \vdots & & \vdots & \ddots & \vdots \\ 0 & 0 & 0 & \cdots & A_{h-1\,h} \\ A_{h1} & 0 & 0 & \cdots & 0 \end{bmatrix}, \tag{2.21}$$

where the zero blocks along the diagonal are square.

Proof (a) Applying Theorem 2.14 with $C = A$ and $\gamma = \lambda_t$, the equality condition implies that

$$A = e^{i\theta_t}D_tAD_t^{-1}, \qquad |D| = I, \qquad t = 0,1,\ldots,h-1. \tag{2.22}$$

Let $Az = rz$, $z \gg 0$ and $y^{(t)} = D_t z$. Then $Ay^{(t)} = \lambda_t y^{(t)}$ so that $y^{(t)}$ is an eigenvector corresponding to the simple eigenvalue λ_t. Therefore $y^{(t)}$ and D_t are determined up to a multiplication by a scalar. We define the matrices $D_0, D_1, \ldots, D_{h-1}$ uniquely by choosing their first diagonal element to be one. From (2.22) it follows that $D_j D_k^{\pm 1} z$ is an eigenvector of A that corresponds to $re^{i(\theta \pm \theta_k)}$. Thus the numbers $e^{i\theta_0} = 1, \ldots, e^{i\theta_{k-1}}$, and the matrices $D_0 = I, \ldots, D_{h-1}$, are isomorphic abelian multiplicative groups of order h. Thus the numbers $e^{i\theta_t}$ are h's roots of unity and $D_t^h = I$.

(b) The spectrum of $e^{i2\pi/h}A$ is $\{e^{i2\pi/h}\lambda_0, \ldots, e^{i2\pi/h}\lambda_{h-1}\}$. Since we saw in (a) that A and $e^{i2\pi/h}A$ are similar, this spectrum is equal to S.

(c) Let $D = D_1$. Since $D^h = I$, one can permute D and A so that

$$PDP^t = \begin{bmatrix} e^{i\delta_0}I_0 & 0 & & \cdots & 0 \\ 0 & e^{i\delta_1}I_1 & & \cdots & 0 \\ \vdots & \vdots & & \ddots & \vdots \\ 0 & 0 & \cdots & & e^{i\delta_{s-1}}I_{s-1} \end{bmatrix}, \tag{2.23}$$

where the I_j are identity matrices of not necessarily the same order, $\delta_j = (2\pi/h)n_j$, $0 < n_0 < \cdots < n_{s-1} < h$ and

$$PAP^t = \begin{bmatrix} A_{11} & A_{12} & \cdots & A_{1s} \\ A_{21} & A_{22} & \cdots & A_{2s} \\ \vdots & \vdots & \ddots & \vdots \\ A_{s1} & A_{s2} & \cdots & A_{ss} \end{bmatrix}, \tag{2.24}$$

where A_{jj} is of the same order as I_{j-1}.

Now $A = e^{i(2\pi/h)}DAD^{-1}$ which can be written as a system of s^2 equations

$$e^{i(2\pi/h)}A_{pq} = \frac{e^{in_{q-1}(2\pi/h)}}{e^{in_{p-1}(2\pi/h)}}A_{pq}, \qquad p,q = 1,2,\ldots,s. \tag{2.25}$$

Thus, $A_{pq} \neq 0$ if and only if $n_q \equiv n_p + 1 \bmod h$. Since A is irreducible, there exists, for every p, a q such that $A_{pq} \neq 0$ and thus $n_q \equiv n_p + 1 \bmod h$, but since $0 < n_0 < \cdots < n_{s-1} < h$, it follows that $s = h$, $n_i = i$, $i = 1, \ldots, s-1$, $n_A = 1$ and $A_{pq} \neq 0$ only when $q \equiv p + 1 \bmod h$. ■

If A is primitive then h in Theorem 2.20 is one.

(2.26) Definition Let $A \geq 0$ be irreducible. The number h of eigenvalues of A of modulus $\rho(A)$ is called the *index of cyclicity* of A. If h is greater than one, A is said to be *cyclic of index h.*

The index of cyclicity of a matrix relates to the coefficients of its characteristic polynomial.

(2.27) Theorem Let $\lambda^n + a_1\lambda^{n_1} + \cdots + a_k\lambda^{n_k}$, where $a_1, \ldots, a_k$ are different from zero and $n > n_1 > \cdots > n_k$, be the characteristic polynomial of an irreducible $A \geq 0$ which is cyclic of index h. Then h is the greatest common divisor of the differences $n - n_1, n_1 - n_2, \ldots, n_{k-1} - n_k$.

Proof Let $m > 1$ be an integer such that A and $e^{i(2\pi/m)}A$ are similar. Then

$$\lambda^n + a_1\lambda^{n_1} + \cdots + a_k\lambda^{n_k} = \lambda^n + a_1\theta^{n-n_1}\lambda^{n_1} + \cdots + a_k\theta^{n-n_k}\lambda^{n_k}$$

where
$$\theta = e^{i(2\pi/m)}.$$

It follows that $a_t = a_t\theta^{n-n_k}$ for $t = 1, \ldots, k$, and therefore m divides each of the differences $n - n_1, n - n_2, \ldots, n - n_k$. Conversely, if m divides each of these differences, A and $e^{i(2\pi/m)}A$ have the same spectrum. By Theorem 2.20, A and $e^{i(2\pi/m)}A$ have the same spectrum for $m = h$ but not for $m > h$. It follows that

$$\begin{aligned} h &= \text{g.c.d.}(n - n_1, n - n_2, \ldots, n - n_k) \\ &= \text{g.c.d.}(n - n_1, n - n_2, \ldots, n_{k-1} - n_k). \quad \blacksquare \end{aligned}$$

Since trace A is the coefficient of λ^{n-h} in $\Delta(\lambda)$, the next corollary follows from (2.21) and Theorem 2.27.

(2.28) Corollary An irreducible matrix is primitive if its trace is positive.

The order of cyclicity of a matrix can be computed by inspection of its directed graph.

(2.29) Definition Let $P = \{P_{i_0}P_{i_1}, P_{i_1}P_{i_2}, \ldots, P_{i_{l-1}}P_{i_l}\}$ be a path in a graph G. Then l is the *length* of P. P is a *simple path* if $P_{i_1}, \ldots, P_{i_l}$ are distinct and a *circuit* if $P_{i_0} = P_{i_l}$.

Notice that if the coefficient of λ^{n-l} in $\Delta(\lambda)$ is not zero, then $G(A)$ contains a simple circuit of length l or disjoint circuits, such that the sum of their lengths is l.

(2.30) Theorem Let $A \geq 0$ be an irreducible matrix of order n. Let S_i be the set of all the lengths m_i of circuits in $G(A)$, through P_i. Let

$$h_i = \underset{m_i \in S_i}{\text{g.c.d.}} \{m_i\}.$$

Then $h_1 = h_2 = \cdots = h_n = h$ and h is the index of cyclicity of A.

Proof S_i is clearly an additive semigroup. As such it contains all but a finite number of multiples of its greatest common divisor h_i.

We can assume that A is in the form (2.21). Then it is clear that only powers of A^h may have positive diagonal elements. In fact (see Exercise 6.10)

$$A^{ph} = \begin{bmatrix} C_1^{(p)} & 0 & \cdots & 0 \\ 0 & C_2^{(p)} & \cdots & 0 \\ \vdots & \vdots & \ddots & \vdots \\ 0 & 0 & \cdots & C_h^{(p)} \end{bmatrix}, \tag{2.31}$$

where the diagonal blocks are square primitive matrices (all with spectral radius equal to $\rho(A)^{ph}$).

Now if $h_i = 1$, then $a_{ii}^{(p)} > 0$ for all p sufficiently large, thus A is primitive, and $a_{jj}^{(p)} > 0$ for all $1 \leq j \leq n$ and p sufficiently large so that $h_1 = h_2 = \cdots = h_n = 1$.

If $h_i > 1$, A cannot be primitive, so it is cyclic of index h. Since the blocks $C_i^{(p)}$ in (2.31) are primitive, they become positive for all p sufficiently large and thus $h_1 = h_2 = \cdots = h_n = h$. ■

(2.32) Examples Let

$$A_1 = \begin{bmatrix} 0 & 1 & 0 & 0 \\ 0 & 0 & 1 & 0 \\ 0 & 0 & 0 & 1 \\ 1 & 0 & 0 & 0 \end{bmatrix}, \qquad A_2 = \begin{bmatrix} 0 & 1 & 0 & 0 \\ 0 & 0 & 1 & 0 \\ 0 & 0 & 0 & 1 \\ 0 & 1 & 0 & 0 \end{bmatrix}$$

and

$$A_3 = \begin{bmatrix} 0 & 1 & 0 & 0 \\ 0 & 0 & 1 & 0 \\ 0 & 0 & 0 & 1 \\ 1 & 1 & 0 & 0 \end{bmatrix}.$$

Then

$$G(A_1) = \begin{array}{ccc} P_1 \odot & \longrightarrow & \odot P_2 \\ \uparrow & & \downarrow \\ P_4 \odot & \longleftarrow & \odot P_3 \end{array}$$

so A_1 is cyclic of index 4,

$$G(A_2) = \begin{array}{ccc} P_1 \odot & \longrightarrow & \odot P_2 \\ & \nearrow & \downarrow \\ P_4 \odot & \longleftarrow & \odot P_3 \end{array}$$

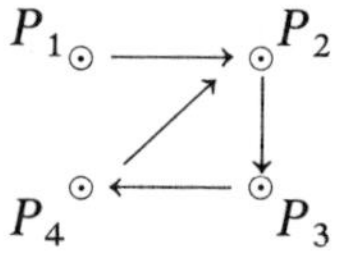

is not strongly connected so A_2 is reducible and

$$G(A_3) = \begin{array}{ccc} P_1 \odot & \longrightarrow & \odot P_2 \\ \uparrow & \nearrow & \downarrow \\ P_4 \odot & \longleftarrow & \odot P_3 \end{array}$$

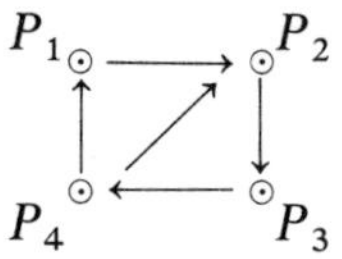

so A_3 is primitive. (Here, for example, $k_4 = \text{g.c.d.}\{3,4,\ldots\} = 1$.)

A partial converse of Theorem (2.20) follows.

(2.33) Theorem Suppose $A \geq 0$ has no zero rows or columns, there exists a permutation matrix P such that

$$PAP^{t} = \begin{bmatrix} 0 & A_1 & 0 & \cdots & 0 \\ 0 & 0 & A_2 & \cdots & 0 \\ \vdots & \vdots & \vdots & \ddots & \vdots \\ 0 & 0 & 0 & \cdots & A_{h-1} \\ A_h & 0 & 0 & \cdots & 0 \end{bmatrix},$$

where the zero blocks on the diagonal are square and $B = A_1 A_2 \cdots A_{h-1} A_h$ is irreducible. Then A is irreducible.

Proof Define a relation R on R^n_+ by writing $x R y$ if $x_i > 0 \leftrightarrow y_i > 0$. By Theorem 1.3 a matrix M is reducible if and only if $Mx\, R x$ for some $x > 0$ which is not positive.

Suppose that $PAP^{t}x\, R\, x$ for $x > 0$. Partition

$$x = \begin{bmatrix} x^{(1)} \\ \vdots \\ x^{(h)} \end{bmatrix}$$

where $x^{(i)}$ has as many rows as A_i does. Then

$$A_i x^{(i+1)}\, R\, x^{(i)}, \qquad i = 1, \ldots, h-1 \quad \text{and} \quad A_h x^{(1)}\, R\, x^{(h)}. \tag{2.34}$$

If $C \geq 0$ and $x R y$, then $Cx\, R\, Cy$. Thus $Bx^{(1)}\, R\, x^{(1)}$. Since B is irreducible, $x^{(1)} = 0$ or $x^{(1)} \gg 0$. If $x^{(1)} = 0$, then $A_1 x^{(2)} = 0$ by (2.34) and hence $x^{(2)} = 0$ because A has no zero columns. By induction $x^{(j)} = 0$ for each $x^{(j)}$ so that $x^{(1)} \gg 0$. If $x^{(1)} > 0$ then a similar argument shows that $x \gg 0$. Therefore A is irreducible. ■

We conclude with bounds for the spectral radius r of a nonnegative irreducible matrix A and for its corresponding positive eigenvectors. Let s_i denote the sum of elements of the ith row of A. Let $S = \max_i s_i$ and $s = \min_i s_i$.

(2.35) Theorem Let $A \geq 0$ be irreducible. Let x be a positive eigenvector and let $\gamma = \max_{i,j}(x_i/x_j)$. Then

$$s \leq r \leq S \tag{2.36}$$

and

$$(S/s)^{1/2} \leq \gamma. \tag{2.37}$$

Moreover, equality holds in either (2.36) or (2.37) if and only if $s = S$.

Proof Expand the determinant of

$$\Delta(r) = \begin{vmatrix} r - a_{11} & -a_{12} & \cdots & -a_{1n} \\ -a_{21} & r - a_{22} & \cdots & -a_{2n} \\ \vdots & \vdots & \ddots & \vdots \\ -a_{n1} & -a_{n2} & \cdots & r - a_{nn} \end{vmatrix}$$

by the last column after adding to it all the preceding columns. The result is

$$\sum_{j=1}^{n} (r - s_j)B_{nj}(r) = 0$$

By Corollary 2.13, $B_{nj}(r) \gg 0$, which proves (2.36) and the condition for equality in (2.36). Let

$$x_M = \max_i x_i \text{ and } \qquad x_m = \min_i x_i.$$

Then

$$rx_i = \sum_{j=1}^{n} a_{ij}x_j \geq \sum_{j=1}^{n} a_{ij}x_m = S_i x_m.$$

Therefore $r/s_i \geq x_m/x_i \geq 1/\gamma$ for all i. In particular,

$$r/S \geq 1/\gamma. \tag{2.38}$$

Similarly,

$$rx_i \leq s_i x_M, \qquad r/s_i \leq x_M/x_i \leq \gamma$$

and in particular

$$r/s \leq \gamma. \tag{2.39}$$

Combining (2.38) and (2.39) proves (2.37).

The equality

$$rx_i = \sum_{j=1}^{n} a_{ij}x_j = r_i x_m$$

holds if and only if all the coordinates of x are equal which implies that the row sums of A must be equal. The converse is obvious. ■

Obviously these inequalities can be stated in terms of column sums.

3 REDUCIBLE MATRICES

Some of the results on irreducible matrices are easily generalized to arbitrary nonnegative matrices by a limiting process since every square $A \geq 0$ can be represented as $A = \lim_{m\to\infty} A_m$ where A_m is positive and thus irreducible.

For example, (2.36) holds for any $A \geq 0$. In general the results obtained by this limiting method are weaker than their irreducible analogues since strict inequalities are not preserved. As an example we generalize Corollary 2.13.

(3.1) Theorem Let r be the spectral radius of $A \geq 0$. Then for $\lambda \geq r$

$$B(\lambda) \geq 0 \quad \text{and} \quad \frac{dB(\lambda)}{d\lambda} \geq 0. \tag{3.2}$$

Proof A passage to the limit gives instead of (2.13), $B(r) \geq 0$. We prove (3.2) by induction on n. For $n = 1$, $B(\lambda) \equiv I$ and $dB(\lambda)/d\lambda \equiv 0$. To prove the inequalities for a matrix of order n we assume they are true for matrices of smaller orders. Let $(\lambda I - A)(k|k)$ be the matrix obtained from $(\lambda I - A)$ by deleting the kth row and the kth column and let $B^{(k)}(\lambda) = \operatorname{adj}[(\lambda I - A)(k|k)]$. Expanding $\Delta(\lambda) = |\lambda I - A|$ by the kth row and the kth column yields

$$\Delta(\lambda) = (\lambda - a_{kk})b_{kk}(\lambda) - \sum_{\substack{i,j=1 \\ i\neq k,\, j\neq k}}^{n} b_{ji}^{(k)}(\lambda)a_{ik}a_{kj}. \tag{3.3}$$

Let r_k denote the maximal nonnegative root of $b_{kk}(\lambda)$. By the induction hypothesis $B^{(k)}(r_k) \geq 0$. Setting $\lambda = r_k$ in (3.3) implies thus that $\Delta(r_k) \leq 0$; so that $r_k \leq r$ for $k = 1, 2, 3, \ldots, n$.

Now,

$$\frac{db_{ij}(\lambda)}{d\lambda} = \sum_{\substack{k=1 \\ k\neq i,\, k\neq j}}^{n} b_{ij}^{(k)}(\lambda).$$

By the induction assumption, $B^{(k)}(\lambda) \geq 0$ for $\lambda \geq \lambda_k$. Thus $dB(\lambda)/d\lambda \geq 0$ for $\lambda \geq r$ and since $B(r) \geq 0$, $B(\lambda) \geq 0$ for $\lambda \geq r$. ■

With the second inequality in (3.2), Corollary 2.13 implies the following.

(3.4) Exercise If $A \geq 0$ is an irreducible matrix with spectral radius r, then for $\lambda \geq r$, $B(\lambda) \gg 0$ and for $\lambda > r$, $(\lambda I - A)^{-1} \gg 0$.

If $A \geq 0$ is reducible then there exists a permutation matrix P_1 so that A can be *reduced* to the form

$$P_1 A P_1^t = \begin{bmatrix} B & 0 \\ C & D \end{bmatrix}, \tag{3.5}$$

where B and D are square matrices. If either B or D are reducible they can be reduced in the manner that A was reduced in (3.5). Finally, by a suitable permutation A can be reduced to triangular block form,

$$PAP^t = \begin{bmatrix} A_{11} & 0 & \cdots & 0 \\ A_{21} & A_{22} & \cdots & 0 \\ \vdots & \vdots & \ddots & \vdots \\ A_{s1} & A_{s2} & & A_{ss} \end{bmatrix}, \tag{3.6}$$

where each block A_{ii} is square and is either irreducible or a 1×1 null matrix. An irreducible matrix is in a triangular block form consisting of only one block.

To study the spectral properties of reducible matrices in greater depth we return to their combinatorial structure. The following terminology will be used in Chapter 8.

(3.7) Definition Let A be a nonnegative square matrix of order n. For $1 \leq i, j \leq n$ we say that i *has an access to* j if in $G(A)$ there is a path from P_i to P_j, and that i and j *communicate* if i has an access to j and j has an access to i.

Communication is an equivalence relation.

(3.8) Definitions The *classes of* $A \geq 0$ are the equivalence classes of the communication relation induced by $G(A)$. *A class* α *has an access to a class* β if for $i \in \alpha$ and $j \in \beta$, i has an access to j. A class is *final* if it has access to no other class. A class α is *basic* if $\rho(A[\alpha]) = \rho(A)$, where $A[\alpha]$ is the submatrix of A based on the indices in α, and *nonbasic* if $\rho(A[\alpha]) < \rho(A)$. (See Corollary 1.6.)

The blocks A_{ii} in the triangular block form (3.6) of A correspond to the classes of A. From (3.6) it is clear that every $A \geq 0$ has at least one basic class and one final class. The class that corresponds to A_{ii} is basic if and only if $\rho(A_{ii}) = \rho(A)$ and final if and only if $A_{ij} = 0, j = 1, \ldots, i-1$. In particular A is irreducible if and only if it has only one (basic and final) class.

(3.9) Example Let

$$A = \begin{bmatrix} 1 & 0 & 0 & 0 & 0 & 0 & 0 & 0 & 0 & 0 \\ 0 & 3 & 0 & 0 & 0 & 0 & 0 & 0 & 0 & 0 \\ 8 & 6 & 2 & 0 & 0 & 0 & 0 & 0 & 0 & 0 \\ 0 & 7 & 0 & 3 & 0 & 0 & 0 & 0 & 0 & 0 \\ 0 & 0 & 1 & 3 & 3 & 0 & 0 & 0 & 0 & 0 \\ 8 & 0 & 0 & 0 & 5 & 3 & 0 & 0 & 0 & 0 \\ 0 & 0 & 0 & 4 & 0 & 0 & 0 & 0 & 0 & 0 \\ 0 & 0 & 0 & 0 & 0 & 0 & 2 & 3 & 0 & 0 \\ 0 & 0 & 0 & 0 & 0 & 0 & 0 & 0 & 1 & 1 \\ 0 & 0 & 0 & 0 & 0 & 0 & 0 & 0 & 1 & 1 \end{bmatrix}.$$

The basic classes of A are $\{2\}$, $\{4\}$, $\{5\}$, $\{6\}$, and $\{8\}$. The nonbasic classes are $\{1\}$, $\{3\}$, $\{7\}$, and $\{9, 10\}$. The final classes are $\{1\}$, $\{2\}$, and $\{9, 10\}$. The accessibility relation between the classes is described by the following diagram in which the basic classes are circled:

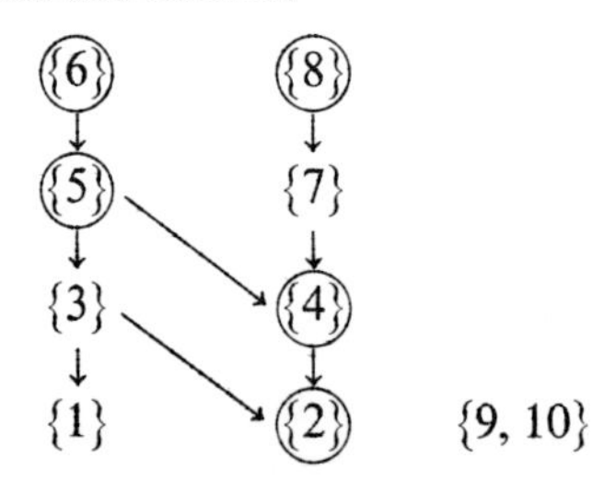

(3.10) Theorem To the spectral radius r of $A \geq 0$ there corresponds a positive eigenvector if and only if the final classes of A are exactly its basic ones.

Proof Only if: We can assume that A is in the triangular block form (3.6). Let z be a positive eigenvector of A that corresponds to r. Partition

$$z = \begin{bmatrix} z^{(1)} \\ z^{(2)} \\ \vdots \\ z^{(s)} \end{bmatrix}$$

in conformity with (3.6). Then for $k = 1, 2, \ldots, s$,

$$A_{k1}z^{(1)} + A_{k2}z^{(2)} + \cdots + A_{k,\,k-1}z^{(k-1)} + A_{kk}z^{(k)} = rz^{(k)}.$$

If the class that corresponds to A_{kk} is final then $A_{kk}z^{(k)} = rz^{(k)}$ so the class is basic. If the class is not final then

$$A_{kk}z^{(k)} < rz^{(k)} \tag{3.11}$$

so

$$\max_i \frac{(A_{kk}z^{(k)})_i}{z_i^{(k)}} \leq r. \tag{3.12}$$

Applying Exercise 2.12 to the irreducible matrix A_{kk} (if $A_{kk} = 0$, the claim of the theorem is obvious) we get

$$\rho(A_{kk}) \leq \max_i \frac{(A_k z^{(k)})_i}{z_i^{(k)}}. \tag{3.13}$$

By the same exercise, equality in (3.12) and (3.13) implies that $A_{kk}^{(k)} = r^{(k)}$, contradicting (3.11).

If: Assume that in the triangular block form (3.6), the diagonal blocks are ordered by their spectral radii so that the blocks $A_{11}, \ldots, A_{gg}$ are those that correspond to the final–basic classes.

Let $z^{(k)}$ be a positive eigenvector of A_{kk} corresponding to r, $k = 1, 2, \ldots, g$, and define

$$z^{(k)} = (rI - A_{kk})^{-1} \sum_{h=1}^{k-1} A_{kh}z^{h}, \qquad k = g + 1, \ldots, s.$$

Since $r > \rho(A_{kk})$ for $k > g$, $(rI - A_{kk})^{-1} \gg 0$, by Exercise 3.4. By induction all the vectors $z^{(k)}$ are positive. Let

$$z = \begin{bmatrix} z^{(1)} \\ z^{(2)} \\ \vdots \\ z^{(s)} \end{bmatrix}.$$

Then $Az = rz$ and $z \gg 0$. ■

(3.14) Theorem To the spectral radius of $A \geq 0$ there corresponds a positive eigenvector of A and a positive eigenvector of A^t if and only if all the classes of A are basic and final. (In other words the triangular block form of A is a direct sum of s matrices having r as a spectral radius.)

Proof If: It follows from the previous theorem.

Only if: By the previous theorem all basic classes are final and if A has nonbasic classes then there is an access from a nonbasic class to a basic one (otherwise a nonbasic class is final). By applying Theorem 3.10 to A^t this implies that there is an access *from* a basic class. This is impossible since the basic classes are final and thus A has only basic classes. ■

(3.15) Corollary A is irreducible if and only if $\rho(A)$ is simple and positive vectors correspond to $\rho(A)$ both for A and for A^t [see Exercise 5.22(iv)].

Recall that the degree of an eigenvalue of A is its multiplicity in the minimal polynomial.

(3.16) Definition The *degree*, $\nu(A)$, of $A \geq 0$ is the degree of $\rho(A)$.

Observe that $\nu(A)$ is the smallest natural number k such that

$$N((\rho(A)I - A)^k) = N((\rho(A)I - A)^{k+1}).$$

(3.17) Definition The null space $N((\rho(A)I - A)^{\nu(A)})$ is called the *algebraic eigenspace* of A and its elements are called *generalized eigenvectors.*

The degree of an irreducible matrix is one. This implies that if all the basic classes of A are final then $\nu(A) = 1$. The algebraic eigenspace of an irreducible matrix is one dimensional and consists of zero and the eigenvectors corresponding to $\rho(A)$. The dimension of the algebraic eigenspace of A is the algebraic multiplicity of $\rho(A)$ and thus equals the number of basic classes of A.

To study the algebraic eigenspace of a reducible matrix A we need more definitions concerning the classes of A.

(3.18) Definition Let $\alpha_1, \alpha_2, \ldots, \alpha_k$ be classes of A. The collection $\{\alpha_1, \alpha_2, \ldots, \alpha_k\}$ is a *chain* from α_1 to α_k if α_i has access to α_{i+1}, $i = 1, \ldots, k-1$. The *length* of a chain is the number of *basic* classes it contains. A class α has access to class β *in m steps* if m is the length of the longest chain from α to β. The *height* of a class β is the length of the longest chain of classes that terminates in β.

(3.19) Example The heights of the classes in Example 3.9 are

Class	Height	Class	Height
{1}	2	{6}	1
{2}	4	{7}	1
{3}	2	{8}	1
{4}	3	{9, 10}	0
{5}	2		

(3.20) Theorem Let $A \geq 0$ have spectral radius r and m basic classes $\alpha_1, \alpha_2, \ldots, \alpha_m$. Then the algebraic eigenspace of A contains nonnegative vectors, $x^{(1)}, \ldots, x^{(m)}$, such that $x_i^{(j)} > 0$ if and only if i has access to α_j, and any such collection is a basis of the algebraic eigenspace of A.

Proof Let α be a basic class of length h. We shall show that the algebraic eigenspace of A contains a nonnegative vector x such that $x_i > 0$ if and only if i has access to α.

Let B_i be the submatrix of A associated with the states having access to α in i steps, $i = 0, 1, \ldots, h$.

By reordering the indices, A may be brought into the form

$$QAQ^t = \begin{bmatrix} B_0 & 0 & \cdots & 0 \\ B_{10} & B_1 & \cdots & 0 \\ \vdots & \vdots & \ddots & \vdots \\ B_{h0} & B_{h1} & \cdots & B_h \end{bmatrix}. \tag{3.21}$$

We may assume that A is in the form (3.21). [The form (3.21) is in general different from (3.6) and the B_i's may be reducible. If A_{kk} is the block in (3.6) corresponding to α, then it is a submatrix of B_1.]

(3.22) Example For A in Example 3.9 and $\alpha = \{2\}$,

$$B_0 = \begin{bmatrix} 1 & 0 & 0 \\ 0 & 1 & 1 \\ 0 & 1 & 1 \end{bmatrix}, \quad B_1 = \begin{bmatrix} 3 & 0 \\ 6 & 2 \end{bmatrix}, \quad B_2 = \begin{bmatrix} 3 & 0 \\ 4 & 0 \end{bmatrix}, \quad B_3 = \begin{bmatrix} 3 & 0 \\ 0 & 3 \end{bmatrix},$$

and $B_4 = [3]$.

For $i = 1, \ldots, h$, let

$$C_{i+1} = \begin{bmatrix} B_1 & \cdots & 0 \\ \vdots & & \vdots \\ B_{i1} & \cdots & B_i \end{bmatrix}$$

and for $i = 2, \ldots, k$,

$$D_i = [B_{i1}, \ldots, B_{i,i-1}]$$

so that for $i \geq 2$,

$$C_{i+1} = \begin{bmatrix} C_i & 0 \\ D_i & B_i \end{bmatrix}$$

Since for $i > 0$, $\rho(B_i) = r$ (one of the classes of B_i is a basic class of A) and since the basic classes of B_i are exactly its final classes, it follows from Theorem 3.10 that there is a positive vector $y^{(i)}$ such that

$$(rI - B_i)y^{(i)} = 0. \tag{3.23}$$

We now show by induction that for $i = 1, \ldots, h$ there is a positive vector $z^{(i)}$ such that

$$(rI - C_{i+1})^i z^{(i)} = 0. \tag{3.24}$$

Choosing $z^{(1)} = y^{(1)}$, (3.24) coincides with (3.23). Suppose (3.24) holds for $i-1$, and consider i. Observe that

$$(rI - C_{i+1})^i = \begin{bmatrix} rI - C_i & 0 \\ D_i & rI - B_i \end{bmatrix}^i = \begin{bmatrix} (rI - C_i)^i & 0 \\ \sum_{k=0}^{i=1} (rI - B_i)^k D_i (rI - C_i)^{i-k-1} & (rI - B_i)^i \end{bmatrix} \tag{3.25}$$

From the induction hypothesis, (3.24), and (3.25) we have for every z

$$(rI - C_{i+1})^i \begin{bmatrix} z^{(i-1)} \\ z \end{bmatrix} = \begin{bmatrix} 0 \\ \sum_{k=1}^{i=1} (rI - B_i)^k D_i (rI - C_i)^{i-1-k} z^{(i-1)} + (rI - B_i)^i z \end{bmatrix}. \tag{3.26}$$

Since all the basic classes of B_i are final, $\nu(B_i) = 1$ and thus the column space of $(rI - B_i)^k$ does not depend on k. Thus there is a vector z such that

$$(rI - C_{i+1})^i \begin{bmatrix} z^{(i-1)} \\ z \end{bmatrix} = 0. \tag{3.27}$$

Choosing $\lambda > 0$ so that $\lambda y^{(i)} + z \gg 0$ and setting

$$z^{(i)} = \begin{bmatrix} z^{(i-1)} \\ \lambda y^{(i)} + z \end{bmatrix},$$

we obtain (3.24) from (3.23), (3.25), and (3.27). This completes the inductive proof.

Letting

$$x = \begin{bmatrix} 0 \\ z^{(h)} \end{bmatrix},$$

we see from (3.22), (3.24), and the positivity of $z^{(h)}$ that

$$(rI - A)^h x = \begin{bmatrix} 0 \\ (rI - C_{h+1})^h z^{(h)} \end{bmatrix} = 0,$$

and $x_i > 0$ if and only if i has access to α.

Let $x^{(j)} \geq 0$ be a vector in the algebraic eigenspace of A such that $x_i^{(j)} > 0$ if and only if i has access to the basic class α_j. The dimension of the algebraic eigenspace of A is equal to the number of the basic classes of A. Thus to show that $x^{(1)}, x^{(2)}, \ldots, x^{(m)}$ are a basis of the algebraic eigenspace, it is enough to show that they are linearly independent.

We can assume that the basic classes are indexed so that α_j does not have access to α_k if $k < j$. Suppose $\sum_{j=1}^{m} a_j x^{(j)} = 0$. Let $x^{(j)}(k)$ be the subvector of $x^{(j)}$ that is based on the indices of α_k. Then for $j < m$, $x^{(j)}(m) = 0$, but since $x^{(m)}(m) > 0$, it follows that $a_m = 0$. Similarly, $a_{m-1} = 0$, and so on. Thus the vectors $x^{(1)}, \ldots, x^{(m)}$ are linearly independent, completing the proof. ■

Theorem 3.18 extends the Perron–Frobenius theorem to reducible matrices. The following (not immediate) exercises complement this extension.

(3.28) Exercise The degree of $A \geq 0$ equals the length of its longest chain (Rothblum [1975]).

(3.29) Exercise Let r be the spectral radius of $A > 0$ and let v be its degree. Then A has a generalized eigenvector having the largest set of positive coordinates among all generalized eigenvectors, and for $k = 1, 2, \ldots, v-1$, $(rI - A)^k x > 0$. Furthermore,

$$((rI - A)^k x)_i > 0,$$

if and only if i has access to some basic class in at least $k + 1$ steps (Rothblum [1975]).

4 PRIMITIVE MATRICES

Let r be the spectral radius of a primitive $A \geq 0$. We shall see now that A/r is semiconvergent, that is, $\lim_{p \to \infty} (A/r)^p$ exists.

(4.1) Theorem Let $A \geq 0$ be primitive and let $\rho(A) = r$. Then $\lim_{p \to \infty} (A/r)^p$ is a positive matrix whose columns are positive eigenvectors corresponding to r. More exactly

$$\lim_{p \to \infty} (A/r)^p = \frac{C(r)}{\psi'(r)}, \tag{4.2}$$

where $\psi(\lambda)$ is the minimal polynomial of A and $C(\lambda) = (\lambda I - A)^{-1}\psi(\lambda)$ is its reduced adjoint matrix.

Proof Let

$$\psi(\lambda) = (\lambda - \lambda_1)^{m_1}(\lambda - \lambda_2)^{m_2} \cdots (\lambda - \lambda_s)^{m_s}, \qquad \text{where} \quad r = \lambda_1 \geq |\lambda_2| \geq \cdots \geq |\lambda_s|.$$

By a well-known formula ([Gantmacher [1959, Vol. I, p. 107]) for functions of matrices

(4.3) $$A^p = \sum_{k=1}^{s} \frac{1}{(m_k - 1)!} B_k,$$

where B_k is the value of the $(m_k - 1)$th derivative of $C(\lambda)\lambda^p(\lambda - \lambda_k)^{m_k}/\psi(\lambda)$, computed at $\lambda = \lambda_k$.

Since A is primitive, $r = \lambda_1 > |\lambda_2|$ and $m_1 = 1$. Thus

$$B_1 = \frac{C(r)r^p(\lambda - r)}{\psi(r)} = \frac{C(r)r^p}{\psi'(r)},$$

while $\lim_{p\to\infty} B_k/r^p = 0$ for $k > 1$, and (4.2) follows from (4.3). The fact that the columns of $C(r)/\psi'(r)$ are positive eigenvectors of A corresponding to r is left to the exercises. (See Exercise 6.8.) ■

Theorem 4.1 offers another proof to the fact (Theorem 1.7) that all but a finite number of powers of a primitive matrix are positive.

(4.4) Definition *The index of primitivity*, $\gamma(A)$, of a primitive matrix A is the smallest positive integer k such that $A^k \gg 0$.

In order to obtain upper bounds for $\gamma(A)$, we introduce the following notation: N denotes the set of indices $\{1,2,\ldots,n\}$. For $L \subseteq N$ and a nonnegative matrix A, $F^0(L) = L$ and $F^h(L)$ is the set of indices i such that there exists in $G(A)$ a path of length h from p_i to p_j for some $j \in L$. $F^h(j)$ denotes $F^h(\{j\})$. Clearly, if A has no zero row, $F(N) = N$ and if A is irreducible and L is a proper subset of N, then $F(L)$ contains some element not in L.

(4.5) Lemma If $A \geq 0$ is irreducible, $j \in N$ and $h \leq n - 1$, then $\bigcup_{l=0}^{h} F^l(j)$ contains at least $h + 1$ elements.

Proof The proof follows by induction on h. ■

(4.6) Lemma Let k be a nonnegative integer, $j \in N$, and $A \geq 0$ be an irreducible matrix of order n. Suppose that for every $l \geq k$, $G(A)$ contains a circuit of length l through p_j. Then $F^{n-1+k}(j) = N$.

Proof By the assumption, $j \in F^{n-1+k-h}(j)$ for every $0 \leq h \leq n - 1$. Thus for $0 \leq h \leq n - 1$, $F^h(j) \subseteq F^{n-1+k}(j)$. By Lemma 4.5,

$$\bigcup_{h=0}^{n-1} F^h(j) = N.$$

Thus

$$N = \bigcup_{h=0}^{n-1} F^h(j) \subseteq F^{n-1+k}(j) \subseteq N. \quad ■$$

(4.7) Theorem Let $A \geq 0$ be an $n \times n$ irreducible matrix. Let k be a nonnegative integer. Suppose that there exist at least d elements in N; $j_1, j_2, \ldots, j_d$, such that for every $l \geq k$ and $i = 1, \ldots, d, A^{(l)}_{j_i j_i} > 0$. Then A is primitive and $\gamma(A) \leq 2n - d - 1 + k$.

Proof We have to show that for every $j \in N$, $F^{2n-d-1+k}(j) = N$. For every $j \in N$ there exist, by Lemma 4.5, $0 \leq h \leq n - d$ and $1 \leq i \leq d$ such that $j_i \in F^h(j)$. Then,

$$N \supseteq F^{2n-d-1+k}(j) = F^{n-d-h}\{F^{n-1+k}[F^h(j)]\} \supseteq F^{n-d-h}(N) = N. \quad \blacksquare$$

(4.8) Corollary Let at least $d > 0$ of the diagonal elements of an irreducible $n \times n$ matrix $A \geq 0$ be positive. Then A is primitive and $\gamma(A) \leq n - d + h(n - 1)$.

Notice that the last result follows for $d = n$ from Theorem 1.3(d).

(4.9) Theorem Let $A \geq 0$ be an $n \times n$ primitive matrix. Suppose that for some positive integer h, $A + A^2 + \cdots + A^h$ has at least $d > 0$ positive diagonal elements. Then $\gamma(A) \leq n - d + h(n - 1)$.

Proof Let $j_1, j_2, \ldots, j_d$ be the d elements such that $j_i \in F^{p_i}(j_i)$, $1 \leq p_i \leq h$, $i = 1, \ldots, d$. Since A is primitive so are the matrices A^{p_i}. Applying Lemma 4.6, with $k = 0$ to A^{p_i} we find that $F^{(n-1)p_i}(j_i) = N$. For arbitrary $j \in N$, Lemma 4.5 implies the existence of $0 \leq l \leq n - d$ and $1 \leq i \leq d$ such that $j_i \in F^l(j)$. Therefore

$$\begin{aligned} N \supseteq F^{n-d+h(n-1)}(j) &= F^{n-d-l+(h-p_i)(n-1)}\{F^{p_i(n-1)}[F^l(j)]\} \\ &\supseteq F^{n-d-l+(h-p_i)(n-1)}N = N, \end{aligned}$$

since $n - d - l + (h - p_i)(n - 1) \geq 0$. $\blacksquare$

(4.10) Corollary Let $A \geq 0$ be an $n \times n$ primitive matrix such that $a_{ij} > 0$ if and only if $a_{ji} > 0$. Then $\gamma(A) \leq 2(n - 1)$.

Proof A^2 has all its diagonal elements positive, so in the upper bound of Theorem 4.9, $d = n$ and $h = 2$. $\blacksquare$

(4.11) Example The primitivity assumption in Theorem 4.9 and its corollary is essential. This is demonstrated by the matrix

$$A = \begin{pmatrix} 0 & 1 \\ 1 & 0 \end{pmatrix}$$

which is cyclic of index 2.

(4.12) Theorem Let $A \geq 0$ be an $n \times n$ primitive matrix and let s be the length of shortest simple circuit in $G(A)$. Then $\gamma(A) \leq h + s(n - 2)$.

Proof We have to show that for every $j \in N$, $F^{n+s(n-2)}(j) = N$. Let $\{p_{j_1},p_{j_2},\ldots,p_{j_s}\}$ be a circuit in $G(A)$ and let m be the smallest nonnegative integer such that $j_i \in F^m(j)$. Then

$$\{j_i\} \subseteq F^s(j_i) \subseteq F^{2s}(j_i) \subseteq \cdots. \tag{4.13}$$

Since A is primitive A^s is irreducible and by Lemma 4.5 the inclusions in (4.13) are proper as long as the sets are proper subsets of N, and thus $F^{(n-1)s}(j_i) = N$. But $F^{(n-1)s}(j_i) \subseteq F^{m+(n-1)s}(j)$, so $F^{m+(n-1)s}(j) = N$. By the definition of m, $m \le n - s$ and thus $\gamma(A) \le n + s(n-2)$. ■

(4.14) Theorem If $A \ge 0$ is an $n \times n$ primitive matrix, then $\gamma(A) \le n^2 - 2n + 2$.

Proof Since A is primitive, $G(A)$ contains at least two simple circuits of different lengths. Thus $s \le n - 1$. ■

(4.15) Example The $n \times n$ matrix

$$A = \begin{bmatrix} 0 & 1 & 0 & \cdots & 0 \\ 0 & 0 & 1 & \cdots & 0 \\ \vdots & \vdots & \vdots & \ddots & \vdots \\ 0 & 0 & 0 & \cdots & 1 \\ 1 & 1 & 0 & \cdots & 0 \end{bmatrix}$$

is primitive and its index of primitivity is $n^2 - 2n + 2$. This shows that the upper bound in Theorem 4.14 is, in general, the best possible.

5 STOCHASTIC MATRICES

Consider n possible states, $s_1,s_2,\ldots,s_n$, of a certain process. Suppose that the probability of the process moving from state s_i to state s_j is time independent and denote this probability by t_{ij}. Such a process is called a *finite homogeneous Markov chain.* These processes will be studied in detail in Chapter 8.

The matrix $T = (t_{ij})$ clearly satisfies

$$t_{ij} \ge 0, \qquad \sum_{j=1}^{n} t_{ij} = 1, \qquad i,j = 1,\ldots,n. \tag{5.1}$$

(5.2) Definition A square matrix T of order n is called (*row*) *stochastic* if it satisfies (5.1) and *doubly stochastic* if, in addition,

$$\sum_{i=1}^{n} t_{ij} = 1, \qquad j = 1,\ldots,n.$$

We shall return to stochastic matrices in Chapter 8 and to doubly stochastic matrices in Chapter 3.

We denote by e the vector all of whose entries are equal to one.

(5.3) Theorem The maximal eigenvalue of a stochastic matrix is one. A nonnegative matrix T is stochastic if and only if e is an eigenvector of T corresponding to the eigenvalue one.

Proof Since (2.36) holds, by the remark in the beginning of Section 3, for any nonnegative matrix, it follows that the spectral radius of a stochastic matrix is one and clearly e is a corresponding eigenvector. Conversely, if $Te = e$, then all the row sums of T are equal to one. ■

The last characterization of stochastic matrices points out that they belong to the class of nonnegative matrices that possess a positive eigenvector that corresponds to the spectral radius, the class which was described in Theorem 3.10. There is a close connection between stochastic matrices and this class.

(5.4) Theorem If $A \geq 0$, $\rho(A) > 0$, $z \gg 0$, and $Az = \rho(A)z$, then $A/\rho(A)$ is similar to a stochastic matrix.

Proof Let D be the diagonal matrix with $d_{ii} = z_i$. Then $P = D^{-1}(A/\rho(A))D$ is stochastic. ■

A linear combination $\sum_{j=1}^{p} \theta_j x_j$ is called a *convex combination* of $x_1, \ldots, x_p$ if the θ_j are nonnegative and $\sum_{j=1}^{p} \theta_j = 1$. The *convex hull* of a set X is the totality of convex combinations of finite sets of elements of X. A point p is an *extreme point* of a convex set C if $p \in C$ but is not a convex combination of other points in C. By a finite-dimensional version of the well-known Krein–Milman theorem, (e.g., Rockafellar [1970]), a bounded convex set is the convex hull of its extreme points. The convex set $\{x; Ax \leq b\}$, where $A \in R^{m \times n}$ and $b \in R^m$, is called a *polyhedron.* A nonempty bounded polyhedron is called a *polytope.* Its extreme points are called *vertices.* The set of row stochastic matrices of order n is clearly a polytope in $R^{n \times n}$. It has n^n vertices, the matrices with exactly one entry 1 in each row. Before we prove the analogous result (of Birkhoff) for Ω_n, the set of all doubly stochastic matrices of order n, we introduce an important theorem of Frobenius and König on the zero–nonzero pattern of a matrix. For this we need the following notation:

If A is an $n \times n$ matrix and σ a permutation on n objects, then the n-tuple $(a_{1\sigma(1)}, a_{2\sigma(2)}, \ldots, a_{n\sigma(n)})$ is called a *diagonal* of A. If k and n are positive integers, $k \leq n$, then $Q_{k,n}$ denotes all $\binom{n}{k}$ increasing sequences $\omega = (\omega_1, \ldots, \omega_k)$, $1 \leq \omega_1 < \omega_2 < \cdots < \omega_k \leq n$. If A is an $m \times n$ matrix, $\alpha \in Q_{h,m}$, and $\beta \in Q_{k,n}$, then $A[\alpha|\beta]$ is the $h \times k$ submatrix of A whose (i,j) entry is $a_{\alpha_i \beta_j}$.

(5.5) Theorem A necessary and sufficient condition that every diagonal of an $n \times n$ matrix A contains a zero is that the matrix contains an $s \times t$ zero submatrix with $s + t = n + 1$.

Proof Let $\alpha \in Q_{s,n}$, $\beta \in Q_{t,n}$, where $s + t = n + 1$. Suppose that $A[\alpha|\beta] = 0$. Then the $s \times n$ matrix $A[\alpha|1, \ldots, n]$ contains at most $n - t = s - 1$ nonzero columns. Hence every diagonal of A must contain at least one zero in rows indexed by α.

Conversely, suppose that A is an $n \times n$ matrix each of whose diagonals contain at least one zero. We use induction on n. If A is the zero matrix there is nothing to prove. Otherwise A must contain a nonzero element a_{hk}. But then every diagonal of the matrix $A(h|k)$, obtained from A by deleting the hth row and the kth column, must contain a zero and by the induction hypothesis $A(h|k)$ contains a $p \times q$ zero submatrix, where $p + q = n$. There exist, therefore, permutation matrices P and Q such that

$$PAQ = \begin{bmatrix} X & 0 \\ Y & Z \end{bmatrix},$$

where X is $p \times p$ and Z is $q \times q$. Clearly every diagonal of PAQ contains a zero. Thus at least one of the two matrices X or Z has the property that every diagonal contains a zero. Suppose that every diagonal of X contains a zero. Then, again by the induction hypothesis, X contains a $u \times v$ zero submatrix where $u + v = p + 1$. But then PAQ contains a $u \times (v + q)$ zero submatrix and

$$u + v + q = p + q + 1 = n + 1.$$

The proof ends in a similar way if every diagonal of Z contains a zero. ∎

With Theorem 5.5 as a lemma we now can prove the following.

(5.6) Theorem The set of all $n \times n$ doubly stochastic matrices, Ω_n, is a convex polyhedron whose vertices are the permutation matrices.

Proof Let $S \in \Omega_n$. We use induction on the number of positive entries in S. If S has exactly n positive entries then S is a permutation matrix so the theorem holds. If S is not a permutation matrix and if it contains an $s \times t$ zero submatrix then $s + t \leq n$. Thus by Theorem 5.5, S contains a diagonal all of whose entries are positive. Let P be the permutation matrix with ones in the positions corresponding to this diagonal and let θ be the least element in that diagonal. Clearly $\theta < 1$ and $T = (1/1 - \theta))(S - \theta P)$ belongs to Ω_n and has less positive entries than S. Hence, by the induction hypothesis, T is a convex combination of permutation matrices and therefore so is $S = \theta P + (1 - \theta)T$. This shows that Ω_n is the convex hull of the permutation matrices. The permutation matrices are the vertices of Ω_n since clearly no permutation matrix is a convex combination of other permutation matrices.

A sharper statement of Theorem 5.6 follows.

(5.7) Exercise Every $S \in \Omega_n$ can be expressed as a convex combination of at most $n^2 - 2n + 2$ permutation matrices (Marcus and Ree [1959]).

We cannot discuss Ω_n without mentioning, at least *en passant*, the celebrated van der Waerden conjecture.

(5.8) Definition The *permanent* of an $n \times n$ matrix A, per A, is defined by

$$\text{per A} = \sum_{\sigma \in S_n} \prod_{i=1}^{n} a_{i\sigma(i)},$$

where S_n is the symmetric group of order n.

(5.9) Example Suppose the n distinct points $c_1, \ldots, c_n$ are connected to one another and that particles q_i are located one at each point c_i $(i = 1, \ldots, n)$. Let the probability that at button push the particle q_i moves from point c_i to point c_j be t_{ij}. Let $T = (t_{ij})$. Then the probability that after the button push there is precisely one particle at each point is per T.

An interesting question is to find bounds for the permanent of doubly stochastic matrices. Obviously,

$$\max_{A \in \Omega_n} \text{per } A = 1,$$

and the maximum is obtained on the permutation matrices which are the vertices of Ω_n. The half century old van der Waerden [1926] conjecture is that

$$\min_{A \in \Omega_n} \text{per } A = \frac{n!}{n^n}$$

and that this minimum is obtained only for the matrix all of whose elements are $1/n$. Hundreds of papers are the offsprings of this very easy looking conjecture which is still unresolved in general. It is known to be true for $n \leq 5$ and for general n if the numerical range of A lies in the angle $-\pi/2n \leq \arg z \leq \pi/2n$.

We conclude the section with an upper bound for eigenvalues of a stochastic matrix which are different from one.

(5.10) Theorem If A is stochastic and $\lambda \neq 1$ is an eigenvalue of A, then

$$|\lambda| \leq \min\left(1 - \sum_j \min_i a_{ij}, \left(\sum_j \max_i a_{ij}\right) - 1\right).$$

Proof Let v be an eigenvector of A^t, corresponding to λ,

$$\lambda v^t = v^t A.$$

From $Ae = e$, it follows that $v^t e = v^t A e = \lambda v^t e$ and thus v and e are orthogonal since $\lambda \neq 1$.

Let $c_j = \min_i a_{ij}$ and consider the matrix $A_c = (a_{ij} - c_j)$. Then

$$\lambda v^t = v^t A = v^t A_c$$

since v and e are orthogonal. Taking absolute values, we have

$$|\lambda|\,|v^t| \leq |v^t| A_c$$

since A_c is nonnegative. By Theorem 1.1

$$|\lambda| \leq \rho(A_c)$$

but all the row sums of A_c are equal to $1 - \sum_{j=1}^{n} c_j$ so

$$\rho(A_c) = 1 - \sum_{j=1}^{n} c_j$$

which proves the first inequality in the theorem. The second is proved in an analogous manner. ■

6 EXERCISES

(6.1) Show that the spectral radius of a nonnegative matrix A is positive if and only if $G(A)$ contains at least one closed path (Ullman [1952]).

(6.2) Prove that if $A \geq 0$ is irreducible then

$$\max_{x \gg 0} \min_{y \gg 0} \frac{y^t A x}{y^t x} = \rho(A) = \min_{y \gg 0} \max_{x \gg 0} \frac{y^t A x}{y^t x}$$

(Birkhoff and Varga [1958]).

(6.3) Let A be a nonnegative matrix of order n and let x be a positive vector. Let $D = \operatorname{diag}\{(Ax_i/x_i\}$. Then the following conditions on the matrix A are equivalent.

(i) A is irreducible.

(ii) $\operatorname{rank}(A - D) = n - 1$ and $A^t z = Dz$ for some positive vector z.

(iii) $\operatorname{rank}(A - D) = n - 1$ and $A^t z = Dz$ for some vector z such that $z_i \neq 0$, $i = 1, \ldots, n$.

(iv) $A^t z = Dz$, $z \neq 0 \rightarrow z_i \neq 0$, $i = 1, \ldots, n$ (Elsner [1976a]).

(6.4) (a) Let x be a positive eigenvector of an irreducible matrix $B \geq 0$. Show that if $A \geq 0$ commutes with B then x is an eigenvector of A.

(b) Prove that the following conditions on $A \geq 0$ are equivalent.

(i) A satisfies the conditions of Theorem 3.14.
(ii) A commutes with a positive matrix.
(iii) A commutes with a nonnegative irreducible matrix.

(6.5) Let A be a square nonnegative matrix. Show that

$$\rho\left(\frac{A + A^{t}}{2}\right) \geq \rho(A)$$

and that equality holds if and only if A and A^{t} have a common eigenvector corresponding to $\rho(A)$ (Levinger [1970]).

(6.6) Let

$$Au = \lambda u \quad \text{where} \quad A \geq 0 \qquad \text{and} \qquad u = \begin{pmatrix} u^{(1)} \\ u^{(2)} \end{pmatrix} \quad \text{where} \quad u^{(1)} \geq 0, \quad u^{(2)} \leq 0.$$

Partition

$$A = \begin{bmatrix} A_{11} & A_{12} \\ A_{21} & A_{22} \end{bmatrix}$$

in conformity with u. Show that $\lambda \leq \min(\rho(A_{11}), \rho(A_{22}))$.

(6.7) (a) Show that if $A \geq 0$ is irreducible then $(\lambda I - A)^{-1}$ is positive for $\lambda > \rho(A)$.

(b) Show that $A \geq 0$ is reducible if and only if $B_{ii}(\rho(A)) = 0$ for some i, where $B(\lambda)$ is the adjoint matrix of A (Corollary 2.13).

(6.8) Let $C(\lambda) = B(\lambda)/D_{n-1}(\lambda)$, where $D_{n-1}(\lambda)$ is the greatest common divisor of the polynomials $B_{ij}(\lambda)$, be the reduced adjoint matrix of A (Theorem 4.1).

(a) Show that if A is nonnegative and $\lambda \geq \rho(A)$, then $C(\lambda)$ is nonnegative.

(b) Show that if in addition A is irreducible then $C(\lambda)$ is positive for $\lambda \geq \rho(A)$.

(c) Complete the proof of Theorem 4.1.

(6.9) Let $A \geq 0$ be an irreducible matrix and let h be the index of cyclicity of A. Show that if some power A^q of A is reducible then A^q is *completely reducible*, i.e., is cogredient to direct sum of d irreducible matrices having the same spectral radius where d is the greatest common divisor of q and h and the spectral radius is $(\rho(A))^q$.

(6.10) Suppose A is a cyclic matrix of index h of the form (2.21).

(a) Show that A^h is the direct sum $B_1 + B_2 + \cdots + B_h$, where $B_j \equiv A_{j,\,j+1}A_{j+1,\,j+2} \cdots A_{j-1,\,j}$ (indices taken modulo h) is a primitive matrix and $\rho(B_j) = (\rho(A))^h$ (Frobenius [1912]).

(b) Moreover, show that each of the matrices B_j has the same nonzero eigenvalues (Sylvester [1883]).

(6.11) Compute the eigenvalues of

$$A = \begin{bmatrix} 0 & B & 0 & 0 \\ 0 & 0 & B & 0 \\ 0 & 0 & 0 & B \\ C & 0 & 0 & 0 \end{bmatrix},$$

where

$$B = \begin{bmatrix} 0 & 1 \\ 1 & 0 \end{bmatrix} \quad \text{and} \quad C = \begin{bmatrix} 6 & 13 \\ 4 & 6 \end{bmatrix}.$$

(6.12) Let A be a 4×4 (0,1) cyclic matrix such that $\sum_{i,j=1}^{4} a_{ij} = 8$. What is the spectral radius of A?

(6.13) (a) Write the triangular block form (3.6) of the matrix

$$A = \begin{bmatrix} 1 & 2 & 0 & 0 & 3 \\ 0 & 0 & 4 & 0 & 0 \\ 0 & 0 & 0 & 2 & 0 \\ 0 & 8 & 0 & 0 & 0 \\ 4 & 0 & 0 & 0 & 0 \end{bmatrix}.$$

(b) Compute $\rho(A)$.

(6.14) Compute a basis of the algebraic eigenspace of the matrix in Example 3.9.

(6.15) Let $H^{(n)}$ be the $n \times n$ (Hilbert) matrix defined by $H_{ij}^{(n)} = 1/(i + j - 1)$. Show that

$$n > m \rightarrow \rho(H^{(n)}) > \rho(H^{(m)}).$$

(6.16) (a) Prove that A is *semiconvergent*, i.e., $\lim_{n\to\infty} A^n$ exists, if and only if $\lambda \in \sigma(A)$ implies that (i) $|\lambda| < 1$ or (ii) $\lambda = 1$ and $\deg \lambda = 1$.

(b) Let $A \geq 0$ be irreducible. Show that $\lim_{n\to\infty}(A/r)^n$ exists if and only if A is primitive.

(6.17) Let P be a stochastic matrix.

(a) Prove that P is semiconvergent if and only if $|\lambda| = 1 \to \lambda = 1$.

(b) Show that in this case $\lim_{n\to\infty} P^n = C(1)/\psi'(1)$ and that if, in addition, $\lambda = 1$ is a simple eigenvalue, then $\lim_{n\to\infty} P^n = B(1)/\Delta'(1)$. Here B, C, Δ, and ψ are the adjoint matrix, the reduced adjoint matrix, the characteristic polynomial, and the minimal polynomial, respectively.

(6.18) Let $A \geq 0$ be irreducible. Prove that $\lim_{m\to\infty}(a_{ij}^{(m)})^{1/m} = \rho(A)$ if and only if A is primitive.

(6.19) Let $A \geq 0$ be an $n \times n$ irreducible matrix and $x^{(0)}$ an arbitrary positive vector. Define $x^{(r)} = Ax^{(r-1)} = A^r x^{(0)}$,

$$\overline{\lambda_r} = \max_{1<i<n} \frac{x_i^{(r+1)}}{x_i^{(r)}} \qquad \text{and } \underline{\lambda_r} = \min_{1<i<n} \frac{x_i^{(r+1)}}{x_i^{(r)}}.$$

(a) Show that

$$\underline{\lambda_0} \leq \underline{\lambda_1} \leq \cdots \leq \underline{\lambda_r} \leq \cdots \leq \rho(A) \leq \cdots \leq \overline{\lambda_r} \cdots \leq \overline{\lambda_1} \leq \overline{\lambda_0}.$$

(b) Prove that the sequences $\{\overline{\lambda_r}\}$ and $\{\underline{\lambda_r}\}$ converge to $\rho(A)$ for an arbitrary initial positive vector $x^{(0)}$ if and only if A is primitive.

(6.20) (a) Prove that a nonnegative irreducible matrix A of order n is primitive if and only if for some q, there is an ordered pair (i,j) such that

$$a_{ij}^{(q)} a_{ij}^{(q+1)} > 0$$

(Lewin [1971a]).

(b) Show that for $n > 4$, q may be taken to be not greater than

$$\frac{(n-2)(n-3)}{2}$$

(Vitek [1975]).

(6.21) A nonnegative matrix is of *doubly stochastic pattern* if there is a doubly stochastic matrix having zeros at precisely the same positions. Show that if A is an $n \times n$ primitive matrix of doubly stochastic pattern then its index of primitivity is

$$\{n^2/4 + 1\} \qquad \text{for} \quad n \equiv 0 \bmod 4 \quad \text{or} \quad n = 5{,}6$$

and

$$\{n^2/4\} \qquad \text{otherwise},$$

where $\{x\}$ denotes the least integer $\geq x$ (Lewin [1974]).

(6.22) A matrix A is *fully indecomposable* if no permutation matrices P and Q exist such that

$$PAQ = \begin{bmatrix} B & 0 \\ C & D \end{bmatrix},$$

where B and C are square submatrices.

(a) Show that A is fully indecomposable if and only if for some permutation matrix P, PA is irreducible and has nonzero entries on the main diagonal (Brualdi *et al.* [1966]).
(b) Show that if $A \geq 0$ is fully indecomposable then it is primitive but that the converse is not true.

(6.23) (a) Find two reducible matrices whose product is positive and two primitive matrices whose product is reducible.
(b) Prove that the product of nonnegative fully indecomposable matrices is fully indecomposable (Lewin [1971b]).

(6.24) Which of the following matrices is fully indecomposable?, primitive?, cyclic? (of what index?), reducible?

$$A_1 = \begin{bmatrix} 0 & 1 \\ 1 & 1 \end{bmatrix}, \quad A_2 = \begin{bmatrix} 1 & 0 \\ 1 & 1 \end{bmatrix},$$

$$A_3 = \begin{bmatrix} 0 & 1 & 0 \\ 0 & 0 & 1 \\ 1 & 0 & 0 \end{bmatrix}, \quad A_4 = \begin{bmatrix} 0 & 0 & 1 \\ 0 & 1 & 0 \\ 1 & 0 & 0 \end{bmatrix}, \quad A_5 = \begin{bmatrix} 0 & 1 & 0 \\ 1 & 1 & 1 \\ 0 & 1 & 0 \end{bmatrix},$$

$$A_6 = \begin{bmatrix} 0 & 1 & 0 & 0 \\ 0 & 0 & 0 & 1 \\ 0 & 0 & 1 & 0 \\ 1 & 0 & 0 & 0 \end{bmatrix}, \quad A_7 = \begin{bmatrix} 1 & 1 & 0 & 0 \\ 0 & 1 & 1 & 0 \\ 0 & 0 & 1 & 1 \\ 1 & 0 & 0 & 1 \end{bmatrix},$$

$$A_8 = \begin{bmatrix} 1 & 0 & 0 & 1 \\ 0 & 1 & 1 & 0 \\ 0 & 1 & 1 & 0 \\ 1 & 0 & 0 & 1 \end{bmatrix}, \quad A_9 = \begin{bmatrix} 0 & 0 & 0 & 1 \\ 0 & 0 & 1 & 1 \\ 1 & 1 & 0 & 0 \\ 1 & 0 & 0 & 0 \end{bmatrix},$$

$$A_{10} = \begin{bmatrix} 0 & 0 & 1 & 0 \\ 0 & 0 & 1 & 1 \\ 1 & 1 & 0 & 0 \\ 0 & 1 & 0 & 0 \end{bmatrix}.$$

(6.25) A matrix is *totally nonnegative* (*totally positive*) if all its minors of any order are nonnegative (positive). A totally nonnegative matrix A is *oscillatory* if there exists a positive number q such that A^q is totally positive.

Show that a totally nonnegative matrix is oscillatory if and only if A is nonsingular and $|i - i| \leq 1 \rightarrow a_{ij} > 0$ (Gantmacher and Krein [1950]).

(6.26) Consider a Jacobi matrix

$$A = \begin{bmatrix} a_1 & b_1 & \cdots & 0 & 0 \\ c_1 & a_2 & b_2 & & 0 \\ \vdots & c_2 & a_3 & & \vdots \\ & & & \ddots & \\ 0 & & & & b_{n-1} \\ 0 & 0 & \cdots & c_{n-1} & a_n \end{bmatrix}.$$

Show that A is totally nonnegative (oscillatory) if all the principal minors of A, the b_i's and the c_i's are nonnegative (positive). (For oscillatory matrices it is enough to check the consecutive principal minors.)

(6.27) Let α and β be disjoint subsets of $N = \{1,2, \ldots ,n\}$ such that $\alpha \cup \beta = N$. Let $A[\alpha|\alpha]$ and $A[\beta|\beta]$ denote the principal submatrices of an $n \times n$ matrix A based on indices in α and β, respectively. Prove (the Hadamard–Fischer inequality) that for a totally nonnegative matrix A

$$\det A \leq \det A[\alpha|\alpha] \det A[\beta|\beta].$$

(Engel and Schneider [1977] and Gantmacher [1959]).

(6.28) (a) Prove that an oscillatory matrix A has n distinct positive eigenvalues

$$\lambda_1 > \lambda_2 > \cdots > \lambda_n.$$

(b) Show that if $u^k = (u_1^k, \ldots ,u_n^k)$ is an eigenvector of A corresponding to λ_k then there are exactly $k - 1$ variations of signs in the coordinates $u_1^k, \ldots ,u_n^k$.

(6.29) Let A be an oscillatory matrix with eigenvalues $\lambda_1 > \lambda_2 > \cdots > \lambda_n$. Let $A(i|i)$ denote the matrix obtained from A by deleting the ith row and the ith column and denote the eigenvalues of $A(i|i)$ by $\lambda_1^{(i)} > \lambda_2^{(i)} > \cdots > \lambda_{n-1}^{(i)}$.

Prove that for $i = 1$ and $i = n$,

$$\lambda_1 > \lambda_1^{(i)} > \lambda_2 > \lambda_2^{(i)} > \cdots > \lambda_{n-1}^{(i)} > \lambda_n$$

and that for every $1 \leq i \leq n$

$$\lambda_1 > \lambda_1^{(i)} > \lambda_2 \qquad \text{and} \qquad \lambda_{n-1}^{(i)} > \lambda_n.$$

(6.30) Prove that if a positive eigenvector corresponds to the spectral radius of a nonnegative matrix A, then the degree of every eigenvalue of A such that $|\lambda| = \rho(A)$ is one. (Hint: Prove it first for stochastic matrices.)

(6.31) Show that if M is a doubly stochastic matrix and $D = \text{diag}\{d_1, \ldots, d_n\}$ is a positive definite diagonal matrix then

$$\rho(DM) \geq \left(\prod_{i=1}^{n} d_i \right)^{1/n}$$

(Friedland and Karlin [1975]).

(6.32) Express

$$A = \begin{pmatrix} a & 1-a & 0 & 0 \\ 0 & a & 1-a & 0 \\ 0 & 0 & b & 1-b \\ 1-a & 0 & a-b & b \end{pmatrix},$$

$1 \geq a \geq b \geq 0$, as a convex combination of permutation matrices.

(6.33) (a) Show that every reducible doubly stochastic matrix is cogredient to a direct sum of irreducible doubly stochastic matrices (Mirsky [1963]).

(b) Show that if an $n \times n$ doubly stochastic matrix is cyclic of index h then h divides n and there exist permutation matrices P and Q such that PAQ is a direct sum of h doubly stochastic matrices of order n/h (Marcus *et al.* [1961]).

(6.34) (a) Prove that if A is a positive square matrix then there is a unique doubly stochastic matrix of the form $D_1 A D_2$ where D_1 and D_2 are positive diagonal matrices.

(b) Show that the matrices D_1 and D_2 are unique up to a scalar factor.

(c) Show that the matrix $D_1 A D_2$ can be obtained as a limit of the sequence of matrices generated by alternately normalizing the rows and columns of A (Sinkhorn [1964]).

(6.35) Let A be a nonnegative matrix of order n.

(a) Prove that a necessary and sufficient condition for the existence of a doubly stochastic matrix $B = D_1 A D_2$ as in Exercise 6.35 is that A has a doubly stochastic pattern, that if B exists then it is unique, and that $\det D_1 D_2 \geq (\rho(A))^{-n}$.

(b) Show that D_1 and D_2 are unique up to a scalar multiplication if and only if A is fully indecomposable.

(c) Prove that a necessary and sufficient condition that the iteration process described in Exercise 6.35 will converge to a doubly stochastic matrix is that A contains a positive diagonal (Sinkhorn and Knopp [1967]; London [1971]).

(6.36) Let $x = (x_i)$ and $y = (y_i)$ be vectors in R^n satisfying

$$x_1 \geq x_2 \geq \cdots \geq x_n, \qquad y_1 \geq y_2 \geq \cdots \geq y_n.$$

Show that the following conditions on x and y are equivalent:

(i) $y = Ax$ for a doubly stochastic matrix A.
(ii) $\sum_{i=1}^{k} x_i \geq \sum_{i=1}^{k} y_i$, $k = 1,2, \ldots ,n - 1$, and $\sum_{i=1}^{n} x_i = \sum_{i=1}^{n} y_i$.
(iii) $\sum_{i=1}^{n} \phi(x_i) \geq \sum_{i=1}^{n} \phi(y_i)$, for all continuous convex functions ϕ (Hardy *et al.* [1952]).
(iv) There exists a symmetric matrix with eigenvalues $x_1, \ldots ,x_n$ and diagonal elements $y_1, \ldots , y_n$ (Horn [1954], Mirsky [1964]).

(6.37) For $A \gg 0$, let $K = \max(a_{ij}a_{kl}a_{il}^{-1}a_{kj}^{-1})^{1/2}$, $M = \max_{i,j} a_{ij}$ and $m = \min_{i,j}, a_{ij}$. Prove that if $\lambda \neq \rho(A)$ is an eigenvalue of A, then

$$|\lambda| \leq (K - 1)/(K + 1)\rho(A) \leq (M - m)/(M + m)\rho(A).$$

Check that for

$$A = \begin{pmatrix} 2 & 1 \\ 1 & 2 \end{pmatrix}, \qquad \lambda_2 = 1 = \tfrac{1}{3}\rho(A)$$

(Hopt [1963], Ostrowski [1963]).

7 NOTES

(7.1) Part (a) of Theorem 1.4 is due to Perron [1907] and part (b) to Frobenius [1912]. There are many proofs of various parts of the theorem. In addition to the references given in Chapter 1 we mention Wielandt [1950], Debreau and Herstein [1953], Brauer [1957b], Fan [1958], Bellman [1960], and Pullman [1971].

(7.2) In a great part of the chapter we follow Gantmacher [1959]. In emphasizing the use of graph theory we follow Varga [1962].

The block triangular form (3.6) is slightly different from the normal form of a reducible matrix as defined by Gantmacher and Varga.

(7.3) The concepts of irreducibility and of full indecomposability and some of the basic results of the chapter, including Theorem 1.4(b) and its corollaries and Theorems 2.20, 2.27, 2.35, and 5.5, are due to Frobenius [1909, 1912].

In a very interesting survey by Schneider [1977], an irreducibility-type condition introduced by Markov [1908] is analyzed and compared with Frobenius' definition of irreducibility.

Theorem 5.5 was reproved by König using graph theory. Schneider's survey contains interesting remarks on Frobenius' [1917] criticism of Konig's work and of the use of graph theory. No doubt that attitudes have changed since 1917.

(7.4) The term indecomposable is also used in the literature, e.g., Marcus and Minc [1964], for irreducible. The following measurement of irreducibility and of full indecomposability were suggested by Hartfiel, e.g., Hartfiel [1975]. Let $A \geq 0$ be $n \times n$ and let

$$u_k(A) = \min_{\substack{R \cap C = \phi \\ |R| + |C| = n-k}} \left(\max_{i \in R,\, j \in C} a_{ij} \right)$$

and

$$U_k(A) = \min_{|R| + |C| = n-k} \left(\max_{i \in R,\, j \in C} a_{ij} \right).$$

Hartfiel shows that $u_k(A)$ is positive if and only if $U_k(A + D)$ is positive for every positive definite diagonal matrix D. Notice that $u_0(A) > 0$ means that A is irreducible and that $U_0(A) > 0$ means that A is fully indecomposable.

A similar measurement of irreducibility was introduced by Fiedler, e.g., Fiedler [1972].

(7.5) Theorem 2.14 and the proof of Theorem 2.20 are due to Wielandt [1950]. The function r_x, used in this proof, and the inequalities (2.11) and (2.12) go back to Collatz [1942] and is known as the Collatz–Wielandt function.

Bounds for the difference of maximal eigenvalues of two irreducible matrices, one of which dominates the other, are given in Marcus *et al.* [1961].

(7.6) Improvements of the bounds in Theorem 2.35 include Lederman [1950], Ostrowski [1952], Brauer [1957a], Ostrowski and Schneider [1960], Ostrowski [1960/61], Lynn and Timlake [1969], Minc [1970], and de Oliveira [1971]. In many practical problems the improvements are not much better than the bounds of Frobenius but are much more complicated computationally.

(7.7) The term imprimitive is used, in Gantmacher [1959] and by other authors, for what we called cyclic. The graph theoretic term cyclic was introduced by Romanovsky [1936] who also proved Theorem 2.30.

(7.8) In studying the index of primitivity we used the technique of Holladay and Varga [1958]. A rough bound for $\gamma(A)$, $2n^2 - 2n$, appears already in Frobenius [1912, p. 463]. The sharp, in general, bound of Theorem 4.14 and the example that follows are Wielandt's (Wielandt [1950]).

The question of bounding $\gamma(A)$ for special classes of primitive matrices is related to the following Diophantine problem of Frobenius:

Let $a_0 < a_1 < \cdots < a_s$ be positive relatively prime integers. Determine the smallest integer $\phi(a_0, a_1, \ldots, a_s)$ such that every integer N, $N \geq \phi(a_0, a_1, \ldots, a_s)$ is expressible as $\sum_{i=0}^{s} \alpha_i a_i$ where α_i are nonnegative integers.

This problem and its application to the study of $\gamma(A)$ are discussed in the Ph.D. dissertation of Vitek [1977].

(7.9) Section 3 is mostly based on Rothblum [1975]. The notation is motivated by the application to Markov chains. However, Theorems 3.1, 3.10, and 3.14 are taken from Gantmacher [1959].

(7.10) In Chapter 5 we shall speak about the index of a square matrix (not to be confused with index of cyclicity or index of primitivity). In terms of this concept, degree A = index $(\rho(A)I - A)$.

(7.11) The proof of Theorem 2.33 is Pullman's (Pullman [1974]). The theorem itself as well as more general versions of it are due to Minc [1974a,b].

Another interesting work of Minc, which was not mentioned in the text, describes the linear transformations which map nonnegative matrices into nonnegative matrices and preserves the spectra of each nonnegative matrix (Minc [1974c]).

(7.12) The set of all points in the unit circle which are eigenvalues of some doubly stochastic matrix were characterized partly by Dmitriev and Dynkin [1945] and completely by Karpelevich [1951]. Barker and Turner [1973] considered a similar problem by extending the concept of a stochastic matrix to matrices in $\pi(K)$.

Stochastic matrices with real eigenvalues were studied by Suleimanova [1949] who found sufficient conditions for n given real numbers to be the eigenvalues of a stochastic matrix. We shall return to this problem in Chapter 4.

(7.13) The original proof of Theorem 5.6 (Birkhoff [1946]) is based on a theorem of Hall [1935] on the number of systems of distinct representatives, and is a slight extension of the Frobenius–König theorem (Theorem 5.5). The proof in the text is taken from Marcus and Minc [1964].

Marcus, *et al.* [1961] study the number of permutation matrices needed to describe a doubly stochastic matrix as a convex combination of permutation matrices. Of the many generalizations of Theorem 5.6 we shall mention only Mirsky [1963] and Cruse [1975a]. We shall return to the latter in Chapter 4.

(7.14) Example 5.9 due to L. H. Harper is borrowed from Marcus and Minc [1964]. The van der Waerden conjecture was proved for $n \leq 5$ by Marcus and Newman [1959], Eberlein and Mudholkar [1969], and Eberlein [1969], and for matrices whose numerical range lies in $-\pi/2n \leq \arg z \leq \pi/2n$ by Friedland [1974]. An encyclopedic reference on permanents and on the history of the conjecture is Minc [1978].

(7.15) Additional works on the diagonal equivalence of a nonnegative matrix to a doubly stochastic one, (Exercises 6.34 and 6.35), include Maxfield and Minc [1962], Brualdi, *et al.* [1966], Menon [1968], and Djokovic [1970].

(7.16) The theorem in Exercise 6.37 (i)–(iii) was first proved by Hardy *et al.* [1929]. Implications (ii) → (iii) is also known as Karamata inequality (Karamata [1932]). Implication (ii) → (i), which is perhaps one of the first results on doubly stochastic matrices, was proved by Schur [1923]. For generalizations of and references on the theorem see Mirsky [1963], Beckenbach and Bellman [1971], and Fischer and Holbrook [1977].

(7.17) References for totally nonnegative matrices include Gantmacher and Krien [1950] and Karlin [1968].

These matrices share many properties with hermitian positive semidefine matrices and with M-matrices, including the Hadamard–Fischer inequality described in Exercise 6.27. See Chapter 6 and Engel and Schneider [1977].

CHAPTER 3

SEMIGROUPS OF NONNEGATIVE MATRICES

1 INTRODUCTION

Since matrix multiplication is associative and the product of two nonnegative matrices is again a nonnegative matrix, the set $\pi(R^n_+)$ of all $n \times n$ nonnegative matrices forms a multiplicative semigroup. The usual semigroup notation, $\mathcal{N}_n$, will be used here to denote $\pi(R^n_+)$. This chapter is devoted to an examination of the algebraic properties of $\mathcal{N}_n$.

In Section 2 some useful ideas from the algebraic theory of semigroups are given. A canonical form for nonnegative idempotent matrices is given in Section 3 and special types of idempotent matrices are considered. The remaining sections consider certain algebraic properties of the semigroup $\mathcal{N}_n$ and its subsemigroup $\mathcal{D}_n$ of doubly stochastic matrices. Particular attention is given to characterizations of the Green's relations on these semigroups and to the characterizations of their maximal subgroups. In the process, nonnegative matrix equations and matrix factorization are considered.

This material is important in our development of the theory and applications of nonnegative matrices in several respects. First, we shall be concerned with convergent and semiconvergent sequences of powers of matrices in $\mathcal{N}_n$ in studying M-matrices (see Chapter 6) and in investigating certain iterative methods for solving associated systems of linear equations (see Chapter 7). Also, powers of stochastic and doubly stochastic matrices in $\mathcal{D}_n$ will be studied in detail, relative to the theory of finite Markov chains (see Chapter 8). Perhaps the most important applications of the material in this chapter involve the solvability of certain nonnegative matrix equations arising in the areas of mathematical economics and mathematical programming (see Chapters 9 and 10). But before turning to the algebraic theory of semigroups of nonnegative matrices, it will be convenient to develop some general definitions and notation.

2 ALGEBRAIC SEMIGROUPS

Let T denote a multiplicative semigroup and let $a,b \in T$. Then the binary relation $\mathscr{R}[\mathscr{L},\mathscr{J}]$ is defined on T by the rule $a\mathscr{R}b[a\mathscr{L}b,a\mathscr{J}b]$ if and only if a and b generate the same principal right [left, two-sided] ideal in T. The binary relation $\mathscr{H}$ is defined to be $\mathscr{L} \cap \mathscr{R}$. Then each of $\mathscr{R}$, $\mathscr{L}$, and $\mathscr{H}$ are equivalence relations on T. The intersection of all the equivalence relations on T containing $\mathscr{L} \cup \mathscr{R}$ is denoted by $\mathscr{D}$. If the semigroup T forms a compact set under some topology, then $\mathscr{D} = \mathscr{J}$ on T. These are known as the *Green's relations* on T and they play a fundamental role in the study of the algebraic structure of semigroups (see Clifford and Preston [1961, Chapter II] for a complete discussion).

In the case where T contains an identity element, as in the case of $\mathscr{N}_n$, the following simple equations establish the Green's relations. Let $a,b \in T$. Then

$$a\mathscr{R}b \Leftrightarrow a = bx, \quad b = ay \qquad \text{for some} \quad x, y \in T,$$

$$a\mathscr{L}b \Leftrightarrow a = xb, \quad b = ya \qquad \text{for some} \quad x, y \in T,$$

$$a\mathscr{J}b \Leftrightarrow a = x_1bx_2, \quad b = y_1ay_2 \qquad \text{for some} \quad x_1,x_2,y_1,y_2 \in T,$$

$$a\mathscr{H}b \Leftrightarrow a\mathscr{R}b \quad \text{and} \quad a\mathscr{L}b,$$

$$a\mathscr{D}b \Leftrightarrow a\mathscr{R}c \quad \text{and} \quad c\mathscr{L}b, \qquad \text{for some} \quad c \in T.$$

Of additional interest is the concept of a generalized inverse in a semigroup T. An element $a \in T$ is said to be *regular* in T if $a = axa$ is solvable for some $x \in T$. If in addition $x = xax$, then a and x are said to be *semi-inverses* of each other. Notice that if $a = axa$ then ax and xa are idempotent elements, that is $(ax)^2 = ax$ and $(xa)^2 = xa$; also, a and xax are semi-inverses. It can be shown that if one element in an equivalence class D of the relation $\mathscr{D}$ is regular then each element of D is regular. Such a class is called a *regular $\mathscr{D}$-class.* Moreover, every $\mathscr{L}$- and $\mathscr{R}$-class contained in a regular $\mathscr{D}$-class D contains an idempotent. (See Exercises 6.1 and 6.3.)

A subgroup G of a semigroup T is called a maximal subgroup of T if it is not properly contained in any other subgroup of T. Associated with each regular $\mathscr{D}$-class D of T there is a maximal subgroup G of T and G is isomorphic to each $\mathscr{H}$-class of T contained in D that contains an idempotent (see Exercises 6.4 and 6.5). Moreover, every maximal subgroup of T is obtained in this way. Idempotent elements then play a fundamental role in algebraic semigroup theory.

3 NONNEGATIVE IDEMPOTENTS

A canonical form for arbitrary idempotents in $\mathscr{N}_n$ is given first.

(3.1) Theorem Let E be a nonnegative idempotent matrix of rank k in $\mathcal{N}_n$. Then there exists a permutation matrix P such that

$$PEP^t = \begin{bmatrix} J & JU & 0 & 0 \\ 0 & 0 & 0 & 0 \\ VJ & VJU & 0 & 0 \\ 0 & 0 & 0 & 0 \end{bmatrix}, \tag{3.2}$$

$$J = \begin{bmatrix} J_1 & & & 0 \\ & J_2 & & \\ & & \ddots & \\ 0 & & & J_k \end{bmatrix}, \tag{3.3}$$

where the J_i are positive idempotent matrices of rank 1; that is, $J_i = x^i(y^i)^t$, where $x^i \gg 0$, $y^i \gg 0$, and $(y^i)^t x^i = 1$. Conversely, every matrix of the form (3.2) where J is given by (3.3), while U and V are arbitrary nonnegative matrices of appropriate sizes, is idempotent and of rank k.

Proof Recall that the change from E to PEP^t means that a certain permutation is simultaneously performed on the rows and columns of E. This operation is an isomorphism for multiplicative semigroups of matrices and thus it preserves idempotents.

Assume first that E has no zero rows and no zero columns. Let e_i denote the ith unit vector for $i = 1,2, \ldots ,n$ and let $E_1,E_2, \ldots ,E_k$ denote the edges of the polyhedral cone K generated by the columns of E. Let M_j denote the collection of unit vectors mapped into the edge E_j for $j = 1,2, \ldots ,k$. That is, M_j is the collection of all unit vectors e_i such that the ith column of E belongs to the edge E_j of K. Let S_j denote the subspace spanned by M_j. Then R^n is the vector space direct sum

$$R^n = S_1 \oplus S_2 \oplus \cdots \oplus S_k.$$

Furthermore ES_j is the edge E_j, so that the restriction of E to S_j has rank 1 for every j. By rearranging the coordinates in such a way that the unit vectors belonging to each S_j are grouped together, we find a matrix

$$PEP^t = J = \begin{bmatrix} J_1 & & & 0 \\ & J_2 & & \\ & & \ddots & \\ 0 & & & J_k \end{bmatrix},$$

where the J_i are idempotents of rank 1. This proves the theorem for the particular case indicated above.

In the general case we begin by grouping the indices $i = 1,2, \ldots ,n$ in four sets according to whether the ith row and column of E are both zero, or the ith row is zero but the ith column is not, and so on. By simultaneously rearranging rows and columns, we find a matrix

$$P_1EP_1^t = E_1 = \begin{bmatrix} A & B & 0 & 0 \\ 0 & 0 & 0 & 0 \\ C & D & 0 & 0 \\ 0 & 0 & 0 & 0 \end{bmatrix},$$

where P_1 is a permutation matrix, and A,B,C are such that A and B have no zero rows in common, and A and C have no zero columns in common. Moreover since E_1 is idempotent, $A^2 = A$, $AB = B$, $CA = C$, and $CB = D$. Since A and $B = AB$ have no zero rows in common, A cannot have a zero row. Similarly, no column of A is zero. By the first part of the proof, a further permutation performed simultaneously on the rows and columns of A produces a matrix $P_2AP_2^t = J$ of the form (3.3). Then setting $U = P_2B$, $V = CP_2^t$ we have $U = JU$, $V = VJ$, and $VJU = VU = BC = D$. Then

$$\begin{bmatrix} P_2 & 0 \\ 0 & 0 \end{bmatrix} P_1EP_1^t \begin{bmatrix} P_2^t & 0 \\ 0 & 0 \end{bmatrix} = PEP^t$$

has the form given by (3.2). ∎

Later, we shall need cannonical forms for some special nonnegative idempotent matrices. The symmetric case is described first. Clearly, if E is a symmetric idempotent, then the matrices U and V in the cannonical form PEP^t given by (3.2) must be zero or missing. This gives the following.

(3.4) Corollary A nonnegative symmetric matrix E of rank k is idempotent if and only if there exists a permutation matrix P such that

$$PEP^t = \begin{bmatrix} J & 0 \\ 0 & 0 \end{bmatrix}, \tag{3.5}$$

where J is given by (3.3) with each $J_i = x^i(x^i)^t$, $x_i \gg 0$, $(x^i)^tx^i = 1$.

The form (3.2) also simplifies somewhat whenever E is stochastic, for then E can have no zero rows.

(3.6) Corollary A stochastic matrix E of rank k is idempotent if and only if there exists a permutation matrix P such that

$$PEP^t = \begin{bmatrix} J & 0 \\ VJ & 0 \end{bmatrix}, \tag{3.7}$$

where J has the form (3.3) with each $J_i = e(y^i)^t$ where $e = (1,1,\ldots,1)^t$, $y^i \gg 0$, and $(y^i)^te = 1$ and where V is a rectangular stochastic matrix of the appropriate size.

The doubly stochastic case is especially simple, since E can have no zero rows or columns and since a doubly stochastic, idempotent, rank 1 matrix or order n is the matrix $(1/n)$.

(3.8) Corollary A doubly stochastic matrix E of rank k is idempotent if and only if there exists a permutation matrix P such that $J = PEP^t$ has the form (3.3), where if J_i is $n_i \times n_i$, then each entry of J_i is $1/n_i$.

4 THE SEMIGROUP $\mathcal{N}_n$

Regularity in $\mathcal{N}_n$ will be investigated first. The primary tool here is the vector space concept of rank.

Clearly not every matrix in the semigroup $\mathcal{N}_n$ of nonnegative matrices is regular (see Exercise 6.9). An $n \times n$ matrix A of rank r will be called *r-monomial* if each column of A contains at most one nonzero entry. If $r = n$ then A will be called *monomial.* The only regular nonsingular matrices in $\mathcal{N}_n$ are the nonnegative monomial matrices.

Now let I_r denote the identity matrix of order r. If A is an $r \times n$ $(n \times r)$ matrix of rank r, then any solution X to $AX = I_r$ $(XA = I_r)$ is called a *right* (left) *inverse* of A. For an $n \times n$ matrix A of rank r, there exist $n \times r$ and $r \times n$ matrices B and G, respectively, such that

$$A = BG. \tag{4.1}$$

In this case (4.1) is called a *rank factorization* of A. If B and G are nonnegative, then (4.1) will be called a *nonnegative rank factorization.* It follows that if B_L is any left inverse of B and G_R is any right inverse of G, then

$$X = G_R B_L$$

is a semi-inverse of A. Conversely, every semi-inverse of A is obtained in this way. This leads to the following lemma.

(4.2) Lemma Let $A \in \mathcal{N}_n$ have rank r and suppose that $A = BG$ is a nonnegative rank factorization. Then A is regular in $\mathcal{N}_n$ if and only if B and G have nonnegative left and right inverses, respectively.

Proof If A is regular and X is a semi-inverse of A in $\mathcal{N}_n$, then $X = G_R B_L$ for some B_L and G_R. In this case $B_L = I_r B_L = G G_R B_L = GX$, which is nonnegative. Similarly, G_R is nonnegative. The converse is immediate. ■

In order to investigate regularity in $\mathcal{N}_n$, it is thus important to know when a nonnegative matrix has a nonnegative right (left) inverse.

(4.3) Lemma Let A be an $r \times n$ nonnegative matrix of rank r. Then A has a nonnegative right inverse if and only if A has a monomial submatrix of order r. In this case A has a nonnegative right inverse with r nonzero entries.

Proof Suppose that X is an $n \times r$ nonnegative matrix such that $AX = I_r$. That is,

$$\sum_{k=1}^{n} a_{ik} a_{kj} = \begin{cases} 0 & \text{if } i \neq j, \\ 1 & \text{if } i = j, \end{cases}$$

for each i,j where $1 \leq i, j \leq r$. Then for each i there exists some k such that $a_{ik} \neq 0$ and $a_{jk} = 0$ for $j \neq i$, $1 \leq j \leq r$. That is, the kth column of A has exactly one nonzero entry and that entry is in the ith row. Since A has rank r and since r columns of A have precisely one nonzero entry, A has a monomial submatrix of order r.

For the converse let P be a permutation matrix of order n such that $AP = (B,C)$ where B is monomial of order r and let

$$Y = \begin{bmatrix} B^{-1} \\ 0 \end{bmatrix}.$$

Then $X = PY$ has the desired properties. ∎

Notice that a result dual to Lemma 4.3 can be stated for left inverses.

Now any nonnegative rank 1 matrix A has a nonnegative rank factorization. In particular there exist nonnegative column n-vectors x and y such that $A = xy^t$. Suppose that A has rank r and has the partitioned row block form

$$A = \begin{bmatrix} H_1 \\ \vdots \\ H_r \\ 0 \end{bmatrix}, \tag{4.4}$$

where each H_i has rank 1 and where the zero block may not appear. Then for A nonnegative there exist nonnegative vectors x^i and y^i such that $H_i = x^i(y^i)^t$, and A has the nonnegative rank factorization

$$A = \begin{bmatrix} x^1 & 0 & \cdots & 0 \\ 0 & x^2 & & \\ \vdots & & \ddots & \vdots \\ 0 & & & x^r \\ \hline & & 0 & \end{bmatrix} (y^1, y^2, \ldots, y^r)^t. \tag{4.5}$$

The main result in this section is given next.

(4.6) Theorem Let A be an $n \times n$ nonnegative matrix of rank r. Then the following statements are equivalent.

(i) A is regular in $\mathcal{N}_n$.
(ii) A has a semi-inverse in $\mathcal{N}_n$ of the form $D_1 A^t D_2$ where D_1 and D_2 are nonnegative diagonal matrices.
(iii) A has a semi-inverse in $\mathcal{N}_n$ which is r-monomial.
(iv) A has a monomial submatrix of order r.

Proof Assume that A is regular in $\mathcal{N}_n$ and let X be a nonnegative semi-inverse of A. For $E = AX$, choose a permutation matrix P such that $K = PEP^t$ has the form (3.2). Then $Y = XP^t$ is a semi-inverse of $C = PA$ in $\mathcal{N}_n$. Next, partition C into the row block form

$$C = \begin{bmatrix} H \\ L \\ M \\ N \end{bmatrix}$$

corresponding to the row block form of K. Now $KC = PEP^tPA = PEA = PA = C$ and thus $JH = H$, $L = 0$, $VJH = VH = M$, and $N = 0$. Thus C has the form

$$C = \begin{bmatrix} H \\ 0 \\ VH \\ 0 \end{bmatrix}. \tag{4.7}$$

Partition the matrix H into the row block form

$$H = \begin{bmatrix} H_1 \\ \vdots \\ H_r \end{bmatrix}$$

corresponding to the partitioned form of J given in (3.2). From $JH = H$ it follows that $J_i H_i = H_i$ for each i. Thus H_i has rank 1. Then H has a nonnegative rank factorization $H = B_1 G$. Let

$$B_2 = \begin{bmatrix} B_1 \\ 0 \\ VB_1 \\ 0 \end{bmatrix}.$$

Then $C = B_2 G$ is a nonnegative rank factorization of C. Moreover for $B = P^t_2 B_2$, $A = BG$ is a nonnegative rank factorization of A. Then by Lemma 4.2,

B and G have nonnegative left and right inverses, respectively. By Lemma 4.3 and a dual result on left inverses, B_L and G_R can be chosen to have exactly r nonzero entries. For this choice, the matrices $D_1 = G_R G_R^t$ and $D_2 = B_L^t B_L$ are $n \times n$ nonnegative diagonal matrices and

$$\begin{aligned} D_1 A^t D_2 &= G_R G_R^t (BG)^t B_L^t B_L = G_R (G G_R)^t (B_L B)^t B_L \\ &= G_R I_r^2 B_L = G_R B_L, \end{aligned}$$

so that $D_1 A^t D_2$ is a nonnegative semi-inverse of A. This establishes (iii). Since in this case $G_R B_L$ is of r-monomial type, (i) also implies (iii).

Next assume (iii) holds and let X be an r-monomial semi-inverse of A. Then there exist permutation matrices P and Q so that $Y = PXQ$ has block form

$$Y = PXQ = \begin{bmatrix} M & 0 \\ 0 & 0 \end{bmatrix}, \tag{4.8}$$

where M is an $r \times r$ monomial matrix. Then Y is a semi-inverse of $B = Q^t A P^t$ in $\mathcal{N}_n$ so that B has the block form

$$B = \begin{bmatrix} B_1 & B_2 \\ B_3 & B_4 \end{bmatrix},$$

where $B_1 = M^{-1}$ and $B_4 = B_3 M B_2$. Thus $A = QBP$ has a monomial submatrix of order r, establishing (iv). The proof that (iv) implies (iii) is obtained by retracing these steps. Since statements (ii) and (iii) each imply (i) trivially, the proof of the theorem is complete. ■

Regular $\mathcal{D}$-classes and maximal subgroups of $\mathcal{N}_n$ are described next. The $\mathcal{D}$-class containing the zero matrix 0 in $\mathcal{N}_n$ is $\{0\}$ while the $\mathcal{D}$-class containing I_n consists of the group of all monomial matrices in $\mathcal{N}_n$. The following result gives a complete description of all the regular $\mathcal{D}$-classes. It will be used to establish the maximal subgroup characterization.

(4.9) Theorem Let $A \in \mathcal{N}_n$ be regular of rank r and let D denote the $\mathcal{D}$-class containing A. Then D contains the canonical idempotent

$$E = \begin{bmatrix} I_r & 0 \\ 0 & 0 \end{bmatrix}. \tag{4.10}$$

Moreover, D consists of all members of $\mathcal{N}_n$ of rank r containing a monomial submatrix of order r.

Proof Since A is regular, it has an r-monomial semi-inverse X in $\mathcal{N}_n$ by Theorem 4.6. It will follow from Exercise 6.3 that $X \in D$. Let P and Q be permutation matrices such that $Y = PXQ$ has the form (4.8), where M is monomial of order r. Then $Y \in D$ so that

$$Z = \begin{bmatrix} M^{-1} & 0 \\ 0 & 0 \end{bmatrix}$$

is a semi-inverse of Y in D. Thus the canonical idempotent $E = YZ$ belongs to D. By this argument, D contains all regular members of $\mathcal{N}_n$ having rank r. By Theorem 4.6 these matrices have a monomial submatrix of order r. Clearly each member of D has this property. ■

Notice that by Theorem 4.9, $\mathcal{N}_n$ has exactly $n + 1$ regular $\mathcal{D}$-classes. Moreover the maximal subgroup of $\mathcal{N}_n$ associated with the $\mathcal{D}$-class D is isomorphic to the $\mathcal{H}$-class, H, containing E. (See Clifford and Preston [1961, Chapter II].) This establishes the following corollary.

(4.11) Corollary The maximal subgroups of $\mathcal{N}_n$ are isomorphic to the complete monomial groups of degree r over the reals, $0 \leq r \leq n$.

We now turn to a description of the Green's relations on the semigroup $\mathcal{N}_n$. The fact that $\mathcal{D} = \mathcal{J}$ on the regular elements is established first, although $\mathcal{N}_n$ is not a compact topological semigroup. We note also that for $A,B \in \mathcal{N}_n$, $A\mathcal{L}B$ if and only if $A^t\mathcal{R}B^t$ so that only one of $\mathcal{L}$ and $\mathcal{R}$ needs to be investigated.

(4.12) Theorem If A,B in $\mathcal{N}_n$ are regular, then $A\mathcal{D}B$ if and only if $A\mathcal{J}B$.

Proof Since $\mathcal{D} \subseteq \mathcal{J}$ in any semigroup, $A\mathcal{D}B$ only if $A\mathcal{J}B$. Conversely, if $A\mathcal{J}B$ then the equations $A = X_1BY_1$ and $B = X_2AY_2$ are solvable for X_1, Y_1 and X_2, Y_2 in $\mathcal{N}_n$. Thus, A and B have the same rank and hence by Theorem 4.9, $A\mathcal{D}B$. ■

From these ideas, it is seen that the tool to be used to characterize Green's relations for regular elements in $\mathcal{N}_n$ would be that of rank. However, this tool is more of a vector space notion and is too sophiscated to characterize Green's relations on the entire semigroup $\mathcal{N}_n$. Here, a tool more concerned with polyhedral cones is necessitated. For this we define a finite nonempty set of vectors $S \subseteq R^n_+$ to be *cone independent* if no vector in S lies in the polyhedral cone generated by the other vectors in S. For $A \in \mathcal{N}_n$, we define a number $d(A)$ by

$$d(A) = \textit{maximum number of cone independent columns of } A.$$

It is immediate that $d(A) \geq \text{rank } A$. Now in general $d(A^t) \neq d(A) \neq \text{rank } A$. Also, for any permutation matrices P and Q, if $B = PAQ$ then $d(B) = d(A)$. Moreover for any $A,B \in \mathcal{N}_n$, $d(AB) \leq d(A)$.

The characterizations of the Green's relations on $\mathcal{N}_n$ are based upon the following lemmas.

(4.13) Lemma Let A, B be in $\mathcal{N}_n$.

(i) If $A\mathcal{R}B$ then $d(A) = d(B)$.
(ii) If $A\mathcal{D}B$ then $d(A) = d(B)$ and $d(A^t) = d(B^t)$.

Proof Note that (i) follows from the definition of $\mathcal{R}$. For (ii), suppose $A\mathcal{R}C$ and $C\mathcal{L}B$. Then $d(A) = d(C)$ from (i). Since $C\mathcal{L}B$, $XC = B$ and $YB = C$ for some X,Y in $\mathcal{N}_n$. But then, $d(B) \leq d(C)$ and $d(C) \leq d(B)$ and consequently $d(A) = d(B)$. Finally, as $A\mathcal{D}B$ if and only if $A^t\mathcal{D}B^t$, $d(A^t) = d(B^t)$. ■

(4.14) Lemma Let A be in $\mathcal{N}_n$ with $d(A) = c$. If A' is any $n \times c$ submatrix of A such that $d(A') = c$ then $A\mathcal{R}[A'0]$. If further, $d(A^t) = r$ and A' is any $r \times c$ submatrix of A with r cone independent rows and c cone independent columns of A then

$$A\mathcal{D}\begin{bmatrix} A'' & 0 \\ 0 & 0 \end{bmatrix}.$$

Proof Without loss of generality suppose the c cone independent columns are in columns $1, \ldots, c$; i.e., $A = [A'A_2]$, where A' is $n \times c$. It is easily verified that $A\mathcal{R}[A'0]$. If further, $d(A^t) = r$ then again without loss of generality, we assume the cone independent columns are columns $1, \ldots, r$. Hence

$$A = \begin{bmatrix} A'' & A_2 \\ A_3 & A_4 \end{bmatrix},$$

where A'' is $r \times c$. As stated previously,

$$A\mathcal{R}\begin{bmatrix} A'' & 0 \\ A_3 & 0 \end{bmatrix} \quad \text{and} \quad \begin{bmatrix} A'' & 0 \\ A_3 & 0 \end{bmatrix}\mathcal{L}\begin{bmatrix} A'' & 0 \\ 0 & 0 \end{bmatrix}.$$

Hence

$$A\mathcal{D}\begin{bmatrix} A'' & 0 \\ 0 & 0 \end{bmatrix}. \quad ■$$

Based on these results, our characterizations of the Green's relations $\mathcal{R}$, $\mathcal{L}$, and $\mathcal{D}$ on $\mathcal{N}_n$ now follows.

(4.15) Theorem Let A,B be in $\mathcal{N}_n$. The following statements are equivalent.

(a) $A\mathcal{R}B$.
(b) (i) $d(A) = d(B) = d$ and (ii) given any $n \times d$ submatrix of cone independent columns of A, say, A', and any $n \times d$ submatrix of cone independent columns of B, say, B' then there is a $d \times d$ monomial matrix X so that $A'X = B'$.

Proof Suppose $d(A) = d$. Let A' be any submatrix of d cone independent columns of A. Now $A\mathscr{R}[A'0]$ by Lemma 4.14. Similarly, if B' is any submatrix of d cone independent columns of B, then $B\mathscr{R}[B'0]$.

Now if $A\mathscr{R}B$, then $d(A) = d(B)$ by Lemma 4.13. Further, from the previous remarks, $A'\mathscr{R}B'$; i.e., $A'X = B'$ and $B'Y = A'$ hold for some X and Y in $\mathcal{N}_d$. Hence $A'(XY) = A'$ and so $XY = I$ from which it follows that X and Y are monomials. Thus, (b) is obtained.

Conversely, if (b) holds, $A'\mathscr{R}B'$. Thus $[A'0]\mathscr{R}[B'0]$ where $[A'0]$ and $[B'0]$ are in $\mathcal{N}_n$. As $A\mathscr{R}[A'0]$ and $B\mathscr{R}[B'0]$, (a) follows. ■

(4.16) Theorem Let A,B be in $\mathcal{N}_n$. The following statements are equivalent

(a) $A\mathscr{D}B$.

(b) (i) $d(A) = d(B) = c$, $d(A^t) = d(B^t) = r$ and (ii) given any $r \times c$ submatrix A' in A and any $r \times c$ submatrix B' in B lying in r cone independent rows and c cone independent columns of A and B, respectively, then there are monomial matrices X in $\mathcal{N}_r$ and Y in $\mathcal{N}_c$ such that $XA'Y = B'$.

Proof The argument is similar to that in Theorem 4.15. ■

Having characterized the Green's relations on $\mathcal{N}_n$ for $\mathscr{L}$, $\mathscr{R}$, and $\mathscr{D}$, our efforts are now turned toward $\mathscr{J}$. Our work rests on the following corollary to Theorem 4.16.

(4.17) Corollary Let A,B be in $\mathcal{N}_n$ and nonsingular. Then $A\mathscr{D}B$ if and only if $XAY = B$ has monomial solutions X and Y in $\mathcal{N}_n$.

Applying this lemma, we can show that for $n \geq 3$, $\mathscr{D} \neq \mathscr{J}$ on $\mathcal{N}_n$. For this consider

$$A = \begin{bmatrix} 1 & 0 & 0 \\ 2 & 1 & 0 \\ 3 & 4 & 1 \end{bmatrix} \quad \text{and} \quad B = \begin{bmatrix} 1 & 0 & 0 \\ 2 & 1 & 0 \\ 6 & 1 & 1 \end{bmatrix}.$$

Then by direct calculation,

$$\begin{bmatrix} 1 & 0 & 0 \\ 0 & 1 & 0 \\ \frac{21}{4} & 0 & \frac{1}{4} \end{bmatrix} \begin{bmatrix} 1 & 0 & 0 \\ 2 & 1 & 0 \\ 3 & 4 & 1 \end{bmatrix} \begin{bmatrix} 1 & 0 & 0 \\ 0 & 1 & 0 \\ 0 & 0 & 4 \end{bmatrix} = \begin{bmatrix} 1 & 0 & 0 \\ 2 & 1 & 0 \\ 6 & 1 & 1 \end{bmatrix}$$

and

$$\begin{bmatrix} 1 & 0 & 0 \\ 0 & 1 & 0 \\ 0 & 0 & \frac{1}{2} \end{bmatrix} \begin{bmatrix} 1 & 0 & 0 \\ 2 & 1 & 0 \\ 6 & 1 & 1 \end{bmatrix} \begin{bmatrix} 1 & 0 & 0 \\ 0 & 1 & 0 \\ 0 & 7 & 2 \end{bmatrix} = \begin{bmatrix} 1 & 0 & 0 \\ 2 & 1 & 0 \\ 3 & 4 & 1 \end{bmatrix}.$$

Hence $A\mathscr{J}B$. But, as there are no monomials D_1 and D_2 so that $D_1AD_2 = B$, it follows that $\mathscr{D} \neq \mathscr{J}$ on $\mathscr{N}_3$.

For $n > 3$, consider

$$\bar{A} = \begin{bmatrix} A & 0 \\ 0 & I_{n-3} \end{bmatrix} \quad \text{and} \quad \bar{B} = \begin{bmatrix} B & 0 \\ 0 & I_{n-3} \end{bmatrix}.$$

From the above calculations, $\bar{A}\mathscr{J}\bar{B}$ yet $\bar{A}\not\mathscr{D}\bar{B}$. Hence $\mathscr{D} \neq \mathscr{J}$ on $\mathscr{N}_n$, $n \geq 3$.

For $n = 2$, the result differs. For this case we show $\mathscr{D} = \mathscr{J}$. In this regard, suppose $A\mathscr{J}B$. We argue cases.

Case 1 A and hence B are singular.

Singularity here implies A and B are regular elements in $\mathscr{N}_2$ and so $A\mathscr{D}B$.

Case 2 A and hence B are nonsingular.

By definition $A\mathscr{J}B$ implies that $X_1AY_1 = B$ and $X_2BY_2 = A$ for some nonsingular X_1, X_2, Y_1, and Y_2 in $\mathscr{N}_2$. Thus, each of X_1, X_2, Y_1, and Y_2 has a positive diagonal. Let $X \prec Y$ denote the property that $x_{ij} > 0$ implies $y_{ij} > 0$ for all i,j. Then there exist permutation matrices P and Q so that $PAQ \prec B$ and permutation matrices R and S so that $RBS \prec A$. Thus, PAQ and B have the same zero pattern. We again argue cases.

Case a A and hence B have one or two zeros.

In this case, by solving equations, diagonal matrices D_1 and D_2 in $\mathscr{N}_2$ may be found so that $D_1PAQD_2 = B$. Hence $A\mathscr{D}B$.

Case b A and hence B are positive.

In this case, as $X_1AY_1 = B$ and $X_2BY_2 = A$, it follows that $(X_2X_1)A(Y_1Y_2) = A$. Set $X = X_2X_1$ and $Y = Y_1Y_2$; i.e., $XAY = A$. As $(cX)A(c^{-1}Y) = A$ for any positive number c, we may assume without loss of generality that $\det X = \det Y = \pm 1$. Suppose $\det X = \det Y = 1$; i.e., $x_{11}x_{22} - x_{12}x_{21} = 1$ and $y_{11}y_{22} - y_{12}y_{21} = 1$. Suppose $\max\{x_{11}, x_{22}\} = x_{11} \geq 1$ and $\max\{y_{11}, y_{12}\} = y_{11} \geq 1$. If either of these two inequalities is strict, the (1,1) entry in XAY is strictly greater than a_{11}, a contradiction. But now $x_{11} = x_{22} = y_{11} = y_{22} = 1$. Further $x_{12} = x_{21} = y_{12} = y_{21} = 0$ so that $X = Y = I$. Considering all other possible cases leads to the conclusion that X and Y are monomials and so X_1, X_2, Y_1, and Y_2 are monomials, hence $A\mathscr{D}B$.

Moreover, as $A\mathscr{J}B$ if and only if the equations $XAY = B$ and $XBY = A$ have solutions X_1, Y_1, X_2, Y_2 in $\mathscr{N}_n$, respectively, and as $\mathscr{D} \neq \mathscr{J}$ on $\mathscr{N}_n$ for $n \geq 3$, we suspect that no further satisfactory characterization of $\mathscr{J}$ exists.

This section is concluded with a discussion of factorizations of nonnegative matrices of order $n \geq 2$.

(4.18) Definition A matrix $P \in \mathcal{N}_n$ is called a *prime* if

(i) P is not monomial, and
(ii) $P = BC$, where $B,C \in \mathcal{N}_n$, implies that either B or C is monomial.

If P is neither a prime nor a monomial matrix it is called *factorizable.*

It will follow from Exercise 6.11 that prime matrices have no nonnegative rank factorizations.

For $A \in \mathcal{N}_n$, a_j denotes the jth column of A. By A^* we denote the (0,1) matrix defined by $a_{ij}^* = 1$ if $a_{ij} > 0$ and $a_{ij}^* = 0$ if $a_{ij} = 0$. The matrix A^* will be called the *incidence matrix* of A. We use the componentwise partial order on $\mathcal{N}_n$ and on the set of column n-tuples. Assume $n \geq 2$.

(4.19) Theorem Let $A \in \mathcal{N}_n$. Let $1 \leq i, k \leq n$, and $i \neq k$. If $a_i^* \geq a_k^*$ then A is factorizable.

Proof By reordering the columns of A, we may assume without loss of generality that $a_1^* \geq a_2^*$. Hence there exists a positive δ such that $b_1 = (a_1 - a_2\delta) \geq 0$ and $b_1^* = a_1^*$. Let $b_i = a_i$, $i = 2, \ldots, n$. Then $B = [b_1, \ldots, b_n] \in \mathcal{N}_n$. We shall prove that B is not monomial.

Either $b_2 = 0$ or $b_2 \neq 0$. If $b_2 = 0$, then B is not monomial. If $b_2 \neq 0$, then there is an r, $1 \leq r \leq n$, such that $b_{r2} > 0$. Since $a_1^* \geq a_2^* = b_2^*$, we have $a_{r1} > 0$ and since $b_1^* = a_1^*$, it follows that $b_{r1} > 0$. Thus in both cases, B is not monomial.

Let

$$C = \begin{bmatrix} 1 & 0 \\ \delta & 1 \end{bmatrix} \oplus I_{n-2},$$

where I_{n-2} is the $(n-2) \times (n-2)$ identity matrix and is missing if $n = 2$. Then C is not monomial. Since $A = BC$, it follows that A is factorizable. ∎

(4.20) Corollary If A is prime, then A^* has a zero and a one in every row and column.

Recall that a matrix $A \in \mathcal{N}_n$ is called *fully indecomposable* if there do *not* exist permutation matrices M,N such that

$$MAN = \begin{bmatrix} A_{11} & A_{12} \\ 0 & A_{22} \end{bmatrix},$$

where A_{11} is square. Otherwise it is *partly decomposable.*

A matrix A is *completely decomposable* if there exist permutation matrices M,N such that $MAN = A_1 \oplus \cdots \oplus A_s$, where A_i is fully indecomposable, $i = 1, \ldots, s$ and $s \geq 1$. (Note that a fully indecomposable matrix is completely decomposable.)

We now state a sufficient condition for A to be prime in $\mathcal{N}_n$.

(4.21) Theorem Let $n > 1$ and let $A \in \mathcal{N}_n$. If
(i) A is fully indecomposable, and
(ii) $(a_i^*)^t a_k^* \leq 1$ for all i,k such that $1 \leq i, k \leq n$, and $i \neq k$, then A is prime.

Proof By (i), A is not monomial.

Let $A = BC$, where $B,C \in \mathcal{N}_n$. Let $\mathscr{Z}_n = \{1, \ldots, n\}$ and let $J = \{j \in \mathscr{Z}_n : b_j$ has at most one positive entry$\}$.

We now assert the following.

(4.22) If $j \in \mathscr{Z}_n/J$, then there is at most one $i \in \mathscr{Z}_n$ such that $c_{ji} > 0$. For suppose that $j \in \mathscr{Z}_n/J$ and that $c_{ji} > 0$, $c_{jk} > 0$, where $i,k \in \mathscr{Z}_n$, $i \neq k$. Then

$$a_i = \sum_{l=1}^{n} b_l c_{li} \geq b_j c_{ji}, \qquad \text{whence} \quad a_i^* \geq b_j^*.$$

Similarly, $a_k^* \geq b_j^*$. Hence $(a_i^*)^t a_k^* \geq 2$, which contradicts (ii). Thus (4.22) is proved.

If E is a set, let $|E|$ denote the number of elements in E.

We shall next show that

$$0 < |J| < n \text{ is impossible.}$$

Let $|J| = q$, and put $I = \{i \in \mathscr{Z}_n : c_{ji} = 0 \text{ for all } j \in \mathscr{Z}_n/J\}$. Suppose that $|I| = r$. Let $d = \sum_{i \in I} a_i$. By (i), d has at least $r + 1$ positive entries. Since for every $i \in I$, we have $a_i = \sum_{j \in J} b_j c_{ji}$ it follows that d has at most q positive entries. Hence $r < q$. Let $I' = \mathscr{Z}_n/I$ and $J' = \mathscr{Z}_n/J$. By definition of I, for each $i \in I'$ there exists a $j \in J'$ such that $c_{ji} > 0$. Since $|I'| = n - r > n - q = |J'|$, there exists a $j \in J'$ such that $c_{ji} > 0$ and $c_{jk} > 0$ for distinct i,k in $\mathscr{Z}_n$. But this contradicts (4.22). Hence $0 < |J| < n$ is impossible.

There are two remaining possibilities.

(a) $|J| = n$.

Then each column of B has at most one positive entry. But by (i), every row of B is nonzero. Hence B is monomial.

(b) $|J| = 0$.

By (4.22), each row of C has at most one positive entry. But by (i), every column of C is nonzero, whence C is monomial. ■

It is clear that Theorem 4.19, Corollary 4.20, and Theorem 4.21 have analogs for rows instead of columns.

(4.23) Theorem Let $A \in \mathcal{N}_n$ and let A be prime. Then there exists an r, $1 \le r \le n$, and a fully indecomposable prime $P \in \mathcal{N}_r$ such that

$$MAN = P \oplus D,$$

where M,N are permutation matrices in $\mathcal{N}_n$ and D is a nonsingular diagonal matrix in $\mathcal{N}_{n-r}$.

Proof The proof is by induction on n. If $n = 1$, the result is trivial, since there are no primes in $\mathcal{N}_1$. So suppose that $n > 1$, and that the theorem holds for $\mathcal{N}_{n-1}$. Let A be a prime in $\mathcal{N}_n$. If A is fully indecomposable, there is no more to prove. So suppose that, for suitable permutation matrices R,S,

$$RAS = \begin{bmatrix} A_{11} & A_{12} \\ 0 & A_{22} \end{bmatrix},$$

where A_{11} is $s \times s$, $0 < s < n$.

We shall show that $A_{12} = 0$.

Suppose $A_{12} \ne 0$, say, $a_{ij} > 0$, $1 \le i \le s$ and $s + 1 \le j \le n$. It follows that A_{11} is not monomial, for otherwise we would have $a_j^* \ge a_k^*$, where $1 \le k \le s$, and by Theorem 4.19 A would not be prime.

Thus

$$RAS = \begin{bmatrix} I_s & A_{12} \\ 0 & A_{22} \end{bmatrix} \begin{bmatrix} A_{11} & 0 \\ 0 & I_{n-s} \end{bmatrix},$$

with neither factor monomial, which is again a contradiction.

Hence $A_{12} = 0$, and

$$RAS = \begin{bmatrix} A_{11} & 0 \\ 0 & A_{22} \end{bmatrix}.$$

If either A_{11} or A_{22} is factorizable, then it is easily seen that RAS is factorizable. Hence since

$$RAS = \begin{bmatrix} A_{11} & 0 \\ 0 & I_{n-s} \end{bmatrix} \begin{bmatrix} I_s & 0 \\ 0 & A_{22} \end{bmatrix},$$

either

(a) A_{22} is monomial and A_{11} is prime, or
(b) A_{11} is monomial and A_{22} is prime.

Suppose (a) holds. By inductive hypothesis we permute the rows and columns of A_{11} to obtain $P \oplus D_1$, where P is a fully indecomposable prime in $\mathcal{N}_r$ where $1 \le r \le s$, and D_1 is a nonsingular diagonal matrix in $\mathcal{N}_{s-r}$. We also permute the rows and columns of A_{22} to obtain a nonsingular diagonal matrix D_2 in $\mathcal{N}_{n-s}$. Thus, for suitable permutation matrices M and N,

$$MAN = P \oplus D,$$

where $D = D_1 \oplus D_2$ is a nonsingular diagonal matrix in $\mathcal{N}_{n-r}$. The proof in case (b) is similar. ■

(4.24) Theorem If P is a prime in $\mathcal{N}_r$ and Q is monomial in $\mathcal{N}_{n-r}$, where $1 \leq r \leq n$, then $P \oplus Q$ is a prime in $\mathcal{N}_n$.

Proof Let $A = P \oplus Q$ and let $A = BC$. Partition

$$B = \begin{bmatrix} B_1 \\ B_2 \end{bmatrix} \quad \text{and} \quad C = [C_1 \quad C_2],$$

where B_1 is $r \times n$ and C_1 is $n \times r$. Replacing B by BN and C by $N^{-1}C$, where N is a permutation matrix, we may suppose that any zero columns of B_1 are at the right. Thus

$$B = \begin{bmatrix} B_{11} & B_{12} \\ B_{21} & B_{22} \end{bmatrix} \quad \text{and} \quad C = \begin{bmatrix} C_{11} & C_{12} \\ C_{21} & C_{22} \end{bmatrix},$$

where C_{11} is $s \times r$, B_{11} is $r \times s$, $B_{12} = 0$, and no column of B_{11} is zero. Clearly $s > 0$, since A has no zero row. We have

$$\begin{bmatrix} P & 0 \\ 0 & Q \end{bmatrix} = A = BC = \begin{bmatrix} B_{11}C_{11} & B_{11}C_{12} \\ B_{21}C_{11} + B_{22}C_{21} & B_{21}C_{12} + B_{22}C_{22} \end{bmatrix},$$

whence $O = B_{11}C_{12}$. Since no column of B_{11} is zero, it follows that $C_{12} = 0$. Hence $s < n$, since A has no zero column. Thus $0 < s < n$.

We now have $P = B_{11}C_{11}$ and $Q = B_{22}C_{22}$.

We next show that $r = s$.

If $s < r$, we have

$$P = B'_{11}C'_{11}, \quad \text{where} \quad B'_{11} = [B_{11} \quad 0] \in \mathcal{N}_r \quad \text{and} \quad C'_{11} = \begin{bmatrix} C_{11} \\ 0 \end{bmatrix} \in \mathcal{N}_r.$$

But this factorization contradicts the fact that P is prime. Similarly, if $s > r$, we obtain $n - r < n - s$, a contradiction to $Q = B_{22}C_{22}$ and that Q is monomial. Hence $r = s$. But

$$A = BC = \begin{bmatrix} B_{11}C_{11} & 0 \\ B_{21}C_{11} + B_{22}C_{21} & B_{22}C_{22} \end{bmatrix},$$

and so $B_{21}C_{11} = 0$ and $B_{22}C_{21} = 0$. Since $P = B_{11}C_{11}$ is a factorization in $\mathcal{N}_r$ it follows that C_{11} is either prime or monomial. Thus C_{11} has no zero row. Hence it follows from $B_{21}C_{11} = 0$ that $B_{21} = 0$. Similarly, we deduce from $B_{22}C_{21} = 0$ and the fact that B_{22} is monomial that $C_{21} = 0$. Hence $B = B_{11} \oplus B_{22}$ and $C = C_{11} \oplus C_{22}$. Since B_{22}, C_{22} are monomial and one of B_{11}, C_{11} is monomial, it follows that either B or C is monomial. ■

(4.25) Theorem Let $A \in \mathcal{N}_n$. Then A is prime if and only if there exists an r, $1 \leq r \leq n$, and an indecomposable prime $P \in \mathcal{N}_r$, such that

$$MAN = P \oplus D,$$

where M,N are permutation matrices in $\mathcal{N}_n$ and D is a nonsingular diagonal matrix in $\mathcal{N}_{n-r}$.

Proof The proof is immediate by Theorems (4.23) and (4.24). ■

Remark Since there are no primes in $\mathcal{N}_1$ and $\mathcal{N}_2$, as will be shown later, we can improve the inequality in Theorems 4.23 and 4.24 to $3 \leq r \leq n$.

(4.26) Corollary Every prime in $\mathcal{N}_n$ is completely decomposable.

We now use Theorems 4.19, 4.21, and 4.25 to classify all primes of orders 2 and 3 and to discuss primes of order 4.

$n = 2$. There are no primes in $\mathcal{N}_2$. This follows from Theorem 4.19
$n = 3$. The matrix $A \in \mathcal{N}_3$ is a prime if and only if

$$MA^*N = \begin{bmatrix} 0 & 1 & 1 \\ 1 & 0 & 1 \\ 1 & 1 & 0 \end{bmatrix}$$

for suitable permutation matrices M and N.

This follows from Theorems 4.19 and 4.21.

We show next that there exist matrices $A \in \mathcal{N}_n$, $n \geq 4$, where A is prime while A^* is not prime. Let

$$A = \begin{bmatrix} 1 & 1 & 5 & 0 \\ 0 & 1 & 1 & 5 \\ 5 & 0 & 1 & 1 \\ 1 & 5 & 0 & 1 \end{bmatrix}.$$

Then

$$A^* = \begin{bmatrix} 1 & 1 & 1 & 0 \\ 0 & 1 & 1 & 1 \\ 1 & 0 & 1 & 1 \\ 1 & 1 & 0 & 1 \end{bmatrix} = \begin{bmatrix} 0 & 1 & 1 & 0 \\ 1 & 0 & 0 & 0 \\ 0 & 0 & 1 & 1 \\ 0 & 1 & 0 & 1 \end{bmatrix} \begin{bmatrix} 0 & 1 & 1 & 1 \\ \frac{1}{2} & 1 & 0 & 0 \\ \frac{1}{2} & 0 & 1 & 0 \\ \frac{1}{2} & 0 & 0 & 1 \end{bmatrix}$$

so that A^* is not prime.

We now establish that A is prime. For this, suppose $A = BC$. We need to show that B or C is a monomial matrix.

Case 1 B is partly decomposable.

In this case there are permutation matrices P and Q so that

$$PBQ = \begin{bmatrix} B_{11} & 0 \\ B_{21} & B_{22} \end{bmatrix},$$

where B_{11} is of order r. Thus, without loss of generality, we assume

$$A = \begin{bmatrix} A_1 \\ A_2 \end{bmatrix} = \begin{bmatrix} B_{11} & 0 \\ B_{21} & B_{22} \end{bmatrix} \begin{bmatrix} C_1 \\ C_2 \end{bmatrix}$$

with A_1 and C_1 being $r \times 4$. By calculation, $A_1 = B_{11}C_1$. Suppose $B_{11} = D_r$, a monomial. Then if $B_{21} \neq 0$, some row of A^* dominates some other row of A^*, a contradiction. Hence, if $B_{11} = D_r$ we must have that $B_{21} = 0$ and so $A_2 = B_{22}C_2$. If $B_{11} = D_r$ and $B_{22} = D_{4-r}$ are monomial, then B is a monomial and we are through. Hence, without loss of generality, we assume B_{11} is not a monomial.

From $A_1 = B_{11}C_1$, note that each column of A_1 is a nonnegative combination of vectors in B_{11}. Of course, $1 < r < 4$; i.e., $r = 2$ or $r = 3$. If $r = 2$ we observe that A_1 has a monomial subpattern and so B_{11} is a monomial, a contradiction. For $r = 3$, A_1 contains a diagonal consisting entirely of zeros and so a subpattern of the form

$$\bar{A}_1 = \begin{bmatrix} X & 0 & X \\ X & X & 0 \\ 0 & X & X \end{bmatrix},$$

which is prime by Theorem 4.21, so $B_{11} = \bar{A}_1 D$, D a monomial. Thus the remaining column, i.e., the positive column, in A_1 must be a nonnegative combination of the columns in $\bar{A}_1$. By checking the four possibilities for A_1, we find that this is not the case.

Hence we conclude that B must be a monomial matrix.

Case 2 C is partly decomposable.

In this case we have $A^t = C^t B^t$. But, as is easily seen, there are permutation matrices P and Q so that $PA^tQ = A$, and so this case reduces to the previous one.

Case 3 B and C are fully indecomposable.

Since B and C are fully indecomposable, each has at least two ones in each row and column. Further, since A has no positive rows, B and C have precisely two ones in each row and column. As $A = BC$ if and only if $A = BQQ^tC$ for any permutation matrix Q, we may assume B has a positive main diagonal. Further, as $A = BC$ if and only if $A = BDD^{-1}C$ for any positive diagonal matrix D, we may assume that B has a main diagonal consisting entirely of ones. Thus, the candidates for B are as follows.

$$\text{(i)} \begin{bmatrix} 1 & a & 0 & 0 \\ 0 & 1 & b & 0 \\ 0 & 0 & 1 & c \\ d & 0 & 0 & 1 \end{bmatrix}, \quad \text{(ii)} \begin{bmatrix} 1 & a & 0 & 0 \\ 0 & 1 & 0 & b \\ c & 0 & 1 & 0 \\ 0 & 0 & d & 1 \end{bmatrix},$$

$$\text{(iii)} \begin{bmatrix} 1 & 0 & a & 0 \\ 0 & 1 & 0 & b \\ 0 & c & 1 & 0 \\ d & 0 & 0 & 1 \end{bmatrix}, \quad \text{(iv)} \begin{bmatrix} 1 & 0 & a & 0 \\ b & 1 & 0 & 0 \\ 0 & 0 & 1 & c \\ 0 & d & 0 & 1 \end{bmatrix},$$

$$\text{(v)} \begin{bmatrix} 1 & 0 & 0 & a \\ b & 1 & 0 & 0 \\ 0 & c & 1 & 0 \\ 0 & 0 & d & 1 \end{bmatrix}, \quad \text{(vi)} \begin{bmatrix} 1 & 0 & 0 & a \\ 0 & 1 & b & 0 \\ c & 0 & 1 & 0 \\ 0 & d & 0 & 1 \end{bmatrix},$$

where a, b, c, and d are positive.

As all arguments are similar, we provide only the argument for pattern (i). If

$$B = \begin{bmatrix} 1 & a & 0 & 0 \\ 0 & 1 & b & 0 \\ 0 & 0 & 1 & c \\ d & 0 & 0 & 1 \end{bmatrix}$$

then as $A = BC$, we must have

$$C = \begin{bmatrix} 1 & w & 0 & 0 \\ 0 & 1 & x & 0 \\ 0 & 0 & 1 & y \\ z & 0 & 0 & 1 \end{bmatrix}.$$

Elementwise comparison of A and BC yields the following equations.

$$a + w = 1 \tag{4.27}$$

$$bw = 5 \tag{4.28}$$

$$b + x = 1 \tag{4.29}$$

$$xc = 5$$

$$c + y = 1$$

$$dy = 5$$

$$d + z = 1$$

$$dw = 5.$$

Equation (4.27) implies $w < 1$ while Eq. (4.28) implies $b > 5$ which contradicts Eq. (4.29). Hence, there are no solutions to these equations and so $A \neq BC$ for B having the (4.27) pattern.

Finally as all possible assumptions of A being factorizable lead to contradictions, we conclude that A *is* prime.

5 THE SEMIGROUP $\mathscr{D}_n$

In this section certain algebraic properties of the semigroup $\mathscr{D}_n$ of doubly stochastic matrices are determined. In particular the Green's relations on $\mathscr{D}_n$ are given, regularity is investigated, and the maximal subgroups are characterized.

Let $\mathscr{P}_n$ denote the set of all $n \times n$ permutation matrices, that is, doubly stochastic matrices $P = (p_{ij})$ with $p_{ij} = 0$ or 1 for each i and j. Then clearly $\mathscr{P}_n$ is a maximal subgroup of $\mathscr{D}_n$. Geometrically, $\mathscr{D}_n$ forms a convex polyhedron with the permutation matrices as vertices (see Chapter 2, where $\mathscr{D}_n$ in denoted by Ω_n).

The Green's relations are determined first. Since $\mathscr{D}_n$ forms a compact semigroup under the natural topology, the relations $\mathscr{D}$ and $\mathscr{J}$ on $\mathscr{D}_n$ are the same (see Hoffman and Mostert [1966]). As before, $A\mathscr{L}B$ if and only if $A^t\mathscr{R}B^t$.

(5.1) Theorem Let $A,B \in \mathscr{D}_n$. Then $A\mathscr{R}B$ if and only if $A = BP$ for some $P \in \mathscr{P}_n$.

Proof Suppose that $A\mathscr{R}B$. Then by Theorem 4.15, $d(A) = d(B) = d$. Moreover any set of d independent columns of A form nonnegative multiples of a set of d independent columns of B. Thus each column of A is a multiple of some column of B and vice versa. Then since $d(A) = d(B)$ and A and B are doubly stochastic, they have the same sets of columns in some ordering.

The converse is immediate. ■

(5.2) Corollary Let $A,B \in \mathscr{D}_n$. Then $A\mathscr{D}B$ if and only if $A = PBQ$ for some $P,Q \in \mathscr{P}_n$.

It follows from Theorem 5.1 and Corollary 5.2 that each $\mathscr{L}$-, $\mathscr{R}$-, $\mathscr{H}$-, and $\mathscr{D}$-equivalence class in $\mathscr{D}_n$ is finite. Thus the maximal subgroups of $\mathscr{D}_n$ are finite. It will be seen that $\mathscr{D}_n$ contains only finitely many regular elements.

(5.3) Lemma An $\mathscr{R}[\mathscr{L}]$-class of $\mathscr{D}_n$ contains at most one idempotent. If a $\mathscr{D}$-class D of $\mathscr{D}_n$ contains an idempotent it contains exactly one idempotent of the form (3.3) where if J_i is $n_i \times n_i$ then $J_i = (1/n_i)$, $n_1 \leq n_2 \leq \cdots \leq n_k$.

Proof Let E,F be idempotent in an $\mathscr{R}$-class R of $\mathscr{D}_n$. Then E and F are left identities for the elements in R so that

$$E = E^t = (FE)^t = E^tF^t = EF = F.$$

The second statement follows from Corollaries 3.8 and 5.2. ■

(5.4) Lemma The semigroup $\mathcal{D}_n$ contains finitely many regular elements. Moreover if $A \in \mathcal{N}_n$ is regular then each component of A is either zero or $1/n_i$ for some $1 \leq n_i \leq n$.

Proof The proof of the first statement is immediate since by Corollary 3.8, $\mathcal{D}_n$ contains finitely many idempotents and by Lemma 5.3, each $\mathcal{D}$-class of $\mathcal{D}_n$ is finite. The second statement follows from Corollary 3.8 and Theorem 5.1. ■

The discussion of regularity in $\mathcal{D}_n$ is concluded with the following result.

(5.5) Theorem If $A \in \mathcal{D}_n$ is regular then A^t is the unique semi-inverse of A in $\mathcal{D}_n$.

Proof Let X be any semi-inverse of A in $\mathcal{D}_n$. Then AX is idempotent and $AX\mathcal{R}A$ so that $AX = AP$ for some $P \in \mathcal{P}_n$ by Theorem 5.1. Since $(AP)^t = AP$, $A^t = PAP$. Then $AA^t = APAP = AP = AX$. Similarly $PA = XA$, so that

$$X = XAX = PAX = PAP = A^t. \quad ■$$

We remark that it will follow from the results in Chapter 5, that if a matrix A in $\mathcal{D}_n$ is regular in $\mathcal{D}_n$ then $A^t = A^+$, the Moore–Penrose generalized inverse of A.

In Section 3 the maximal subgroups of $\mathcal{N}_n$ were shown to be isomorphic to complete monomial groups over the reals. Our final result provides a similar result for $\mathcal{D}_n$. Its proof is immediate from Lemma 5.3 and Theorem 5.5.

(5.6) Corollary The semigroups $\mathcal{D}_n$ contain finitely many maximal subgroups, each of which is isomorphic to a finite direct product of full symmetric groups.

6 EXERCISES

(6.1) Show that an element a in a semigroup T is regular if and only if the $\mathcal{R}[\mathcal{L}]$ class of T containing a contains an idempotent element.

(6.2) Show that two elements a,b of a semigroup T are group inverses of each other within some subgroup of T if and only if they satisfy the equations $a = aba$, $b = bab$, and $ab = ba$.

(6.3) Show that if a $\mathcal{D}$-class D of a semigroup T contains a regular element a then every element of D is regular and moreover every semi-inverse of a in T is contained in D.

(6.4) Let H be an $\mathscr{H}$-class of a semigroup T. Show that if a,b and ab all belong to H then H is subgroup of T. Conclude that an $\mathscr{H}$-class is a subgroup if and only if it contains an idempotent element.

(6.5) Use Exercise 6.4 to show that the maximal subgroups of a semigroup T are precisely the $\mathscr{H}$-classes of T containing idempotents.

(6.6) Prove that every rank one nonnegative idempotent matrix has the form

$$E = xy^{t}$$

for some $x, y \in R^n_+$ with $y'x = 1$.

(6.7) Show that if A is nonnegative and nonsingular then A^{-1} is nonnegative if and only if A is monomial. (Hint: Use Theorem 4.6.)

(6.8) Characterize all nonnegative nonsingular matrices A with $A = A^{-1}$ (Harary and Minc [1976]).

(6.9) Show that not every matrix in $\mathscr{N}_n$ is regular for $n \geq 2$.

(6.10) Consider the following matrices in $\mathscr{N}_3$.

$$A = \begin{bmatrix} 2 & 0 & 1 \\ 4 & 2 & 0 \\ 4 & 1 & 1 \end{bmatrix}, \qquad B = \begin{bmatrix} 4 & 0 & 2 \\ 2 & 1 & 0 \\ 5 & \frac{1}{2} & 2 \end{bmatrix}.$$

Show that A and B are regular and that they belong to the same subgroup of $\mathscr{N}_3$. (Hint: For the last statement it suffices to show that $A \mathscr{H} B$.)

(6.11) Show that a prime matrix in $\mathscr{N}_n$ can have no nonnegative rank factorization (Berman and Plemmons [1974b]). Does every factorizable matrix have a nonnegative rank factorization?

(6.12) Show that

$$A = \begin{bmatrix} 1 & 1 & 0 & 0 \\ 0 & 1 & 1 & 0 \\ 0 & 0 & 1 & 1 \\ 1 & 0 & 0 & 1 \end{bmatrix}$$

is prime in $\mathscr{N}_4$. (Hint: Use Theorem 4.21.)

(6.13) Let

$$A = \begin{bmatrix} \frac{1}{2} & 0 & \frac{1}{3} & \frac{1}{6} \\ \frac{1}{2} & 0 & \frac{1}{3} & \frac{1}{6} \\ 0 & \frac{1}{2} & \frac{1}{6} & \frac{1}{3} \\ 0 & \frac{1}{2} & \frac{1}{6} & \frac{1}{3} \end{bmatrix}, \qquad B = \begin{bmatrix} \frac{1}{2} & 0 & \frac{1}{8} & \frac{3}{8} \\ \frac{1}{2} & 0 & \frac{1}{8} & \frac{3}{8} \\ 0 & \frac{1}{2} & \frac{3}{8} & \frac{1}{8} \\ 0 & \frac{1}{2} & \frac{3}{8} & \frac{1}{8} \end{bmatrix}.$$

Show that the convex polyhedrons generated by the columns of A and B are the same. In particular, show that this convex polyhedron consists of all nonnegative vectors of the form

$$\begin{bmatrix} a \\ a \\ b \\ b \end{bmatrix}, \qquad a + b = \tfrac{1}{2}.$$

Note however, that A and B are not $\mathcal{R}$-equivalent in $\mathcal{D}_4$.

(6.14) Let $\|A\|$ denote the spectral norm of A. Show that $A \in \mathcal{D}_n$ is regular in $\mathcal{D}_n$ if and only if $\|A\| = 1$.

(6.15) A matrix $A = (a_{ij})$ is said to be orthostochastic if there is a unitary matrix $U = (u_{ij})$ with $a_{ij} = |u_{ij}|^2$ for every i and j. Show that every regular doubly stochastic matrix is orthostochastic. Is the converse of this statement true?

(6.16) Let S_n denote the set of all $n \times n$ nonnegative matrices having row and column sums at most one. Show then that S_n is a multiplicative semigroup with $\mathcal{D}_n$ as a subsemigroup. Show that Theorems 5.1 and 5.5 and Corollary 5.6 in fact hold with $\mathcal{D}_n$ replaced by S_n.

7 NOTES

(7.1) The notation and terminology given in Section 2 for algebraic semigroups follow that of Clifford and Preston [1961]. Properties of topological semigroups can be found in Hoffman and Mostert [1966].

(7.2) The characterizations of nonnegative idempotents in Section 3 is due to Flor [1969]. Stochastic idempotents were first characterized by Doob [1942]. De Marr [1974] has also studied nonnegative idempotents.

(7.3) The material in Section 4 on regularity in $\mathcal{N}_n$ is due to Plemmons [1973]. The characterizations of the Green's relations on $\mathcal{N}_n$ are due to Hartfiel *et al.* [1976], while the maximal subgroups of $\mathcal{N}_n$ were first characterized by Flor [1969]. Conditions under which $A = A^{-1} \geq 0$ were established by Harary and Minc [1976] (see Exercise 6.8). Their result was extended by Berman [1974b] to the case where $A = A^{+} \geq 0$, where A^{+} is the Moore–Penrose inverse of A (see Theorem 5.5.6). Compact topological groups of nonnegative matrices were shown to be finite by Brown [1964]. Prime elements and nonnegative matrix equations were investigated by Richman and Schneider [1974] and some open questions in their paper were answered in the negative by Borosh *et al.* [1976].

(7.4) From Exercise 6.12, it follows that not every nonnegative matrix has a nonnegative rank factorization. Thomas [1974] has given a geometric characterization of all nonnegative matrices having a nonnegative rank factorization.

(7.5) The semigroup of doubly stochastic matrices has been investigated by many authors. Theorems 5.1 and 5.5 are due to Montague and Plemmons [1973]. The maximal subgroups in $\mathcal{D}_n$ were first characterized by Schwarz [1967]. Farahat [1966] independently obtained Corollary 5.6.

(7.6) The algebraic properties of the semigroup of row-stochastic matrices were not given here since the theory is not completely developed. Stochastic matrices having certain types of stochastic generalized inverses have been studied by Hartfiel [1974], Rao [1973], and Wall [1975].

(7.7) Finally, factorizations of $A \in \mathcal{N}_n$ into $A = LU$ or $A = UL$, where $L \geq 0$ and $U \geq 0$ are lower and upper triangular matrices, respectively, have been studied by Markham [1972]. His primary tool in these studies is the theory of determinants.

CHAPTER 4

SYMMETRIC NONNEGATIVE MATRICES

1 INTRODUCTION

The original outline of this book included a chapter on "Miscellaneous." It did not take us long to find out that several books can be written on these miscellaneous topics. Still, it was decided to touch upon two of them, the theory of which is particularly nice in the case of symmetric matrices. These topics are described in this chapter.

First, we consider the inverse eigenvalue problem of finding necessary and sufficient conditions for a set $\{\lambda_1, \ldots, \lambda_n\}$ to be the set of eigenvalues of some nonnegative $n \times n$ matrix.

In the most general sufficient condition available at the writing of this chapter $\lambda_1, \ldots, \lambda_n$ are real, so it is natural to ask whether they are the eigenvalues of a symmetric nonnegative matrix (Kellog [1971]). This question was asked and answered affirmatively by Fiedler [1974a]. This and related results are described in Section 2.

In Section 3 we accumulate several results on polytopes of nonnegative matrices with given row sums and column sums and, in particular, on polytopes of symmetric nonnegative matrices. As a compromise between the wish to include results that, in our opinion, are interesting and useful and space limitations imposed by the scope of the book, the results are stated without proofs in Section 3.

Several additional results on symmetric nonnegative matrices are mentioned in the exercises and in the notes.

2 INVERSE EIGENVALUE PROBLEMS

If $\{\lambda_1, \ldots, \lambda_n\}$ is the spectrum of an $n \times n$ nonnegative matrix A, then for every positive integer k,

$$s_k \equiv \sum_{i=1}^{n} \lambda_i^k \geq 0. \tag{2.1}$$

In fact,

$$\sum_{i=1}^{n} \lambda_i^k \geq \sum_{i=1}^{n} a_{ii}^k,$$

since the sum of the eigenvalues is the trace and for a nonnegative matrix A,

$$\text{trace}(A^k) \geq \sum_{i=1}^{n} a_{ii}^k.$$

A less immediate necessary condition, which can be proved by Holder's inequality, is (Exercise 4.1):

$$(2.2) \qquad (s_k)^m \leq n^{m-1} s_{km}, \qquad k,m = 1,2,\ldots.$$

The inequalities (2.2) are sharp since equality holds in them for the identity matrix.

For $n \leq 3$ (and $n = 4$, if $\lambda_1, \ldots, \lambda_n$ are assumed to be real), conditions (2.1) and (2.2) are also sufficient. (In fact, weaker conditions are sufficient, see Exercises 4.2 and 4.3.) In general, however, they are not sufficient. The set $(\sqrt{2},\sqrt{2},i,-i)$ satisfies the conditions. If it were the spectrum of a nonnegative matrix A then, by the Perron–Frobenius theorem, so would be $(\sqrt{2},i,-i)$, but this is impossible since $(\sqrt{2},i,-i)$ does not satisfy condition (2.2) for $k = 1$ and $m = 2$. Similarly, the set $(1,1,-\frac{2}{3},-\frac{2}{3},-\frac{2}{3})$ satisfies conditions (2.1) and (2.2) but, again by the Perron–Frobenius theorem, cannot be the spectrum of a nonnegative matrix.

The problem of obtaining sufficient conditions for a set of complex numbers to be the spectrum of a nonnegative matrix seems to be a very difficult one. We now present sufficient conditions for a set of real numbers to be the spectrum of a nonnegative matrix, in fact, of a symmetric one. For this we need the following lemma.

(2.3) Lemma Let A be a symmetric $m \times m$ matrix with eigenvalues $\alpha_1, \ldots, \alpha_n$. Let u, $\|u\| = 1$, be an eigenvector corresponding to α_1. Let B be a symmetric $n \times n$ matrix with eigenvalues $\beta_1, \ldots, \beta_n$ and v, $\|v\| = 1$, be an eigenvector corresponding to β_1. Then for any ρ, the matrix

$$C = \begin{bmatrix} A & \rho u v^t \\ \rho v u^t & B \end{bmatrix}$$

has eigenvalues $\alpha_2, \ldots, \alpha_m, \beta_2, \ldots, \beta_n, \gamma_1, \gamma_2$ where γ_1 and γ_2 are the eigenvalues of

$$\hat{C} = \begin{bmatrix} \alpha_1 & \rho \\ \rho & \beta_1 \end{bmatrix}.$$

Proof Let $u, u^{(2)}, \ldots, u^{(m)}$ form an orthonormal system of eigenvectors of A.

$$Au^{(i)} = \alpha_i u^{(i)}, \qquad i = 2, \ldots, m.$$

Then the vectors (of order $m + n$)

$$\begin{bmatrix} u^{(i)} \\ 0 \end{bmatrix}, \qquad i = 2, \ldots, m,$$

are eigenvectors of C corresponding to α_i, $i = 2, \ldots, m$.

Similarly, the vectors

$$\begin{bmatrix} 0 \\ v^{(i)} \end{bmatrix}, \qquad i = 2, \ldots, m,$$

are eigenvectors of C corresponding to β_i, $i = 2, \ldots, m$, where $v, v^{(2)}, \ldots, v^{(m)}$ form an orthonormal system of eigenvectors of B.

Let

$$\begin{bmatrix} r_1 \\ s_1 \end{bmatrix} \quad \text{and} \quad \begin{bmatrix} r_2 \\ s_2 \end{bmatrix}$$

be eigenvectors of $\hat{C}$ that correspond to γ_1 and γ_2, respectively, and form an orthnormal system. Then

$$\begin{bmatrix} r_i u \\ s_i v \end{bmatrix}$$

is an eigenvector corresponding to γ_i, $i = 1,2$. Since the eigenvectors just mentioned form an orthonormal set of $m + n$ vectors, the corresponding eigenvalues form a complete set of eigenvalues for C. ■

We now introduce the following notation.

(2.4) Definition A matrix A is said to *realize* a set of n numbers $(\lambda_1; \lambda_2, \ldots, \lambda_n)$, where $\lambda_2, \ldots, \lambda_n$ are considered unordered, if $\lambda_1, \ldots, \lambda_n$ are the eigenvalues of A and if $\rho(A) = \lambda_1$. Denote by S_n $(\hat{S}_n)$, the collection of all n-tuples, of real numbers, which are realized by a symmetric nonnegative (positive) matrix.

(2.5) Theorem If

$$(\alpha_1; \alpha_2, \ldots, \alpha_m) \in S_m, \qquad (\beta_1; \beta_2, \ldots, \beta_n) \in S_n, \qquad \alpha_1 \geq \beta_1, \qquad \text{and} \qquad \sigma \geq 0,$$

then

$$(\alpha_1 + \sigma, \beta_1 - \sigma, \alpha_2, \ldots, \alpha_m, \beta_2, \ldots, \beta_n) \in S_{m+n}.$$

Proof Let A be a symmetric nonnegative matrix which realizes $(\alpha_1; \alpha_2, \ldots, \alpha_m)$ and let $u \geq 0$, $\|u\| = 1$, be an eigenvector corresponding to $\rho(A)$. Similarly, let B be a nonnegative symmetric matrix that realizes $(\beta_1; \beta_2, \ldots, \beta_n)$ and let $Bv = \beta_1 v$, $v \geq 0$, $\|v\| = 1$. Then by Lemma 2.3, $(\alpha_1 + \sigma; \beta_1 - \sigma, \alpha_2, \ldots, \alpha_m, \beta_2, \ldots, \beta_n)$ is realized by the nonnegative symmetric matrix

$$\begin{bmatrix} A & \rho u v^t \\ \rho v u^t & B \end{bmatrix},$$

where $\rho = (\sigma(\alpha_1 - \beta_1 + \sigma))^{1/2}$. ■

The first sufficient condition for a set to be in S_n is the following.

(2.6) Theorem If

$$\lambda_1 \geq 0 \geq \lambda_2 \geq \cdots \geq \lambda_n \qquad \text{and} \qquad \sum_{i=1}^{n} \lambda_i \geq 0,$$

then

$$(\lambda_1; \lambda_2, \ldots, \lambda_n) \in S_n.$$

Proof The proof is by induction on n. For $n = 1$ the theorem is clear. If $n = 2$, the matrix

$$\begin{bmatrix} 0 & (-\lambda_1\lambda_2)^{1/2} \\ (-\lambda_1\lambda_2)^{1/2} & \lambda_1 + \lambda_2 \end{bmatrix}$$

satisfies the conditions.

To prove the theorem for $n > 2$, assume that the result is true for all smaller sets. The set $\lambda_1' = \lambda_1 + \lambda_2$, $\lambda_2' = \lambda_3, \ldots, \lambda_{n-1}' = \lambda_n$, clearly satisfies the assumptions, so, by the induction hypothesis, $(\lambda_1'; \lambda_2', \ldots, \lambda_{n-1}') \in S_{n-1}$. Applying Theorem 2.5 with $\sigma = |\lambda_2|$, to this and to $(0) \in S_1$, we obtain the desired result. ■

For $\lambda_1 \geq \lambda_2 \geq \cdots \geq \lambda_n$, let

$$K = \left\{ i \in \left\{2, \ldots, \left[\frac{n+1}{2}\right]\right\}; \lambda_i \geq 0 \text{ and } \lambda_i + \lambda_{n+2-i} < 0 \right\}$$

and let m be the greatest index j for which $\lambda_j \geq 0$. With this notation we are able to prove the main result of this section.

(2.7) Theorem If $\lambda_1, \ldots, \lambda_n$ satisfy the conditions

$$\lambda_1 + \sum_{\substack{i \in K \\ i < k}} (\lambda_i + \lambda_{n+2-i}) + \lambda_{n+2-k} \geq 0, \qquad \text{for all} \quad k \in K \tag{2.8}$$

and

$$(2.9) \qquad \lambda_1 + \sum_{i \in K} (\lambda_i + \lambda_{n+2-i}) + \sum_{j=m+1}^{n+1-m} \lambda_j \geq 0,$$

then $(\lambda_1; \lambda_2, \ldots, \lambda_n) \in S_n$.

Proof Here, too, we use induction with respect to n. If $n = 1$, K is void, $m = 1$, and the theorem is true. If $n = 2$, K is void as well. For $m = 2$ the result is trivially true and for $m = 1$ it follows from Theorem 2.6.

For $n > 2$, suppose the theorem is true for all sets of fewer than n numbers. If $m > [(n+1)/2]$, then λ_m does not appear in conditions (2.8) and (2.9), since $m \notin K$ and $\lambda_{n+2-m} \geq \lambda_m$ imply $n + 2 - m \notin K$. Thus, by omitting λ_m from $\{\lambda_1, \ldots, \lambda_n\}$ one obtains a smaller set that satisfies conditions (2.8) and (2.9) and, by the induction hypothesis, belongs to S_{n-1}. This, combined with $(\lambda_m) \in S_1$, implies, by Theorem 2.5, that $(\lambda_1; \lambda_2, \ldots, \lambda_n) \in S_n$.

If $m = 1$, the result follows from Theorem 2.6. Thus, let $2 \leq m \leq [(n+1)/2]$. If $K \neq \{2, \ldots, m\}$, let j be an integer, $2 \leq j \leq m$, such that $j \notin K$. Thus $\lambda_j \geq 0$ and $\lambda_j + \lambda_{n+2-j} \geq 0$. Since neither λ_j, nor λ_{n+2-j}, occur in (2.8) and (2.9), one obtains by omitting them from $\{\lambda_1, \ldots, \lambda_n\}$, a set that satisfies the conditions and thus, by the induction assumption, belongs to S_{n-2}. Since $(\lambda_j; \lambda_{n+2-j}) \in S_2$, we have, again by Theorem 2.5, $(\lambda_1; \lambda_2, \ldots, \lambda_n) \in S_n$.

It remains to consider the case when $2 \leq m \leq [(n+1)/2]$ and $K = \{2, \ldots, m\}$. In this case the conditions in the theorem and the definition of K become

$$(2.10) \qquad \sum_{j=1}^{k} \lambda_j + \sum_{j=1}^{k} \lambda_{n+1-j} \geq 0, \qquad k = 1, \ldots, m-1,$$

$$(2.11) \qquad \sum_{i=1}^{n} \lambda_i \geq 0,$$

$$(2.12) \qquad \lambda_i \geq 0, \qquad i = 1, \ldots, m,$$

and

$$(2.13) \qquad \lambda_i + \lambda_{n+2-i} < 0, \qquad i = 2, \ldots, m.$$

Define $\varepsilon_1 = \lambda_1 + \lambda_n$ and $\varepsilon_2 = -(\lambda_2 + \lambda_n)$. By (2.10) and (2.13), $\varepsilon_1 \geq 0$ and $\varepsilon_2 > 0$. Define

$$\varepsilon = \min(\varepsilon_1, \varepsilon_2), \quad \lambda_1' = \lambda_1 - \varepsilon, \quad \lambda_2' = \lambda_2 + \varepsilon, \quad \lambda_1' = \lambda_i, \quad i = 3, \ldots, n,$$

Since $\lambda_1 - \lambda_2 = \varepsilon_1 + \varepsilon_2 \geq 2\varepsilon$, it follows that

$$(2.14) \qquad \lambda_1' \geq \lambda_2'$$

If $\varepsilon = \varepsilon_1$, then $(\lambda_1'\,; \lambda_2', \ldots, \lambda_{n-1}') \in S_{n-1}$, by the induction hypothesis, (2.10), and (2.11).

Since $\lambda_1' \geq \lambda_n'$ and $\lambda_1' + \lambda_n' = 0$ (by the definition of ε_1), $(\lambda_1'\,; \lambda_n') \in S_2$. Thus, by (2.14) and Theorem 2.5, applied to $\sigma = \varepsilon$, $(\lambda_1\,; \lambda_2, \ldots, \lambda_n) \in S_n$.

If, on the other hand, $\varepsilon = \varepsilon_2$, then one can show, in a similar way that

$$(\lambda_1'\,; \lambda_3', \ldots, \lambda_{n-1}') \in S_{n-2}, \qquad (\lambda_2'\,; \lambda_n') \in S_2,$$

so, here too, $(\lambda_1\,; \lambda_2, \ldots, \lambda_n) \in S_n$, ■

In order to derive similar conditions for positive matrices we prove the following.

(2.15) Theorem If $(\lambda_1\,; \lambda_2, \ldots, \lambda_n) \in S_n$ and if $\varepsilon > 0$, then

$$(\lambda_1 + \varepsilon; \lambda_2, \ldots, \lambda_n) \in \hat{S}_n.$$

Proof Let

$$A = \begin{bmatrix} A_1 & 0 & \cdots & 0 \\ 0 & A_2 & \cdots & 0 \\ \vdots & \vdots & & \vdots \\ 0 & 0 & \cdots & A_k \end{bmatrix} \geq 0$$

be a symmetric matrix that realizes $(\lambda_1\,; \lambda_2, \ldots, \lambda_n)$, where A_i, $i = 1, \ldots, k$, are irreducible and $\lambda_1 = \rho(A_1)$. We proceed by induction with respect to k. If A is irreducible, that is $k = 1$, let u, $\|u\| = 1$, be a positive eigenvector of A, corresponding to $\rho(A)$. Then $\tilde{A} \equiv A + \varepsilon u u^{\mathrm{t}} \gg 0$ and realizes $(\lambda_1 + \varepsilon; \lambda_2, \ldots, \lambda_n)$.

If $k \geq 2$, assume that the assertion is true if A has less than k irreducible diagonal blocks. We can assume that $(\lambda_2, \ldots, \lambda_n)$ are ordered in such a way that

$$\begin{bmatrix} A_1 & 0 & \cdots & 0 \\ 0 & A_2 & \cdots & 0 \\ \vdots & \vdots & & \vdots \\ 0 & 0 & \cdots & A_{k-1} \end{bmatrix}$$

realizes $(\lambda_1\,; \lambda_2, \ldots, \lambda_t) \in S_t$ while A_k realizes $(\lambda_{t+1}\,; \lambda_{t+2}, \ldots, \lambda_n) \in S_{n-t}$. Choose ε' and ε'' so that

$$\varepsilon' > \varepsilon'' > 0 \qquad \text{and} \qquad \varepsilon' + \varepsilon'' = \varepsilon.$$

By the induction hypothesis, there exists a matrix $\hat{A} \gg 0$ which realizes $(\lambda_1 + \varepsilon'; \lambda_2, \ldots, \lambda_t) \in \hat{S}_t$ and a matrix $\hat{B} \gg 0$ which realizes $(\lambda_{t+1} + \varepsilon''; \lambda_{t+2}, \ldots, \lambda_n) \in \hat{S}_{n-t}$. Let $c = (\varepsilon''(\lambda_1 + \varepsilon' - \lambda_t))^{1/2}$. Then c is a positive number and $\lambda_1 + \varepsilon$ and λ_t are the eigenvalues of

$$\begin{bmatrix} \lambda_1 + \varepsilon' & c \\ c & \lambda_t + \varepsilon'' \end{bmatrix}.$$

Let $u \gg 0$, $\|u\| = 1$, be an eigenvector of $\hat{A}$ corresponding to $\lambda_1 + \varepsilon'$ and $v \gg 0$, $\|v\| = 1$, be the eigenvector of $\hat{B}$, corresponding to $\lambda_t + \varepsilon''$. Then, by Lemma 2.3, the matrix

$$\begin{bmatrix} \hat{A} & cuv^t \\ cvu^t & \hat{B} \end{bmatrix}$$

realizes $(\lambda_1 + \varepsilon; \lambda_2, \ldots, \lambda_n)$. ■

Using the notation of Theorem 2.7 we now have the following.

(2.16) Theorem Let $\lambda_1 > \lambda_2 \geq \lambda_3 \geq \cdots \geq \lambda_n$ satisfy conditions (2.8) and (2.9) with strict inequalities; namely,

$$\lambda_1 + \sum_{\substack{i \in K \\ i < k}} (\lambda_i + \lambda_{n+2-i}) + \lambda_{n+2-k} > 0, \qquad \text{for all } k \in K, \tag{2.17}$$

$$\lambda_1 + \sum_{i \in K} (\lambda_i + \lambda_{n+2-i}) + \sum_{j=m+1}^{n+1-m} \lambda_j > 0. \tag{2.18}$$

Then $(\lambda_1; \lambda_2, \ldots, \lambda_n) \in \hat{S}_n$.

Proof Under the assumptions of the theorem there is an $\varepsilon > 0$ such that $\lambda_1 - \varepsilon \geq \lambda_2 \geq \cdots \geq \lambda_n$ satisfy conditions (2.8) and (2.9). Thus

$$(\lambda_1 - \varepsilon; \lambda_2, \ldots, \lambda_n) \in S_n \qquad \text{and} \qquad (\lambda_1; \lambda_2, \ldots, \lambda_n) \in \hat{S}_n. \quad ■$$

Another result on sets realized by positive symmetric matrices is the following "converse" of Theorem 2.15.

(2.19) Theorem Let $(\lambda_1; \lambda_2, \ldots, \lambda_n) \in \hat{S}_n$. Then there exists an $\varepsilon > 0$ such that for all $\sigma \leq \varepsilon$, $(\lambda_1 - \sigma; \lambda_2, \ldots, \lambda_n) \in \hat{S}_n$.

Proof Let $(\lambda_1; \lambda_2, \ldots, \lambda_n)$ be realized by a positive symmetric $n \times n$ matrix A. Let $Au = \lambda_1 u$, $u \gg 0$, $\|u\| = 1$. Then there exists a positive number ε such that $A - \varepsilon uu^t \gg 0$. Consequently, whenever $\sigma \leq \varepsilon$, $B = A - \sigma uu^t \gg 0$ and B is easily seen to realize $(\lambda_1 - \sigma; \lambda_2, \ldots, \lambda_n)$. ■

We now consider the question of the existence of a symmetric nonnegative matrix with prescribed eigenvalues and diagonal elements.

(2.20) Definition A set $(\lambda_1; \lambda_2, \ldots, \lambda_n | a_1, a_2, \ldots, a_n)$, where the numbers $\lambda_2, \ldots, \lambda_n$, as well as the numbers $a_1, \ldots, a_n$, are considered unordered, is said to be *realized* by a matrix A if $a_{ii} = a_i$ and the eigenvalues of A are

$\lambda_1, \ldots, \lambda_n$ and $\lambda_1 = \rho(A)$. We denote by S_n^* the set of all $2n$-tuples $(\lambda_1; \lambda_2, \ldots, \lambda_n | a_1, \ldots, a_n)$ which are realized by some nonnegative symmetric $n \times n$ matrix.

A sufficient condition for a set to be in S_2^* is

(2.21) Lemma If $\lambda_1 \geq \max(a_1, a_2)$ and $\lambda_1 + \lambda_2 = a_1 + a_2$, then

$$(\lambda_1; \lambda_2 | a_1, a_2) \in S_2^*.$$

Proof The set is realized by the matrix

$$\begin{bmatrix} a_1 & t \\ t & a_2 \end{bmatrix} \qquad \text{where} \quad t = (\lambda_1 - a_1)^{1/2}(\lambda_1 - a_2)^{1/2}.$$

The following result extends Theorem 2.6.

(2.22) Theorem If

$$a_{ii} \geq 0, \qquad i = 1, \ldots, n,$$

$$a_1 = \max_i a_i,$$

$$\lambda_j \leq a_j, \qquad j = 2, \ldots, n,$$

and

$$\sum_{i=1}^{n} \lambda_i = \sum_{i=1}^{n} a_i,$$

then

$$(\lambda_1; \lambda_2, \ldots, \lambda_n | a_1, \ldots, a_n) \in S_n^*.$$

Proof The proof is by induction on n. The assertion is trivial for $n = 1$ and follows from the lemma for $n = 2$. Let $n > 2$ and assume that the assertion is true for all $2k$ – tuples with $k < n$.

Define

$$\lambda_1' = \lambda_1 + \lambda_n - a_n,$$

$$\lambda_i' = \lambda_i, \qquad i = 2, \ldots, n-1.$$

The numbers $\lambda_1', \ldots, \lambda_{n-1}', a_1, \ldots, a_{n-1}$ satisfy the conditions of the theorem so, by the induction hypothesis, there exists a nonnegative symmetric matrix $\tilde{A}$ realizing $(\lambda_1'; \lambda_2', \ldots, \lambda_{n-1}' | a_1, \ldots, a_{n-1})$.

According to Lemma 2.21 and the definition of λ_1', there exists a nonnegative symmetric matrix

$$\begin{bmatrix} \lambda_1' & \sigma \\ \sigma & a_n \end{bmatrix} \qquad \text{realizing} \quad (\lambda_1; \lambda_n | \lambda_1', a_n).$$

Let $\tilde{A}u = \lambda_1' u$, $u \geq 0$, $\|u\| = 1$. By Lemma 2.3 the matrix

$$\begin{bmatrix} \tilde{A} & \sigma u \\ \sigma u^t & a_n \end{bmatrix}$$

realizes $(\lambda_1; \lambda_2, \ldots, \lambda_n | a_1, \ldots, a_n)$. ■

Theorem 2.6 follows from Theorem 2.22 by choosing $a_1 = \sum_{j=1}^n \lambda_j$ and $a_j = 0, j = 2, \ldots, n$.

Let $\lambda_1 \geq \lambda_2 \geq \cdots \geq \lambda_n$ and $a_1 \geq a_2 \geq \cdots \geq a_n$. Recall (Exercise 2.6.36) that

$$\sum_{i=1}^{s} \lambda_i \geq \sum_{i=1}^{s} a_i, \qquad s = 1, \ldots, n-1 \tag{2.23}$$

and

$$\sum_{i=1}^{n} \lambda_i = \sum_{i=1}^{n} a_i \tag{2.24}$$

are necessary and sufficient conditions for the existence of a symmetric (not necessarily nonnegative) matrix with eigenvalues λ_i, $i = 1, \ldots, n$ and diagonal elements a_i, $i = 1, \ldots, n$.

(2.25) Theorem If $\lambda_1 \geq \lambda_2 \geq \cdots \geq \lambda_n$ and $a_1 \geq a_2 \geq \cdots \geq a_n$ satisfy $\lambda_k \leq a_{k-1}$, $k = 2, \ldots, n-1$, in addition to conditions (2.23) and (2.24), then

$$(\lambda_1; \lambda_2, \ldots, \lambda_n | a_1, \ldots, a_n) \in S_n^*.$$

Proof Here, too, we use induction on n. The assertion is trivial for $n = 1$ and follows from Lemma 2.21 for $n = 2$.

Let $n \geq 3$ and suppose the assertion is true for $2k$-tuples with $k < n$. Define

$$\lambda_2' = \lambda_1 + \lambda_2 - a_1.$$

Applying the induction hypothesis to $(\lambda_2'; \lambda_3, \ldots, \lambda_n | a_2, \ldots, a_n)$, we see that it is realized by some nonnegative symmetric matrix $\tilde{A}$, of order $n - 1$.

Since $\lambda_1 \geq \lambda_2', \lambda_1 \geq a_1$, and $\lambda_1 + \lambda_2 = \lambda_2' + a_1$, there exists, by Lemma 2.21, a nonnegative 2×2 matrix

$$\begin{bmatrix} \lambda_2' & \sigma \\ \sigma & a_1 \end{bmatrix}$$

with eigenvalues λ_1 and λ_2. Thus, by Lemma 2.3, $(\lambda_1; \lambda_2, \ldots, \lambda_n | a_1, \ldots, a_n)$ is realized by the nonnegative symmetric matrix

$$\begin{bmatrix} \tilde{A} & \sigma u \\ \sigma u^t & a_1 \end{bmatrix},$$

where $\tilde{A}u = \lambda_2' u$, $u \geq 0$, and $\|u\| = 1$. ■

The last theorem of this section gives necessary conditions. In proving it we shall use the following lemma.

(2.26) Lemma Let $\lambda_1 \geq \cdots \geq \lambda_n$ and $a_1 \geq \cdots \geq a_n$, $n \geq 2$. If

$$(\lambda_1; \lambda_2, \ldots, \lambda_n | a_1, \ldots, a_n) \in S_n^*,$$

then $\lambda_1 + \lambda_n \geq a_{n-1} + a_n$.

Proof Let A be a symmetric nonnegative matrix with eigenvalues $\lambda_1, \ldots, \lambda_n$ and diagonal entries $a_1, \ldots, a_n$. Assume first that A is irreducible. Then λ_1 corresponds to a positive eigenvector u of A and λ_n corresponds to an eigenvector v of A which cannot be nonnegative (and not nonpositive). Without loss of generality, we can assume that the positive coordinates of v are its first m coordinates. Let the corresponding partitioning of A, u and, v be

$$A = \begin{bmatrix} A_{11} & A_{12} \\ A_{12}^{t} & A_{22} \end{bmatrix}, \qquad u = \begin{bmatrix} u^{(1)} \\ u^{(2)} \end{bmatrix}, \qquad v = \begin{bmatrix} v^{(1)} \\ v^{(2)} \end{bmatrix},$$

where $u^{(1)} \gg 0$, $u^{(2)} \gg 0$, $v^{(1)} \gg 0$, and $-v^{(2)} > 0$.

Since $\lambda_1 \neq \lambda_n$ (otherwise A is a scalar matrix and thus reducible), it follows that u and v are orthogonal; i.e.,

$$v^{(1)t}u^{(1)} = -v^{(2)t}u^{(2)}. \tag{2.27}$$

From $Au = \lambda_1 u$ follows

$$A_{11}u^{(1)} + A_{12}u^{(2)} = \lambda_1 u^{(1)}. \tag{2.28}$$

From $Av = \lambda_n v$ follows

$$A_{12}^{t}v^{(1)} + A_{22}v^{(2)} = \lambda_n v^{(2)}. \tag{2.29}$$

Multiplying (2.28) by $v^{(1)t}$, (2.29) by $-u^{(2)t}$, and adding, we obtain

$$v^{(1)t}A_{11}u^{(1)} - u^{(2)t}A_{22}v^{(2)} = \lambda_1 v^{(1)t}u^{(1)} - \lambda_n u^{(2)t}v^{(2)}.$$

According to (2.27), this can be written in the form

$$(\lambda_1 + \lambda_n)\, v^{(1)t}u^{(1)} = v^{(1)t}A_{11}u^{(2)} + u^{(2)t}A_{22}(-v^{(2)}). \tag{2.30}$$

Let σ_1 and σ_2 be the minimal diagonal elements of A_{11} and A_{22}, respectively. Then $A_{11} \geq \sigma_1 I$ and $A_{22} \geq \sigma_2 I$, so the expression in (2.30) is greater than or equal to

$$\sigma_1(v^{(1)t}u^{(1)}) + \sigma_2(u^{(2)t}(-v^{(2)})) = (\sigma_1 + \sigma_2)v^{(1)t}u^{(1)}.$$

Thus, $\lambda_1 + \lambda_n \geq \sigma_1 + \sigma_2 \geq a_{n-1} + a_n$.

If A is reducible, it is cogredient to

$$\begin{bmatrix} A_1 & 0 \\ 0 & A_2 \end{bmatrix},$$

where A_1 is that irreducible block which has eigenvalue λ_n, If A_1 is a 1×1 block, then $\lambda_n = a_i \geq a_n$, for some i (in fact, $\lambda_n = a_n$ by conditions (2.23) and (2.24)). Also, $\lambda_1 \geq a_1$, by condition (2.23), so $\lambda_1 + \lambda_n \geq a_{n-1} + a_n$.

If the order of A_1 is greater than one, let $\lambda_1' = \rho(A_1)$ and a_{n-1}' and a_n' be the smallest diagonal elements of A_1. By the proof for irreducible matrices, $\lambda_1' + \lambda_n \geq a_{n-1}' + a_n'$ and by Corollary 2.1.6, $\lambda_1 + \lambda_n \geq a_{n-1} + a_n$. ■

Using the lemma we can prove the following.

(2.31) Theorem If $\lambda_1 \geq \cdots \geq \lambda_n$ and $a_1 \geq \cdots \geq a_n$ are eigenvalues and diagonal elements of a nonnegative symmetric matrix, then

$$\lambda_1 \geq a_1, \qquad \sum_{i=1}^{n} \lambda_i = \sum_{i=1}^{n} a_i$$

and

$$\sum_{i=1}^{s} \lambda_i + \lambda_k \geq \sum_{i=1}^{s-1} a_i + a_{k-1} + a_k \qquad \text{for all } s \text{ and } k, \quad 1 \leq s < k \leq n. \tag{2.32}$$

If $n = 2$ or 3, these conditions, together with $a_n \geq 0$, are also sufficient.

Proof The first two conditions are conditions (2.23) ($s = 1$) and (2.24). We prove (2.32) by induction on n. For $k = n$ and $s = 1$ the condition follows from Lemma 2.26, so let $n \geq 3$ and assume all the conditions (2.32) are fulfilled for matrices of order $n - 1$.

First let $s < k < n$. Let $\hat{A}$ be the principal submatrix of A obtained by deleting the row and column of A containing the diagonal entry a_n.

Let $\hat{\lambda}_1 \geq \hat{\lambda}_2 \geq \cdots \geq \hat{\lambda}_{n-1}$ be the eigenvalues of $\hat{A}$. Then, by the induction hypothesis

$$\sum_{i=1}^{s} \hat{\lambda}_i + \hat{\lambda}_n \geq \sum_{i=1}^{s-1} a_i + a_{k-1} + a_k.$$

By the well-known Cauchy inequalities for eigenvalues of symmetric matrices

$$\lambda_i \geq \hat{\lambda}_i \geq \lambda_{i+1}, \qquad i = 1, \ldots, n-1,$$

so

$$\sum_{i=1}^{s} \lambda_i + \lambda_k \geq \sum_{i=1}^{s-1} a_i + a_{k-1} + a_k.$$

The only remaining cases are $k = n$ and $s > 1$. This time, let $\hat{A}$ be the submatrix of A obtained by deleting the row and column containing a_1. The eigenvalues $\hat{\lambda}_i$ of $\hat{A}$ satisfy, by the induction hypothesis,

$$\sum_{i=1}^{s-1} \hat{\lambda}_i + \hat{\lambda}_{n-1} \geq \sum_{i=2}^{s-1} a_i + a_{n-1} + a_n.$$

Since

$$\sum_{i=1}^{n-1} \hat{\lambda}_i = \sum_{i=2}^{n} a_i,$$

this is equivalent to

$$\sum_{i=s}^{n-2} \hat{\lambda}_i \leq \sum_{i=s}^{n-2} a_i.$$

By the interlacing property of the eigenvalues of A and $\hat{A}$, this implies

$$\sum_{i=s+1}^{n-1} \lambda_i \leq \sum_{i=s}^{n-2} a_i,$$

which is equivalent to

$$\sum_{i=1}^{s} \lambda_i + \lambda_n \geq \sum_{i=1}^{s-1} a_i + a_{n-1} + a_n.$$

Sufficiency is clear for $n = 1$, follows from Lemma 2.21 for $n = 2$, and is stated for $n = 3$ in Theorem 2.25. ■

The example $\lambda_1 = \lambda_2 = 4$, $\lambda_3 = -1$, $\lambda_4 = -3$, $a_1 = a_2 = 2$, $a_3 = a_4 = 0$, demonstrates that the conditions of Theorem 2.31 are not sufficient for $n = 4$.

3 NONNEGATIVE MATRICES WITH GIVEN SUMS

In Chapter 2 we introduced the polytope Ω_n (denoted $\mathscr{D}_n$ in Chapter 3) of doubly stochastic matrices of order n. We saw that the vertices of Ω_n are the $n \times n$ permutation matrices (Birkhoff's theorem, Theorem 2.5.6) and considered diagonal equivalence of a nonnegative matrix to a doubly stochastic one (Exercises 2.6.34 and 2.6.35).

These results can be extened by replacing Ω_n by classes of nonnegative matrices with given row sums and column sums.

For a matrix A, let $r(A)$ denote the vector of row sums of A and let $c(A)$ denote the vector of its column sums.

Let $r^t = [r_1 \cdots r_m]$ and $c^t = [c_1 \cdots c_n]$ be nonnegative vectors such that

$$\sum_{i=1}^{m} r_i = \sum_{j=1}^{n} c_j = s > 0,$$

and let

$$U(r,c) = \{A \in R^{m \times n};\ A \geq 0,\ r(A) = r, \text{ and } c(A) = c\}$$

be the set of *nonnegative matrices with row sums vector r and column sums vector c*. $U(r,c)$ is the domain of the transportation problem of Hitchcock

[1941]. An example of a matrix in $U(r,c)$ is $A = r_i c_j / s$. Clearly $U(r,c)$ is a polytope in $R^{m \times n}$. To describe its vertices we need some basic graph theoretical concepts.

(3.1) Definitions Let G be a (not necessarily directed) graph. A *chain* from a vertex x to a vertex y in G is a sequence of edges $(u_1, \ldots, u_q)$, where $u_i = (x_i, x_{i+1})$ is the edge between x_i and $x_{i+1}, x_1 = x$ and $x_{q+1} = y$. The *length* of a chain is the number of the edges in it. A *cycle* is a chain from a vertex to itself such that no edge appears twice in the sequence.

Let V denote the set of vertices of G. If A is a subset of V, then the graph with A as its vertex set and with all the edges in G that have their endpoints in A is called a *subgraph* of G.

A *connected graph* is a graph that contains a chain from x to y for each pair x,y of distinct vertices. The relation $[x = y$ or there exists a chain in G from x to $y]$ is an equivalence relation on V. The classes of this equivalence relation partition V into connected subgraphs called the *connected components* of G. A *tree* is a connected graph without cycles.

We remark that the graphs associated with the matrices in Chapter 2 are directed and that a path (used to define strong connectedness) is a chain of a special kind.

(3.2) Example The graph

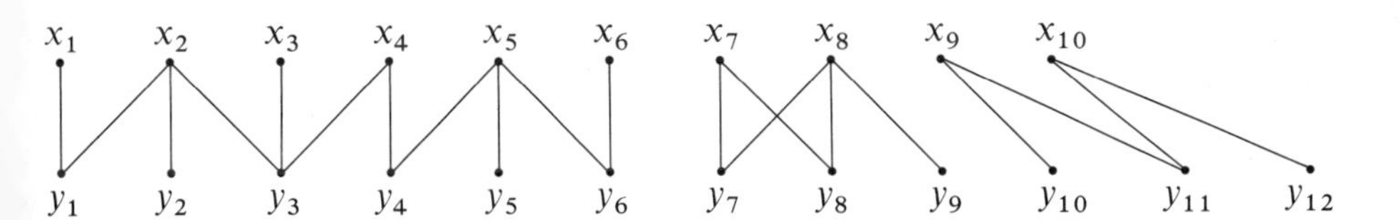

has three connected components. Two are trees but the subgraph with vertices x_7, x_8, y_7, y_8, and y_9 is not a tree since it contains the cycle $((x_7,y_7), (y_7,x_8), (x_8,y_8), (y_8,x_7))$.

Note that there is no edge between a vertex in $\{x_1, \ldots, x_{10}\}$ and a vertex in $\{y_1, \ldots, y_{12}\}$, so G is what is called a *bipartite graph.*

With a nonnegative $m \times n$ matrix A we associate a (nondirected bipartite) graph $BG(A)$ with vertices $x_1, \ldots, x_m, y_1, \ldots, y_n$ where there is no edge between x_{i_1} and $x_{i_2}, i_1, i_2 \in \{1, \ldots, m\}$, or between y_{j_1} and $y_{j_2}, j_1, j_2 \in \{1, \ldots, n\}$, and there is an edge between x_i and y_j if and only if $a_{ij} > 0$.

A graph theoretical characterization of the vertices of $U(r,c)$ is the following.

(3.3) Theorem A matrix A in $U(r,c)$ is an extreme point of $U(r,c)$ if and only if the connected components of $BG(A)$ are trees.

As mentioned in the introduction, the results in this section will be stated without proofs. The reader should consult the notes for references to the original proofs. The following example demonstrates the theorem.

(3.4) Example Thc matrix

$$A = \begin{bmatrix} 2 & 0 & 0 & 0 & 0 & 0 & 0 & 0 & 0 & 0 & 0 & 0 \\ 1 & 2 & 3 & 0 & 0 & 0 & 0 & 0 & 0 & 0 & 0 & 0 \\ 0 & 0 & 1 & 0 & 0 & 0 & 0 & 0 & 0 & 0 & 0 & 0 \\ 0 & 0 & 1 & 2 & 0 & 0 & 0 & 0 & 0 & 0 & 0 & 0 \\ 0 & 0 & 0 & 2 & 1 & 2 & 0 & 0 & 0 & 0 & 0 & 0 \\ 0 & 0 & 0 & 0 & 0 & 2 & 0 & 0 & 0 & 0 & 0 & 0 \\ 0 & 0 & 0 & 0 & 0 & 0 & 1 & 2 & 0 & 0 & 0 & 0 \\ 0 & 0 & 0 & 0 & 0 & 0 & 2 & 1 & 2 & 0 & 0 & 0 \\ 0 & 0 & 0 & 0 & 0 & 0 & 0 & 0 & 0 & 2 & 3 & 0 \\ 0 & 0 & 0 & 0 & 0 & 0 & 0 & 0 & 0 & 0 & 2 & 3 \end{bmatrix}$$

belongs to $U(r,c)$, where $r^t = [2\ 6\ 1\ 3\ 5\ 2\ 3\ 5\ 5\ 5]$ and $u^t = [3\ 2\ 5\ 4\ 1\ 4\ 3\ 3\ 2\ 2\ 5\ 3]$, but is not an extreme point of the set since $BG(A)$, which is the graph in Example 3.2, has a connected component which is not a trec. However its submatrix

$$B = \begin{bmatrix} 2 & 0 & 0 & 0 & 0 & 0 & 0 & 0 & 0 \\ 1 & 2 & 3 & 0 & 0 & 0 & 0 & 0 & 0 \\ 0 & 0 & 1 & 0 & 0 & 0 & 0 & 0 & 0 \\ 0 & 0 & 1 & 2 & 0 & 0 & 0 & 0 & 0 \\ 0 & 0 & 0 & 2 & 1 & 2 & 0 & 0 & 0 \\ 0 & 0 & 0 & 0 & 0 & 2 & 0 & 0 & 0 \\ 0 & 0 & 0 & 0 & 0 & 0 & 2 & 3 & 0 \\ 0 & 0 & 0 & 0 & 0 & 0 & 0 & 2 & 3 \end{bmatrix}$$

is an extreme point of the polytope $U(r,c)$ that contains it.

If $c = r$ and $r_i = k$, $i = 1, \ldots, m$, then the trees corrcsponding to the connected components of the bipartite graph of an extreme point can only have a single edge. Birkhoff's theorem follows in the special case when $k = 1$.

Now consider the convex set of *symmetric nonnegative matrices with row sums vector r*

$$U(r) = \{A \in R^{m \times m};\ A = A^t \geq 0,\ r(A) = r\}.$$

An example of a matrix in $U(r)$ is $A = (r_i r_j / s)$.

With every $m \times m$ symmetric matrix A we associate a graph $G(A)$ which has m vertices, $x_1, \ldots, x_m$, and there is an edge between x_i and x_j if and only if $a_{ij} > 0$. (This is compatible with the definition of the directed graph $G(A)$ in Chapter 2, since A is symmetric.)

As in the case of $U(r,c)$, we use the graphs associated with matrices in $U(r)$ to state a combinatorial classification of the extreme points of the polytope. For this we introduce additional graphical concepts.

(3.5) Definitions A cycle $((x_1x_2),(x_2x_3), \ldots ,(x_qx_1))$ is *elementary* if the q vertices $x_1,x_2, \ldots ,x_q$ are distinct. A *chord* of an elementary cycle is an edge that joins two nonconsecutive vertices of the cycle. A *simple cactus* consists of an elementary cycle without chords along with a tree (possibly without edges) rooted at each vertex of the elementary cycle, where these trees are pairwise vertex disjoint and have no other vertices in common with the elementary cycle. A *simple odd cactus* is a simple cactus whose unique cycle has odd length.

(3.6) Examples Consider the graphs.

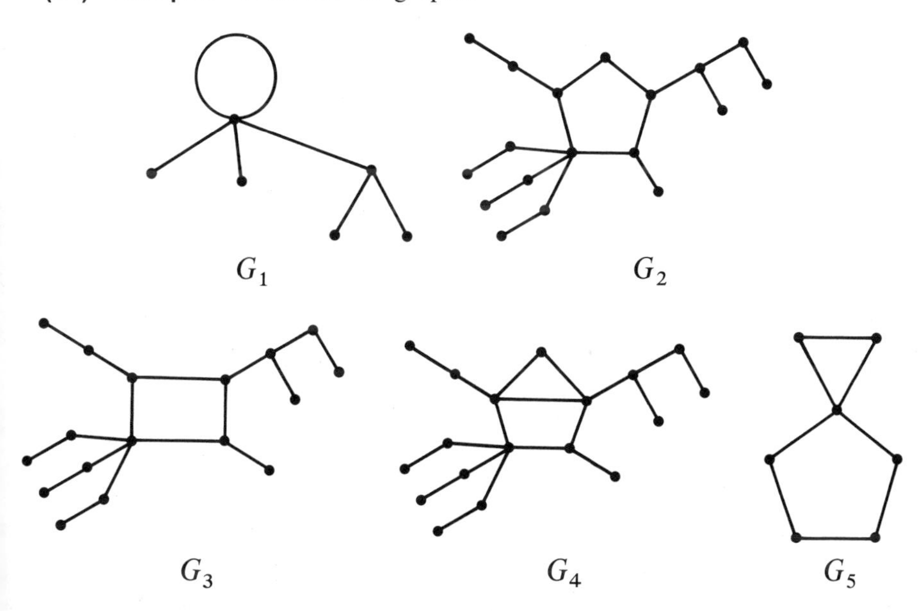

The first three are simple cacti but only G_1 and G_2 are simple odd cacti. G_4 is not a simple cactus because its unique cycle has a chord. The cycle G_5 is not elementary.

(3.7) Theorem A matrix A in $U(r)$ is an extreme point of $U(r)$ if and only if the connected components of $G(A)$ are trees or simple odd cacti.

(3.8) Example Let $r^t = [6\ 1\ 2\ 4\ 2\ 1]$.
The matrix

$$A = \begin{bmatrix} 2 & 1 & 2 & 1 & 0 & 0 \\ 1 & 0 & 0 & 0 & 0 & 0 \\ 2 & 0 & 0 & 0 & 0 & 0 \\ 1 & 0 & 0 & 0 & 2 & 1 \\ 0 & 0 & 0 & 2 & 0 & 0 \\ 0 & 0 & 0 & 1 & 0 & 0 \end{bmatrix}$$

is a vertex of $U(r)$, since $G(A)$ ($=G_1$ of Example 3.6.) is a simple odd cactus.

For $r = e$, the vector of ones, Theorem 3.7 becomes the following.

(3.9) Theorem The extreme points of the polytope of $n \times n$ doubly stochastic symmetric matrices are those matrices which are cogredient to direct sums of (some of) the following three types of matrices:

(i) $[1]$, 1×1 matrix,

(ii)

$$\begin{bmatrix} 0 & 1 \\ 1 & 0 \end{bmatrix}, \quad 2 \times 2 \text{ matrix,}$$

(iii)

$$\begin{bmatrix} 0 & \frac{1}{2} & 0 & 0 & \frac{1}{2} \\ \frac{1}{2} & 0 & \frac{1}{2} & 0 & 0 \\ 0 & \frac{1}{2} & 0 & \frac{1}{2} & 0 \\ 0 & 0 & \frac{1}{2} & 0 & \frac{1}{2} \\ \frac{1}{2} & 0 & 0 & \frac{1}{2} & 0 \end{bmatrix}, \quad k \times k \text{ matrices for any odd } k \geq 3.$$

The graphs $BG(A)$ and $G(A)$, obviously depend only on the zero pattern of A. Given an $m \times n$ nonnegative matrix A it is natural to ask whether there is an $X \in U(r,c)$ with the same zero pattern as A; that is, $a_{ij} > 0$ if and only if $x_{ij} > 0$. Similarly, if A is a symmetric nonnegative matrix of order m, one can inquire if there is an $X \in U(r)$ with the same zero pattern as A. The answers to these questions follow.

(3.10) Theorem Let A be an $m \times n$ nonnegative matrix. The polytope $U(r,c)$ contains a matrix with the same zero pattern as A if and only if

(3.11) for all partitions α_1, α_2 of $\{1, \ldots, m\}$ and β_1, β_2 of $\{1, \ldots, n\}$ into nonempty sets such that $A[\alpha_1 | \beta_2]$ (the submatrix of A based on rows with indices in α_1 and columns with indices in β_2) is a zero matrix,

$$\sum_{j \in \beta_1} c_i \geq \sum_{i \in a_1} r_i,$$

with equality holding if and only if $A[\alpha_2 | \beta_1]$ is also a zero matrix.

(3.12) Theorem Let A be an $m \times m$ symmetric nonnegative matrix. The polytope $U(r)$ contains a matrix with the same zero pattern as A if and only if the following is true.

(3.13) For all partitions α,β,γ of $\{1, \ldots, m\}$, such that $A[\beta \cup \gamma | \gamma]$ is a zero matrix,

$$\sum_{i \in \alpha} r_i \geq \sum_{j \in \gamma} r_j,$$

with equality holding if and only if $A[\alpha | \alpha \cup \beta]$ is also a zero matrix.

Conditions (3.11) and (3.13) are useful in extending and complementing Exercises 2.6.34 and 2.6.35.

(3.14) Theorem Let $A \in R^{m \times n}$ be nonnegative. There exist positive diagonal matrices D_1 and D_2 such that $D_1 A D_2 \in U(r,c)$ if and only if A satisfies condition (3.11).

The diagonal matrices D_1 and D_2 are unique up to a scalar factor unless there exist permutation matrices P and Q such that A can be permuted to a direct sum

$$PAQ = \begin{bmatrix} B & 0 \\ 0 & C \end{bmatrix}.$$

(3.15) Theorem Let $A \in R^{m \times m}$ be nonnegative and symmetric. There exists a positive diagonal matrix D such that $DAD \in U(r)$ if and only if A satisfies condition (3.13).

(3.16) Corollary Let A be a nonnegative symmetric matrix. There exists a positive diagonal matrix D such that DAD is symmetric (doubly) stochastic if and only if there exists a symmetric stochastic matrix having the same zero pattern as A.

For r and c defined as previously we now consider the polytopes

$$U(\leq r, \leq c) = \{A \in R^{m \times n};\ A \geq 0,\ r(A) \leq r,\ c(A) \leq c\}$$

and

$$U(\leq r) = \{A \in R^{m \times m};\ A = A^t \geq 0,\ r(A) \leq r\}.$$

For a matrix A in any of these sets we shall say that the ith row (column) sum of A is *attained* if it is equal to $r_i(c_i)$ and *unattained* if it is strictly less than $r_i(c_i)$.

The extreme points of $U(r,c)$ are also extreme points of $U(\leq r, \leq c)$. The other extreme points of $U(\leq r, \leq c)$ can be constructed from those of $U(r,c)$.

(3.17) Theorem The extreme points of $U(\leq r, \leq c)$ are precisely those matrices obtained as follows: Take an extreme point A of $U(r,c)$ and in each of the trees of $BG(A)$ which are connected components delete a set (possibly empty) of edges of a subtree (a subgraph of a tree which itself is a tree). Replace by zero the positive entries of A which correspond to the edges of $BG(A)$ that were deleted. To be more specific, a matrix A in $U(\leq r, \leq c)$ is an extreme point if and only if the connected components of $BG(A)$ are trees where at most one vertex of each tree corresponds to a row or column of A whose sum is unattained.

(3.18) Example Let

$$r^t = [2\ \ 6\ \ 1\ \ 3\ \ 5\ \ 2\ \ 5\ \ 5] \quad \text{and} \quad c^t = [3\ \ 2\ \ 5\ \ 4\ \ 1\ \ 4\ \ 2\ \ 5\ \ 3].$$

The matrix

$$A = \begin{bmatrix} 2 & 0 & 0 & 0 & 0 & 0 & 0 & 0 & 0 \\ 1 & 2 & 0 & 0 & 0 & 0 & 0 & 0 & 0 \\ 0 & 0 & 1 & 0 & 0 & 0 & 0 & 0 & 0 \\ 0 & 0 & 0 & 0 & 0 & 0 & 0 & 0 & 0 \\ 0 & 0 & 0 & 2 & 1 & 2 & 0 & 0 & 0 \\ 0 & 0 & 0 & 0 & 0 & 2 & 0 & 0 & 0 \\ 0 & 0 & 0 & 0 & 0 & 0 & 0 & 0 & 0 \\ 0 & 0 & 0 & 0 & 0 & 0 & 0 & 0 & 3 \end{bmatrix}$$

is an extreme point of $U(\leq r, \leq c)$ since its graph is

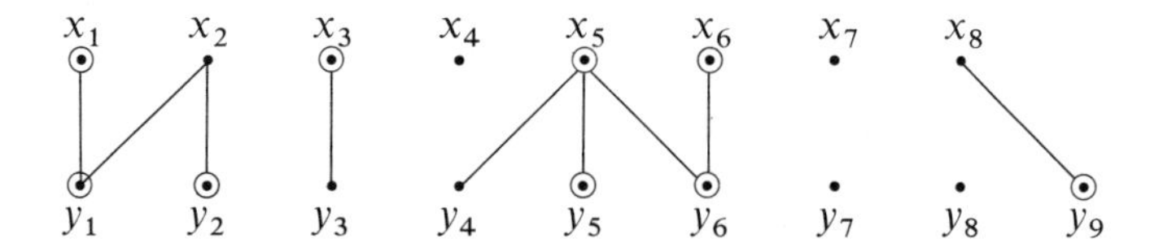

The vertices of the graph that correspond to row sums and column sums that are attained are circled.

Note that A is obtained from the matrix B of Example 3.4 by the rule described in the theorem.

A *subpermutation* matrix is a square (0,1) matrix with at most one 1 in each row and each column. A *doubly substochastic* matrix is a square nonnegative matrix with row sums and column sums at most one.

(3.19) Corollary The subpermutation matrices of order n are the vertices of the polytope of the $n \times n$ doubly substochastic matrices.

The analog of Theorem 3.17 in the symmetric case is the following.

(3.20) Theorem The extreme points of $U(\leq r)$ are precisely those matrices obtained as follows: Take an extreme point A of $U(r)$. In each of the connected components of $G(A)$ which are trees delete a set (possibly empty) of edges of a subtree. In each of the components of $G(A)$ which are simple odd cacti delete a (possibly empty) set of edges of a connected subgraph such that if the set of edges deleted is nonempty it contains all the edges of the cycle. Replace by zero the positive entries of A which correspond to the edges of $G(A)$ that were deleted.

In other words, $A \in U(\leq r)$ is an extreme point if and only if the connected components of $G(A)$ are trees with at most one vertex corresponding to a row whose sum is unattained or simple odd cacti all of whose vertices correspond to rows whose sum is attained.

(3.21) Example The graph $G(A)$

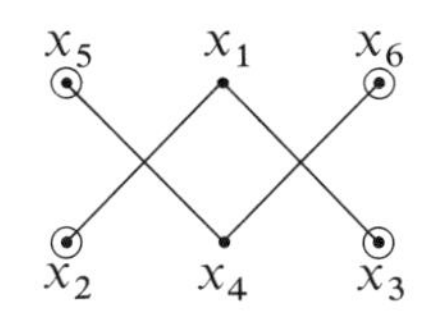

is obtained from the graph G_1 in Example 3.6, by the rule of the theorem. Thus the corresponding matrix

$$A = \begin{bmatrix} 0 & 1 & 2 & 0 & 0 & 0 \\ 1 & 0 & 0 & 0 & 0 & 0 \\ 2 & 0 & 0 & 0 & 0 & 0 \\ 0 & 0 & 0 & 0 & 2 & 1 \\ 0 & 0 & 0 & 2 & 0 & 0 \\ 0 & 0 & 0 & 1 & 0 & 0 \end{bmatrix}$$

is an extreme point of $U(\leq r)$, for the vector r of Example 3.8. The vertices of $G(A)$ that correspond to rows whose sums are attained are circled.

(3.22) Corollary The extreme points of the polytope of symmetric doubly substochastic matrices of order n are those matrices A that are cogredient to a direct sum

$$PAP^{t} = \begin{bmatrix} B & 0 \\ 0 & 0 \end{bmatrix},$$

where B is an extreme point of the set of symmetric doubly stochastic matrices of order k, $k \leq n$.

Let $s(A)$ denote the sum of the elements of A. The concluding result of the section will be an upper bound for the sum of the elements of a power of a symmetric matrix. It will be a special case of the following result.

For $x \in R^m$ let $\hat{x}_i$ be the sum of the i largest elements of x (if $x_1 \geq \cdots \geq x_m$, then $\hat{x}_i = x_1 + \cdots + x_i$), and let $\hat{x} = (\hat{x}_i)$.

(3.23) Theorem Let

$$x_1 \geq \cdots \geq x_m, \qquad y_1 \geq \cdots \geq y_m, \quad \text{and} \qquad z_i = x_i y_i, \quad i = 1, \ldots, m.$$

If

$$A \geq 0, \quad B \geq 0, \quad \hat{r}(A) \leq \hat{x}, \quad \hat{c}(A) \leq \hat{x}, \quad \hat{r}(B) \leq \hat{y}, \quad \text{and} \quad \hat{c}(B) \leq \hat{y},$$

then $\hat{r}(AB) \leq \hat{z}$ and $\hat{c}(AB) \leq \hat{z}$.

The desired upper bound is as follows.

(3.24) Corollary If $A \in U(r)$, then

$$s(A^k) \leq \sum_{i=1}^{m} r_i^k, \qquad k = 1,2, \ldots .$$

4 EXERCISES

(4.1) Prove that if $\lambda_1, \ldots, \lambda_n$ are the eigenvalues of a $n \times n$ nonnegative matrix and if s_k denotes the kth moment $\sum_{i=1}^n \lambda_i^k$, then $s_k^m \leq n^{m-1} s_{km}$, $k,m = 1,2, \ldots$ (Loewy and London [1978]).

(4.2) Prove that if λ_2 and λ_3 are real numbers or conjugate complex numbers and if λ_1 satisfies

$$\lambda_1 \geq \max\{|\lambda_2|, |\lambda_3|\}, \qquad \lambda_1 + \lambda_2 + \lambda_3 \geq 0,$$

and

$$(\lambda_1 + \lambda_2 + \lambda_3)^2 \leq 3(\lambda_1^2 + \lambda_2^2 + \lambda_3^2),$$

then λ_1, λ_2, and λ_3 are the eigenvalues of a nonnegative matrix (Loewy and London [1978]).

(4.3) Let λ_1, λ_2, λ_3, and λ_4 be real numbers such that

$$\lambda_1 = \max_{1 \leq i \leq 4} |\lambda_i| \qquad \text{and} \qquad \lambda_1 + \lambda_2 + \lambda_3 + \lambda_4 \geq 0.$$

Show that they are eigenvalues of a nonnegative matrix (Perfect [1953], Loewy and London [1978]).

(4.4) The real numbers $\lambda_1 \geq \lambda_2 \geq \cdots \geq \lambda_n$ are the eigenvalues of an $n \times n$ Jacobi matrix (i.e., $a_{ij} = 0$ for $|i - j| > 1$) if and only if $\lambda_i + \lambda_{n-i+1} \geq 0$ for all i (Friedland and Melkman [a]).

(4.5) Determine the vertices of $U(r,c)$, $U(r,r)$, $U(r)$, $U(\leq r, \leq c)$, $U(\leq r, \leq r)$, and $U(\leq r)$, for

$$r = \begin{bmatrix} 2 \\ 4 \end{bmatrix} \quad \text{and} \quad c = \begin{bmatrix} 2 \\ 2 \\ 2 \end{bmatrix}.$$

(4.6) Show that

$$A = \begin{bmatrix} 0 & \frac{1}{2} & 0 & \frac{1}{2} \\ \frac{1}{2} & 0 & \frac{1}{2} & 0 \\ 0 & \frac{1}{2} & 0 & \frac{1}{2} \\ \frac{1}{2} & 0 & \frac{1}{2} & 0 \end{bmatrix}$$

is not an extreme point of

$$U\begin{bmatrix} 1 \\ 1 \\ 1 \\ 1 \end{bmatrix}$$

and express it as a convex combination of extreme points.

(4.7) Determine the number of extreme points of $U([1\ 1\ 1\ 1\ 1\ 1]^t)$ and $U([1\ 1\ 1\ 1\ 1]^t)$. Find all the extreme points of $U([1\ 1\ 1\ 1]^t)$ and of $U(\leq [1\ 1\ 1\ 1]^t)$ (Katz [1970]).

(4.8) Let r, c, and s be defined as in Section 3. Prove that $X \in R^{m \times n}$ can be written as $X = A + B$, where $A \in U(r,c)$ and B is nonnegative, if and only if X is nonnegative and

$$\sum_{i \in I} \sum_{j \in J} x_{ij} \geq \sum_{i \in I} r_i + \sum_{j \in J} c_j - s$$

for $I \subseteq \{1, \ldots, m\}$ and $J \subseteq \{1, \ldots, n\}$ (Cruse [1975b], Aharoni [1979]).

(4.9) Let $0 \leq a_i \leq 1$, $i = 1, \ldots, n$. Prove that a necessary and sufficient condition for the existence of a doubly stochastic matrix A such that $a_{ii} = a_i$ is that

$$\sum_{i=1}^{n} a_i \leq n - 2 + 2 \min_{1 \leq j \leq n} a_j$$

(Horn [1954], Brualdi [1968], Erdös and Minc [1973]).

(4.10) Let $0 \leq a_i \leq r_i$, $i = 1, \ldots, m$. Prove that there exists a matrix A in $U(r)$ such that $a_{ii} = a_i$ if and only if

$$\sum_{i=1}^{m} r_i - \sum_{i=1}^{m} a_i \geq 2 \max_{1 \leq j \leq m} (r_j - a_j)$$

(Brualdi [1976]).

(4.11) Show that the convex hull of the symmetric permutation matrices of order m consists of all the symmetric doubly stochastic matrices X that satisfy

$$\sum_{i \in S} \sum_{j \in S - \{i\}} x_{ij} \leq 2k$$

for each subset S of $\{1, \ldots, m\}$ which has $2k + 1$ elements, $k > 0$ (Cruse 1975a]).

(4.12) Let A be an $m \times m$ symmetric nonnegative matrix and let x be a nonnegative vector.

(a) Prove that for every natural number k

$$(x^t A x)^k \leq (x^t x)^{k-1} x^t A^k x.$$

(b) Show that equality holds if x is an eigenvector of A and that if equality holds and x is positive then x is an eigenvector of A.

(c) As a special case, deduce that

$$(s(A))^k \leq m^{k-1} s(A^k)$$

(Mulholland and Smith [1959], London [1966a]).

(4.13) Prove that if $A \geq 0$ is an $m \times m$ symmetric matrix then

$$a_k \equiv (s(A^k)/m)^{1/k} \leq \rho(A), \quad k = 1,2, \ldots, \qquad \text{and} \qquad \lim_{k \to \infty} a_k = \rho(A).$$

Show that $a_k = \rho(A)$ if and only if all the row sums of A^k are equal (London [1966a]).

(4.14) Prove that if $A \geq 0$ is symmetric and A^2 is irreducible, then $(s(A^{2k})/s(A^{2k-2}))^{1/2}$ converges to $\rho(A)$ monotonically (Marcus and Newman [1962]).

(4.15) Let $A \geq 0$ be a symmetric matrix with row sums $r_1, \ldots, r_m$ and let D be the diagonal matrix $D = \text{diag}\{r_1, \ldots, r_m\}$. Prove that A is irreducible if and only if $\text{rank}(A - D) = m - 1$ (Chakravarti [1975]).

(4.16) Let $\lambda_1 \geq \lambda_2 \geq \cdots \geq \lambda_m$ be the eigenvalues of an $m \times m$ irreducible nonnegative symmetric matrix A. Prove that A is either primitive (if and only if $\lambda_1 + \lambda_n > 0$) or cyclic of index 2 (if and only if $\lambda_k + \lambda_{n+1-k} = 0$, $k = 1, \ldots, n$).

(4.17) Determine all 4×4 nonnegative symmetric matrices which are not stochastic but that have a stochastic power (London [1964]).

5 NOTES

(5.1) Most of Section 2 is based on Fiedler [1974a]. Suleimanova [1953] proved, that if

$$\lambda_1 \geq \lambda_2 \geq \cdots \geq \lambda_p \geq 0 > \lambda_{p+1} \geq \cdots \geq \lambda_n \qquad \text{and} \qquad \lambda_1 + \sum_{i=p+1}^{n} \lambda_i \geq 0$$

(which are essentially the conditions of Theorem 2.6) then $\{\lambda_1, \ldots, \lambda_n\}$ is the spectrum of a nonnegative matrix. The same conclusion holds, by a result of Kellog [1971], under the conditions of Theorem 2.7. The reader should observe that in the conclusion of Theorems 2.6 and 2.7 (due to Fiedler), $\{\lambda_1, \ldots, \lambda_n\}$ is the spectrum of a nonnegative *symmetric* matrix.

An excellent survey of inverse eigenvalue problems is included in a thesis by Hershkowits [1978], a student of D. London. The following, open and seemingly very difficult, questions are introduced in the thesis:

(a) Is every real spectrum of a nonnegative matrix the spectrum of a symmetric nonnegative matrix?

(b) Is every (not necessarily real) spectrum of a nonnegative matrix the spectrum of a normal nonnegative one?

(5.2) Additional references on inverse eigenvalue problems include Mirsky [1964], Salzmann [1972], Friedland [1977, 1978, a], Friedland and Karlin [1975], Friedland and Melkman [a], and Loewy and London [1978], from which condition (2.2) is taken.

(5.3) Let P_n be the set of all polynomials $p(\lambda)$ such that $p(A)$ is nonnegative for all $n \times n$ nonnegative matrices A. All polynomials with nonnegative coefficients belong to P_n, but P_n includes also polynomials with negative coefficients. The problem of characterizing the set P_n is suggested in Loewy and London [1978].

A somewhat similar problem is considered by Micchelli and Willoughly [1979] who give conditions on a function f so that if A is symmetric and nonnegative so is $f(A)$. They use these conditions to characterize functions which preserve the class of Stieltjes matrices (see Chapter 6).

(5.4) The transportation problem mentioned in the beginning of Section 3 is as follows:

A homogeneous product is to be shipped in the amounts $r_1, \ldots, r_m$, respectively, from each of m shipping origins and received in amounts $c_1, \ldots, c_n$, respectively, by each of n shipping destinations. The cost of shipping a unit amount from the ith origin to the jth destination is a_{ij}. The problem is to determine the amounts x_{ij} to be shipped from the ith origin to the jth destination so as to minimize the total cost of transportation.

Since a negative shipment has no valid interpretation for the problem as stated, it can be stated as

$$\text{minimize} \sum_{i=1}^{m} \sum_{j=1}^{n} a_{ij} x_{ij}$$

subject to $X \in U(r,c)$.

For details and methods of solution the reader is referred to Hitchcock [1941], Koopmans [1949], Dantzig [1951], and Fulkerson [1956].

(5.5) Theorem 3.3 is taken from Brualdi [1968]. Similar characterizations are given in Fulkerson [1956], Dantzig [1963], and Jurkat and Ryser [1967].

(5.6) Theorem 3.7 is taken from Brualdi [1976]. Similar characterizations are given in Converse and Katz [1975] and Lewin [1977]. Corollary 3.9 was proved by Katz [1970] and by Cruse [1975a].

Inequalities for the eigenvalues of symmetric stochastic matrices are given in Fiedler [1974b].

(5.7) Theorem 3.10 is due to Brualdi [1968] and Theorem 3.12 to Brualdi [1974].

(5.8) Theorem 3.14 is due to Menon and Schneider [1969]. See also Sinkhorn [1974] and Brualdi [1974]. Theorem 3.15 is due to Brualdi [1974] and Corollary 3.16 to Csima and Datta [1972]. See also Maxfield and Minc [1962] and Marcus and Newman [1965].

(5.9) Theorems 3.17 and 3.20 are taken from Brualdi [1976]. Corollary 3.19 is due to Mirsky [1959] and Corollary 3.22 to Katz [1972].

(5.10) The *term rank* of a nonnegative matrix A is defined to be the maximum cardinality of a set of positive entries of A, no two lying in the same row or column. The *symmetric term rank* of a symmetric matrix A is the maximum cardinality of a symmetric set of positive entries of A with no two positive entries in the same row or column. For example, the term rank of

$$A = \begin{bmatrix} 0 & 1 & 1 \\ 1 & 0 & 1 \\ 1 & 1 & 0 \end{bmatrix}$$

is three and its symmetric term rank is two. A formula for the minimum term rank of matrices in $U(r,c)$ was derived by Jurkat and Ryser [1966]. Brualdi [1976] derived an analogous formula for the minimum symmetric term rank of matrices in $U(r)$.

(5.11) Corollary 3.24 was conjectured by London [1966b], who proved it for small k and all m and for small m and all k, and was completely proved by Hoffman [1967] who proved Theorem 3.23.

(5.12) For the graph theoretical concepts used in Section 3, the reader is referred to Berge [1976] and Harary [1969].

We did not mention any of the applications of symmetric nonnegative matrices to graph theory. Examples of such applications are given in Hoffman [1972] and Fiedler [1975].

CHAPTER 5

GENERALIZED INVERSE-POSITIVITY

1 INTRODUCTION

Let K be a proper cone, i.e., a closed, pointed, convex subset of R^n. A matrix $A \in R^{n \times n}$ is called *K-monotone* if it satisfies

(1.1) $$Ax \in K \rightarrow x \in K.$$

This is equivalent to A being nonsingular and

(1.2) $$A^{-1} \in \pi(K), \quad \text{i.e.,} \quad A^{-1}K \subseteq K.$$

We call matrices satisfying (1.2), *K-inverse-positive.* An example of K-monotone matrices is matrices of the form $A = \alpha I - B$ where $B \in \pi(K)$ and $\alpha > \rho(B)$. Such matrices, which are called *K-nonsingular M-matrices,* are of particular interest and will be studied in detail in Chapter 6 for $K = R^n_+$.

In the following chapters we shall explore several applications of K-monotonicity and of extensions of this concept. These extensions allow A to be singular or even rectangular and involve either replacing K in (1.1) by nonempty sets P and Q so that

(1.3) $$Ax \in Q \rightarrow x \in P$$

or replacing the inverse by a generalized inverse. Cone monotonicity is discussed in Section 2, while the more general set monotonicity is the topic of Section 6. The topic of Section 3 is the concept of irreducible monotonicity. Preliminaries on generalized inverses are given in Section 4 and are used there to define generalized inverse positivity. Matrices which are both nonnegative and monotone in some extended sense are studied in Section 5. The chapter is concluded with exercises and bibliographical notes.

2 CONE MONOTONICITY

(2.1) Definition Let K be a proper cone. A matrix A is called *K-monotone* if

(2.2) $$Ax \in K \rightarrow x \in K.$$

(2.3) Theorem A matrix A is K-monotone if and only if it is nonsingular and $A^{-1} \in \pi(K)$.

Proof For a nonsingular matrix A of order n and nonempty sets P and Q in R^n it is obvious that the more general equivalence

$$Q \subseteq AP \leftrightarrow A^{-1}Q \subseteq P \tag{2.4}$$

holds. We have to show that (2.2) implies that A is nonsingular. Let $x \in N(A)$, the null space of A. Then, by (2.2), $x \in K$ and $-x \in K$, and by the pointedness of K, $x = 0$. ∎

The two properties in statement (2.4) will be generalized in the case where A is singular or even rectangular, in Section 4.

To motivate the interest in cone monotonicity, we observe the following simple application to bounds for solutions of linear equations.

(2.5) Theorem Let K be a proper cone and let A be K-monotone. Let u_1 and u_2 be vectors such that $Au_1 \overset{K}{\geq} b \overset{K}{\geq} Au_2$. Then $u_1 \overset{K}{\geq} A^{-1}b \overset{K}{\geq} u_2$.

More sophisticated applications of monotonicity will be studied in forthcoming chapters.

An important class of K-monotone matrices is supplied by the following theorem. These results are strengthened in Chapter 6 for the cone R^n_+.

(2.6) Theorem Let

$$A = \alpha I - B, \qquad B \in \pi(K). \tag{2.7}$$

Then the following are equivalent:

(i) The matrix A is K-monotone.
(ii) The spectral radius of B is smaller than α.
(iii) The matrix A is positive stable; i.e., if λ is an eigenvalue of A, then $\operatorname{Re}\lambda > 0$.

Proof (i) → (ii): Let $x \in K$ be an eigenvector of B, corresponding to $\rho(B)$. Then $Ax = (\alpha - \rho(B))x$.

Since A is nonsingular, being K-monotone, $\alpha \neq \rho(B)$ and $A^{-1}x = (\alpha - \rho(B))^{-1}x$. By K-monotonicity, $A^{-1}x \in K$ and thus $\alpha > \rho(B)$.

(ii) → (i): A is clearly nonsingular, since $\alpha^{-1}B$ is convergent,

$$A^{-1} = \alpha^{-1} \sum_{i=0}^{\infty} (\alpha^{-1}B)^i,$$

so A is K-monotone.

(ii) → (iii): Let λ be an eigenvalue of A. Then $\lambda = \alpha - \mu$, where μ is an eigenvalue of B. Thus (iii) follows from $\alpha > \rho(B) \geq |\mu|$.

(iii) → (ii): Let $x \in K$ be an eigenvector of B corresponding to $\rho(B)$. Then $\lambda = \alpha - \rho(B)$ is an eigenvalue of A and $\operatorname{Re}\lambda > 0$ implies $\alpha > \rho(B)$. ∎

(2.8) Definitions A matrix A is

(i) a *K-nonsingular* M-*matrix* if it can be expressed in the form (2.7) and $\alpha > \rho(B)$ (i.e., if it satisfies any of the equivalent conditions of Theorem 2.6).
(ii) a *K-singular* M-*matrix* if it can be expressed in the form (2.7) where $\alpha = \rho(B)$.

Singular and nonsingular M-matrices (with respect to $K = R^n_+$) will be studied in detail in Chapter 6.

Two corollaries of Theorem 2.6, when B is symmetric and when B is K-irreducible, follow now.

(2.9) Corollary If $B \in \pi(K)$ is symmetric, then $A = \alpha I - B$ is a K-nonsingular M-matrix if and only if A is positive definite.

(2.10) Corollary Let $A = \alpha I - B$, $B \in \pi(K)$, be a K-nonsingular M-matrix. Then $A^{-1} \in \operatorname{int} \pi(K)$ if and only if B is K-irreducible.

Proof If: Let $0 \neq x \in K$. Let $A^{-1}x = y$. Clearly $0 \neq y \in K$ since A is K-monotone. Also $\alpha y \overset{K}{\geq} By$ and by Theorem 1.3.14 and the K-irreducibility of B, $y \in \operatorname{int} K$.

Only if: Suppose B is K-reducible and let x be an eigenvector of B on $\operatorname{bd} K$, say, $Bx = x$. Then, $0 \neq Ax = (\alpha - \beta)x \in \operatorname{bd} K$. But $A^{-1}y \notin \operatorname{int} K$. ■

Recall that A is K-semipositive if $A(\operatorname{int} K) \cap \operatorname{int} K \neq \varnothing$. Rephrasing Theorems 1.3.7, 1.3.14, and 1.3.24 we have, for $A = \alpha I - B \in \pi(K)$:

(2.11) A is a K-nonsingular M-matrix if and only if it is K-semipositive.

(2.12) If B is K-irreducible, then A is a K-nonsingular M-matrix if and only if $A(\operatorname{int} K) \cap (K - \{0\}) \neq \varnothing$.

(2.13) $A(\operatorname{int} K) \cap K \neq \varnothing$ implies that A is a K-(possibly singular) M-matrix. This is an "if and only if" statement if B is K-irreducible.

Statement (2.11) shows that K-monotonicity and K-semipositivity are equivalent for matrices of the form $A = \alpha I - B$, $B \in \pi(K)$. In general, K-monotonicity implies K-semipositivity and, obviously, the converse is false. The relation between the two properties is described by the following results:

(2.14) Theorem The following are equivalent:

(i) A matrix A is K-monotone.
(ii) There exists a K-monotone matrix M such that

$$M^{-1}(M - A) \in \pi(K) \qquad \text{and} \qquad A \text{ is } K\text{-semipositive.} \tag{2.15}$$

(iii) There exists a K-monotone matrix M such that

(2.16) $$M^{-1}(M-A) \in \pi(K) \qquad \text{and} \qquad \rho(M^{-1}(M-A)) < 1.$$

Proof Assuming (i), one can choose $M = A$ in (ii) and (iii). Conversely, either (2.15) or (2.16) means that $M^{-1}A = I - M^{-1}(M-A)$ is a K-nonsingular M-matrix and thus $A = MM^{-1}A$ is K-monotone, being the product of two K-monotone matrices. ■

(2.17) Corollary The following are equivalent:

(i) A matrix A is K-monotone.

(ii) There exists a K-monotone matrix M such that $M \overset{\pi(K)}{\geq} A$ and A is K-semipositive.

(iii) There exists a K-monotone matrix M such that $M \overset{\pi(K)}{\geq} A$ and $\rho(M^{-1}(M-A)) < 1$.

We conclude with a proof of another corollary.

(2.18) Corollary If M_1 and M_2 are K-monotone and $M_2 \overset{\pi(K)}{\geq} A \overset{\pi(K)}{\geq} M_1$, then A, too, is K-monotone.

Proof Let $x_1 = M_1^{-1}y_1$, where $y_1 \in \operatorname{int} K$. Then $0 \neq x \in \operatorname{bd} K$. By a continuity argument there exists $y \in \operatorname{int} K$ such that $x = M_1^{-1}y \in \operatorname{int} K$. Then $Ax \overset{K}{\geq} M_1 x$, so that $Ax \in \operatorname{int} K$, proving that A is K-semipositive. This and $M_2 \overset{\pi(K)}{\geq} A$ satisfy condition (ii) of the previous corollary. ■

3 IRREDUCIBLE MONOTONICITY

In this section we characterize matrices whose inverse is K-positive (not just K-nonnegative). This is done by introducing a broader class of (rectangular) matrices which are called irreducible monotone. The name and definition of these matrices are motivated by Theorem 1.3.14 and Corollary 2.10.

Let K_1 and K_2 be proper cones in R^n and R^m, respectively. Recall that $A \in S_0(K_1, K_2)$ if there exists a nonzero vector in K_1 whose image lies in K_2.

(3.1) Lemma Let $A \in S_0(K_1, K_2)$ be such that

$$Ax \in K_2, \qquad 0 \neq x \in K_1 \rightarrow x \in \operatorname{int} K_1.$$

Let $z \in \operatorname{int} K_1$, $Az \in K_2$, $y \notin K_1$, $Ay \in K_2$. Then

$$y = \alpha z \quad (\alpha \text{ negative}) \qquad \text{and} \qquad Az = Ay = 0.$$

Proof Let $\lambda > 0$ be such that $(\lambda z + y) \in \operatorname{bd} K_1$. Then $A(\lambda z + y) \in K_2$ and, since $(\lambda z + y) \notin \operatorname{int} K_1$, it follows that $\lambda z + y = 0$ and $Az = Ay = 0$. ■

(3.2) Theorem Let $A \in S_0(K_1,K_2)$. Then the following are equivalent:

(i) $Ax \in K_2, 0 \neq x \in K_1 \to x \in \operatorname{int} K_1$.
(ii) $Ax \in K_2, x \neq 0 \to x \in \operatorname{int} K_1$ or $-x \in \operatorname{int} K_1$ and $Az = 0$.

Proof Statement (ii) clearly implies (i). The converse is obvious if $x \in K_1$ and follows from Lemma 3.1 if $x \notin K_1$. ■

(3.3) Definition A matrix in $S_0(K_1,K_2)$ which satisfies the conditions of Theorem 3.2 is called (K_1,K_2)-*irreducible monotone*. The set of (K_1,K_2)-irreducible monotone matrices will be denoted by $M(K_1,K_2)$.

(3.4) Lemma Let $A \in M(K_1,K_2)$. Let $x \in \operatorname{int} K_1$, $Ax \in K_2$, $Az = 0$. Then $z = \alpha x$.

Proof The proof is clear if $z = 0$. If $0 \neq z \in K_1$, the proof follows from Lemma 3.1 with $y = -z$. If $z \notin K_1$, it follows from the lemma with $y = z$. ■

An irreducible monotone matrix has full column rank or almost full column rank as follows.

(3.5) Theorem Let $K_1 \subseteq R^n$, $K_2 \subseteq R^m$, and $A \in M(K_1,K_2)$. Let h be the rank of A. Then $h = n$ or $h = n - 1$.

Proof Suppose $h \leq n - 2$; then there are two nonzero linearly independent vectors y_1 and y_2 such that $Ay_1 = Ay_2 = 0$. Since $A \in M(K_1,K_2)$, there is a $y \in \operatorname{int} K_1$ such that $Ay \in K_2$. Then by Lemma 3.4, $y = \alpha_1 y_1 = \alpha_2 y_2$, contradicting the independence of y_1 and y_2. ■

As a corollary of Theorem 3.5 and Lemma 3.4 we have the following.

(3.6) Corollary Let $A \in M(K_1,K_2)$ be an $m \times n$ matrix of rank h. Then

(i) If $h = n$, then there is $y \in \operatorname{int} K_1$ such that

$$0 \neq Ay \in K_2, \qquad Ax = 0 \to x = 0, \qquad \text{and} \qquad 0 \neq Ax \in K_2 \to x \in \operatorname{int} K_1.$$

(ii) If $h = n - 1$, then $N(A)$ is a line passing through the interior of K_1 and $Ax \in K_2 \to Ax = 0$.

Now we can characterize a matrix whose inverse is K-positive.

(3.7) Theorem Let K_1 and K_2 be (possibly different) proper cones in R^n. Then the following are equivalent:

(i) The matrix A is nonsingular and $A \in M(K_1,K_2)$,
(ii) The matrix A^{-1} maps $K_2 - \{0\}$ into $\operatorname{int} K_1$.

In particular, if K is a proper cone in R^n, then A^{-1} is K-positive if and only if A is a nonsingular matrix in $M(K,K)$.

Proof (i) $\rightarrow$ (ii): Let $x = A^{-1}y$, where $0 \neq y \in K_2$. Then $x \in \operatorname{int} K_1$ by part (i) of Corollary 3.6. (ii) $\rightarrow$ (i): Clearly, $A \in S_0(K_1,K_2)$. Let $0 \neq Ax \in K_2$. Then $x = A^{-1}Ax \in \operatorname{int} K$. ∎

The next, last in this section, corollary, follows from Theorem 3.7 and Corollary 2.10.

(3.8) Corollary Let A be a K-irreducible nonsingular M-matrix. Then $A \in M(K,K)$.

4 GENERALIZED INVERSE-POSITIVITY

To extend cone monotonicity to singular or rectangular matrices we use the concept of generalized inverses.

We start this section with the definition of the basic generalized inverses. If A is nonsingular, then A and $X = A^{-1}$ satisfy

$$AXA = A \tag{4.1}$$

$$XAX = X \tag{4.2}$$

$$AX = (AX)^{\mathrm{t}} \tag{4.3}$$

$$XA = (XA)^{\mathrm{t}} \tag{4.4}$$

$$AX = XA \tag{4.5}$$

$$A^k = XA^{k+1}, \qquad k = 0, 1, 2, \cdots. \tag{4.6}$$

These conditions are used in the following definition.

(4.7) Definition Let λ be a subset of $\{1,2,3,4\}$ which contains 1 and let $A \in R^{m \times n}$. Then a *λ-inverse* of A is a matrix X which satisfies condition (4.i) for each $i \in \lambda$.

The assumption $1 \in \lambda$ assures that the zero matrix is a λ-inverse of A if and only if $A = 0$. Recall that a $\{1,2\}$-inverse is called a *semi-inverse* in Chapter 3.

(4.8) Definition A $\{1,2,3,4\}$-inverse is called the *Moore–Penrose generalized inverse.*

(4.9) Definition The *index* of a square matrix A is the smallest nonnegative integer k such that rank $A^{k+1} = \text{rank } A^k$. (Recall Note 2.7.10.).

(4.10) Definition Let A be a square matrix of index k. The *Drazin inverse* of A is a matrix X which satisfies conditions (4.2), (4.5), and (4.6).

(4.11) Definition A matrix X is a *group inverse* of a (square) matrix A if it satisfies conditions (4.1), (4.2), and (4.5).

(4.12) Definition A matrix Y is a *left* (right) *inverse* of an $m \times n$ matrix A if $YA = I_n$ $(AY = I_m)$, the identity matrix of order n (m).

(4.13) Definition A matrix Y is called a *generalized left inverse* of a (square) matrix A if

$$YAx = x \qquad \text{for all} \quad x \in \bigcap_{m=0}^{\infty} R(A^m).$$

Similarly, Z is called a *generalized right inverse* of A if

$$x^t AZ = x^t \qquad \text{for all} \quad x \in \bigcap_{m=0}^{\infty} R[(A^t)^m].$$

Notice that the $\{1,2\}$-inverses include the left inverses, the right inverses, the Moore–Penrose inverses, and the group inverses but not, in general, the Drazin inverses. For nonsingular matrices all the generalized inverses are A^{-1}.

We now describe, without proof, some properties of these inverses.

Two subspaces L and M of R^n, the n-dimensional real space, are called *complementary* if $L + M = R^n$ and $L \cap M = \{0\}$. In this case every $x \in R^n$ can be expressed uniquely as a sum $x = y + z$, $y \in L$, $z \in M$; and y is called the *projection of* x *on* L *along* M. The linear transformation that carries any $x \in C^n$ into its projection on L along M is called the *projector on* L *along* M.

If A and X are $\{1,2\}$-inverses of each other, then AX is the projector on $R(A)$ along $N(X)$ and XA is the projector on $R(X)$ along $N(A)$.

The Drazin inverse of a matrix A of index k is unique. It is denoted by A^{D} and $A^{\mathrm{D}}A$ is a projector on $R(A^k)$ along $N(A^k)$.

The Moore–Penrose inverse of A is unique and is denoted by A^+. The projectors AA^+ and A^+A are orthogonal. The group inverse of A is denoted by $A^\#$. It is unique if it exists. Let A be an $m \times n$ matrix of rank r factorized as $A = BG$ where B is $m \times r$ and G is $r \times n$. The existence of $A^{\#}$ is equivalent to any of the following conditions:

(a) $R(A)$ and $N(A)$ are complementary subspaces.
(b) The elementary divisors of zero, if any, are linear.
(c) GB is nonsingular.

In this case

$$A^{\#} = B(GB)^{-2}G$$

The group inverse A is the Drazin inverse if the index of A is zero or one and is the same as the Moore–Penrose inverse if and only if $R(A) = R(A^{\mathrm{t}})$. The proof of the following lemma relating generalized left inverses to Eq. (4.6) is obvious and will be omitted.

(4.14) Lemma Let A and Y be square matrices of order n. Then the following statements are equivalent:

(a) Y is a generalized left inverse of A.
(b) Y satisfies Eq. (4.6).
(c) $YA^{m+1} = A^m$ for each $m \geq$ index A.
(d) $YA^{m+1} = A^m$ for some $m \geq 0$.

Consider the system.

$$Ax = b. \tag{4.15}$$

If the system is consistent and G is any $\{1\}$-inverse of A, then Gb is a solution of (4.15). If G is any $\{1,4\}$-inverse at A, the Gb is a solution of (4.15) of minimum Euclidean norm [and $GA = A^+A$ is a projector on $R(A^{\mathrm{t}})$ along $N(A)$]. If (4.15) is inconsistent, a $\{1,3\}$-inverse G of A provides a least-squares solution to (4.15); that is, Gb minimizes the Euclidean norm $\|Ax - b\|$. In general A^+b is the unique least-squares solution having minimum norm. Drazin inverses have applications to differential equations. In this book we shall need them in Chapters 6 and 8.

We now have the necessary background to generalize the concept of inverse-positivity. Monotonicity in this section is with respect to R^n_+.

(4.16) Definitions A is *rectangular monotone* if it has a nonnegative left inverse. A *is λ-monotone* if it has a nonnegative λ-inverse. In particular, A is *semimonotone* if $A^+ \geq 0$ (not if it is $\{1,2\}$-monotone). A is *group monotone* if $A^{\#}$ exists and is nonnegative and D-*monotone* if $A^{\mathrm{D}} \geq 0$.

(4.17) Definition Let A be an $m \times n$ real matrix and let S be a subset of R^n. Then A is said to be *nonnegative* on S if $x \in (R^n_+ \cap S) \rightarrow Ax \in R^n_+$ and *monotone on* S if $Ax \geq 0, x \in S \rightarrow x \geq 0$.

In the study of regular splittings (see Note 8.5) of singular matrices we shall encounter the following classes of matrices.

(4.18) Definition A is *T-monotone* if T is a complementary subspace of $N(A)$ and A is monotone on T. A is *almost monotone* if it is T-monotone for every complementary subspace T of $N(A)$ and *weak monotone* if

$$Ax \geq 0 \rightarrow x \in R_+^n + N(A).$$

We now characterize some of the types of monotonicity just defined, suggesting some of the characterizations as exercises.

(4.19) Theorem Let A be a square matrix of order n. Then A is group monotone if and only if

$$Ax \in R_+^n + N(A), \qquad x \in R(A) \rightarrow x \geq 0. \tag{4.20}$$

Proof Suppose $A^\# \geq 0$, $x \in R(A)$, and $Ax = u + v$, $u \geq 0$, $Av = 0$. Then $x = A^\# Ax = A^\# u + A^\# v = A^\# u \geq 0$. Conversely assume that (4.20) holds. We show first that $A^\#$ exists. Suppose rank $A^2 >$ rank A. Then there exists a vector y such that $A^2 y = 0$ and

$$x = Ay \neq 0. \tag{4.21}$$

Now $Ax = 0 \in R_+^n + N(A)$, $x \in R(A)$, and so by (4.20), $x = Ay \geq 0$. Similarly, $-x \geq 0$, contradicting (4.21). To show that $A^\# \geq 0$, let $w \geq 0$ and decompose w into $w = u + v$, where $u \in R(A)$ and $Av = 0$. Such u and v can be chosen since $A^\#$ exists. Then, since $u \geq -v$, $u \in R_+^n + N(A)$. Thus

$$AA^\# u = u \in R_+^n + N(A), \qquad A^\# u \in R(A).$$

By (4.20) then, $x = A^\# w - A^\# u \geq 0$. ∎

(4.22) Exercise Let A be a square matrix of order n of index k. Prove that $A^{\mathrm{D}} \geq 0$ if and only if

$$Ax \in R_+^n + N(A^k), \qquad x \in R(A^k) \rightarrow x \geq 0.$$

(4.23) Exercise Let A be an $m \times n$ matrix. Prove that $A^+ \geq 0$ if and only if

$$Ax \in R_+^n + N(A^{\mathrm{t}}), \qquad x \in R(A^{\mathrm{t}}) \rightarrow x \geq 0.$$

Notice that if $R(A^{\mathrm{t}}) = R(A)$, then the last condition coincides with (4.20).

The next theorem relates the existence of a nonnegative generalized left inverse of a matrix A to some important monotonicity conditions on A.

(4.24) Theorem Let A be a square matrix of order n and let

$$S = \bigcap_{m=0}^{\infty} R(A^m)$$

then the following statements are equivalent.

(a) A is *generalized left inverse-positive*. That is, A has a generalized left inverse Y with

$$Y \geq 0.$$

(b) A has a generalized left inverse Y which is nonnegative on S.
(c) Every generalized left inverse of A is nonnegative on S. In particular, A^{D} is nonnegative on S.
(d) A is monotone on S.

Proof That (a) → (b) is trivial. Suppose (b) holds and that L is any generalized left inverse of A. Suppose that $x \geq 0$ and $x \in S$. Then $x = A^{k+1}z$ for some z, where $k = \text{index } A$, so that

$$A^k z = YA^{k+1}z = Yx \geq 0$$

by (b). Then

$$Lx = LA^{k+1}z = A^k z \geq 0$$

and thus (c) holds. Now assuming (c), suppose that $Ax \geq 0$ and $x \in S$. Let L be any generalized left inverse of A. Then $Ax \in S$ so that

$$x = LAx \geq 0,$$

since $Ax \geq 0$. This establishes (d). Finally, it will be shown that (d) → (a) by using the theory of the alternative. Let $k = \text{index } A$ and let $(A^k)_i$ denote the ith row of A^k for $i = 1, \ldots, n$. Suppose $A^{k+1}x \geq 0$ and $x = u + v$, $u \in S$, and $A^k v = 0$. Then

$$A^k(Au) = A^{k+1}u = A^{k+1}x \geq 0 \qquad \text{and} \qquad A^k u \in S,$$

so that $A^k u \geq 0$ by (d). Then

$$(A^k)_i x = (A^k)_i u \geq 0.$$

Thus

$$A^{k+1}x \geq 0 \rightarrow (A^k)_i x \geq 0$$

and so by the theory of the alternative it follows that

$$y^{\mathrm{t}} A^{k+1} = (A^k)_i, \qquad y \geq 0,$$

is consistent for $i = 1, \ldots, n$. Then

$$YA^{k+1} = A^k, \qquad Y \geq 0,$$

is consistent so that (a) holds by Lemma 4.14. ■

(4.25) Exercise Prove that $A^{\#}$ exists and is nonnegative on $R(A)$ if and only if A is monotone on $R(A)$.

(4.26) Theorem Let A be a matrix with $N(A) \neq \{0\}$. Then the following statements are equivalent.

(a) A is almost monotone.
(b) $Ax \geq 0 \rightarrow Ax = 0$.

Proof It is clear that (b) implies (a). Conversely, suppose $Ax \geq 0$ and A is almost monotone. Then $x \in N(A)$ or $0 \neq x \in R_+^n$. The latter is imposible since $A(x + \lambda y) = Ax$ for every real λ and $y \in N(A)$ and because of the pointedness of the nonnegative orthant. ■

We conclude by pointing out a relation between λ-monotonicity and T-monotonicity.

(4.27) Exercise A is $R(A^t)$-monotone if and only if it is $\{1,4\}$-monotone.

5 GENERALIZED MONOMIAL MATRICES

In this section we study matrices which are K-nonnegative and K-monotone. For $K = R_+{}^n$ these are the monomial matrices, i.e., products of a permutation matrix and a positive diagonal matrix, studied in Chapter 3. Then, restricting our attention to positivity and monotonicity with respect to the nonnegative orthant we consider nonnegative matrices which are monotone in some sense and finally, we use the Perron–Frobenius theory of Chapter 2 to characterize nonnegative matrices which are equal to their Moore–Penrose generalized inverse. Let $AK = K$ where K is a proper cone. Then clearly A is nonsingular, $A \in \pi(K)$, and $A^{-1} \in \pi(K)$.

(5.1) Theorem Let K be a proper cone.

(a) If A is K-irreducible and $AK = K$, then A^{-1} is also K-irreducible and $A(\text{bd } K) = A^{-1}(\text{bd } K) = \text{bd } K$ (i.e., A is not K-primitive).

(b) Let $A \in \pi(K)$ be K-semipositive (which is the case if A is K-irreducible). Then $AK = K$.

Proof (a) $Ax = \lambda x$ $(x \neq 0)$ if and only if $A^{-1}x = \lambda^{-1}x$, so A^{-1} has one (up to multiplication by a scalar) eigenvector in K and it lies in int K.

(b) By the continuity of A^{-1} it is enough to show that $A^{-1}(\operatorname{int} K) \subseteq K$. Suppose this is not the case and there exists $y \in \operatorname{int} K$ with $A^{-1}y \notin K$. By the semipositivity of A, there exists $x \in \operatorname{int} K$ such that $Ax \in \operatorname{int} K$. Thus for some α between 0 and 1, $z = \alpha y + (1 - \alpha)Ax \in \operatorname{int} K$ and $A^{-1}z \in \operatorname{bd} K$. But this contradicts the assumption that $A(\operatorname{bd} K) = \operatorname{bd} K$. ■

Consider the system (4.15). In many problems, the data in A and b is nonnegative. Then one is naturally interested in nonnegative (least-squares) solutions of (4.15). The existence of such solutions is guaranteed if A has an appropriate nonnegative generalized inverse. We prove the related theorem on A^+ and $A^\#$ and leave the corresponding results on other inverses as exercises.

(5.2) Theorem Let A be an $m \times n$ nonnegative matrix of rank r. Then the following are equivalent.

(a) $A^+ \geq 0$.
(b) There exists a permutation matrix P, such that PA has the form

$$PA = \begin{pmatrix} B_1 \\ \vdots \\ B_r \\ 0 \end{pmatrix}, \tag{5.3}$$

where each B_i has rank 1 and the rows of B_i are orthogonal to the rows of B_j, whenever $i \neq j$. (The zero matrix may be absent.)

(c) $A^+ = DA^t$ for some positive diagonal matrix D.
(d) A is $\{1,3\}$-monotone and $\{1,4\}$-monotone.

Proof (a) → (b): Since $E = AA^+$ is a nonnegative symmetric idempotent, there exists a permutation matrix P so that

$$L = PEP^t = \begin{pmatrix} J_1 & & & 0 \\ & \ddots & & \\ & & J_r & \\ 0 & & & 0 \end{pmatrix} \tag{5.4}$$

where each J_i is a nonnegative idempotent matrix of rank 1 by Theorem 3.3.1. Let $B = PA$. Then $B^+ = A^+P^t$, $BB^+ = L$, $LB = B$, and $B^+L = B^+$. Now B can be partitioned into the form (5.3), where r is the rank of A and where each B_i, $1 \leq i \leq r$, is a $\lambda_i \times n$ matrix with no zero rows. It remains to show that each B_i has rank 1 and $B_iB_j^t = 0$ for $i \neq j$. Let $C = B^+$. Then C can be partitioned into the form

$$C = (C_1, \ldots, C_r, 0),$$

where, for $1 \le i \le r$, C_i is an $n \times \lambda_i$ matrix with no zero column. Moreover, since CB is symmetric, a column of B is nonzero if and only if the corresponding row of C is nonzero. Now $LB = B$ implies that the rows of B_i are orthogonal to the rows of B_j, for $i \ne j$. Since $BC = L$ has the form (5.4),

$$B_i C_j = \begin{cases} J_i & \text{if } i = j, \\ 0 & \text{if } i \ne j, \end{cases}$$

for $1 \le i, j \le r$. Suppose the lth column of B_i is nonzero. Then $B_i C_k = 0$ for $k \ne i$ implies that the lth row of C_k is zero. However, since the lth row of C is nonzero, the lth row of C_i is nonzero. In this case, the lth column of B_k is zero for all $k \ne i$, since $B_k C_i = 0$. Thus $B_i B_i^t = 0$ for all $i \ne j$, proving (b).
(b) → (c): Let $B = PA$ have the form (5.3). Then for $1 \le i \le r$, there exist column vectors x_i, y_i such that $B_i = x_i y_i^t$ so that B_i^+ is the nonnegative matrix

$$B_i^+ = (\|x_i\|^2 \|y_i\|^2)^{-1} B_i^t$$

and moreover $B^+ = (B_1^+, \ldots, B_t^+, 0)$, since $B_i B_j^t = 0$ for $i \ne j$. In particular then, $B^+ = DB^t$ where D is a positive diagonal matrix and thus $A^+ = DA^t$, yielding (c).

(c) → (a): Trivial.

(a) → (d): Trivial.

(d) → (a): Let Y and X be nonnegative matrices such that $YA = A^+A$ and $AX = AA^+$. Thus $A^+ = A^+AA^+ = YAA^+ = YAX \ge 0$. ■

Recall from Chapter 3 that a nonnegative rank factorization of a nonnegative $m \times n$ matrix A of rank r is a factorization $A = BG$ where B is $m \times r$, G is $r \times n$ and B and G are nonnegative. Recall that not every nonnegative matrix has a nonnegative rank factorization, but a nonnegative $\{1\}$-monotone matrix does have such a factorization.

(5.5) Theorem Let A be nonnegative. Then A is group monotone if and only if A has a nonnegative rank factorization $A = BG$, where GB is monomial. In this case, every nonnegative rank factorization of A has this property.

Proof Suppose $A^\# \ge 0$. Then A is $\{1\}$-monotone and has a nonnegative rank factorization. Let $A = BG$ be any such factorization. Then GB is nonsingular and also, $A^\# = B(GB)^{-2}G$. Now by Lemma 3.4.3, B has a nonnegative left inverse B_L and G has a nonnegative right inverse G_R, and thus $(GB)^{-1} = GB(GB)^{-2} = GBB_L A^\# G_R \ge 0$ and since GB is nonnegative, it is monomial. Suppose $A = BG$ is a nonnegative rank factorization such that GB is monomial. Then $(GB)^{-2} \ge 0$ and thus $A^\#$ exists and is nonnegative. ■

It is not difficult to see (Exercise 3.6.7) that a nonnegative matrix is equal to its inverse if and only if A is cogredient to a direct sum of 1×1 matrices $[1]$ and 2×2 matrices of the form

$$\begin{bmatrix} 0 & a \\ 1/a & 0 \end{bmatrix}, \qquad a > 0.$$

We conclude the section with a characterization of matrices which are equal to their Moore–Penrose inverses.

(5.6) Theorem Let A be a nonnegative matrix. Then $A = A^+$ if and only if A is square and there exists a permutation matrix P such that PAP^t is a direct sum of square matrices of the following (not necessarily all) three types:

(i) xx^t, where x is a positive vector such that $x^tx = 1$.

(ii)

$$\begin{pmatrix} 0 & xy^t \\ dyx^t & 0 \end{pmatrix},$$

where x and y are positive vectors (not necessarily of the same order), $d > 0$, $dx^txy^ty = 1$, and the 0's stand for square matrices of the appropriate sizes.

(iii) A zero matrix.

Proof If $A = A^+$, it has to be square. From (4.1) and (4.3) it follows that $A = A^+$ if and only if

$$A^3 = A \tag{5.7}$$

and

$$A^2 \text{ is symmetric.} \tag{5.8}$$

By Theorem 5.2 a necessary condition for $A = A^+ \geq 0$ is

$$A = DA^t, \qquad D = \{\operatorname{diag} d_i\}, \qquad d_i > 0. \tag{5.9}$$

Since A is nonnegative, it can be reduced by a suitable permutation to the triangular block form (2.3.6)

$$PAP^t = \begin{bmatrix} A_{11} & 0 & \cdots & 0 \\ A_{21} & A_{22} & \cdots & 0 \\ \vdots & \vdots & \ddots & \vdots \\ A_{s1} & A_{s2} & \cdots & A_{ss} \end{bmatrix},$$

where the diagonal blocks are square irreducible or zero matrices.

The cogredient permutation is an isomorphism for matrix multiplication and symmetry is invariant by it, thus $A = A^+$ if and only if $(PAP^t)^+ = PAP^t$, (P being a permutation matrix) and PAP^t has to satisfy (5.9). Thus the reducible normal form of A is a diagonal block matrix:

$$PAP^t = \begin{bmatrix} A^1 & 0 & \cdots & 0 \\ 0 & A_2 & & \\ \vdots & \vdots & \ddots & \vdots \\ 0 & 0 & \cdots & A_S \end{bmatrix},$$

where the matrices A_i are irreducible or zero matrices.

We now show that the matrices A_i are of the types given in the theorem. From (5.7) it follows that the minimal polynomials of the matrices A_i (and of A) divide $\lambda^3 - \lambda$. Thus A and the matrices A_i are similar to diagonal matrices and their eigenvalues can be 0, 1, or -1.

The matrix A_i is thus either a zero matrix [type (iii)] or by Theorem 2.1.4 has one as a simple majoring eigenvalue. This is possible in two cases: (i) A_i is primitive or (ii) its index of cyclicity is two.

Suppose A_i is primitive. Then one is a simple eigenvalue and all other eigenvalues (if any) are zeros. Thus trace $A_i = 1$ and since A_i is diagonalizable, rank $A_i = 1$ which means that

$$A_i = xy^t, \tag{5.10}$$

where x and y are positive vectors of the same size. (If x or y had a zero coordinate, A_i would have a zero row (or column) contradicting its irreducibility.) Now by (5.9),

$$A_i = DA_i^t, \qquad D = \{\operatorname{diag} d_k\}, \qquad d_k > 0$$

but

$$A_i \gg 0 \rightarrow (A_i)_{kk} > 0 \rightarrow D = I \rightarrow A_i = A_i^t.$$

The symmetry of A_i follows also from its being idempotent, which follows from

$$A_i = S^{-1} \operatorname{diag}\{1,0,\ldots,0\}S,$$

and from (5.8), and it implies that y in (5.10) may be taken as x. The conditions that trace $A_i = 1$ means that $x^t x = 1$ and thus A_i is of type (i).

Suppose now that the index of cyclicity of A_i is two. Then the eigenvalues of A_i are 1, -1, and possibly zero and the matrix A_i is cogredient to

$$\begin{bmatrix} 0 & B \\ C & 0 \end{bmatrix}, \tag{5.11}$$

where the zero blocks are square matrices (of possibly different orders). By (5.9)

$$\begin{bmatrix} 0 & B \\ C & 0 \end{bmatrix} = \begin{bmatrix} D_1 & 0 \\ 0 & D_2 \end{bmatrix} \begin{bmatrix} 0 & C^t \\ B^t & 0 \end{bmatrix} = \begin{bmatrix} 0 & D_1 C^t \\ D_2 B^t & 0 \end{bmatrix},$$

where $D_1 = \{\operatorname{diag} e_l\}$, $e_l > 0$, $D_2 = \{\operatorname{diag} d_m\}$, $d_m > 0$ are diagonal matrices of the appropriate sizes.

Thus

$$B = D_1 C^t, \tag{5.12}$$

$$C = D_2 B^t. \tag{5.13}$$

Since A_i is diagonalizable, its rank is 2 and rank B = rank $C = 1$. Thus B and C are positive matrices (otherwise A_i would have a zero row or column contradicting its irreducibility), and this, together with (5.12) and (5.13), implies that D_1 and D_2 are scalar matrices: $e_l = 1/d$, $d_m = d$.

Thus (5.11) reduces to the form

$$\begin{bmatrix} 0 & xy^t/d \\ dyx^t & 0 \end{bmatrix},$$

where x and y are positive vectors of the appropriate sizes.

In order that the nonzero eigenvalues of A_i be 1 and -1, the sum of its principal minors of order 2 has to be one, but this sum is $-dx^t xy^t y$, which shows that A_i is of type (ii), completing the "only if" part of the theorem.

To prove the "if" part, recall the remark that $(PAP^t)^+ = PAP^t$ if and only if $A^+ = A$ and check (5.7) and (5.8) for each of the three types of the theorem. They follow from the idempotency and symmetry of the matrices of types (i) and (iii) and can be easily checked for matrices of type (ii). ■

6 SET MONOTONICITY

In the last sections we mainly considered monotonicity with respect to nonnegative orthants. We now generalize this idea and conclude the main part of the chapter with an extension of (1.3), (2.4), and Exercise 4.23.

Let S be a complementary subspace of $R(A)$ and let T be a complementary subspace of $N(A)$. Then A has a unique $\{1,2\}$-inverse, denoted by A^{12}_{ST}, having range T and null space S. In particular

$$A^{12}_{N(A^t)R(A^t)} = A^+.$$

(6.1) Theorem Let A be an $m \times n$ real matrix. Let S be a complementary subspace of $R(A)$. Let T be a complementary subspace of $N(A)$. Let P and Q

be nonempty sets in R^n and R^m, respectively. Then the following statements are equivalent:

(a) $A^{12}_{ST}Q \subseteq P$,
(b) $Ax \in AA^{12}_{ST}Q \rightarrow A^{12}_{ST}Ax \in P$,
(c) $Ax \in Q + S \rightarrow A^{12}_{ST}Ax \in P$,
(d) $Ax \in AA^{12}_{ST}Q,\ x \in T \rightarrow x \in P$,
(e) $Ax \in Q + S,\ x \in T \rightarrow x \in P$.

Proof (a) → (b): Let $Ax = AA^{12}_{ST}u$ for some $u \in Q$. Then $A^{12}_{ST}u$ for some $u \in Q$. Then $A^{12}_{ST}u \in A^{12}_{ST}Q \subseteq P$.

(b) → (c): Let $Ax = u + v$, $u \in Q$, $v \in S$. Then $Ax = AA^{12}_{ST}u \in AA^{12}_{ST}u \in AA^{12}_{ST}Q$ and thus $A^{12}_{ST}Ax \in P$.

(c) → (a): Let $u \in Q$, $u = Ax - v$, $v \in S$, (i.e., $Ax \in Q + S$). Then $A^{12}_{ST}u = A^{12}_{ST}Ax \in P$ so that $A^{12}_{ST}Q \subseteq P$.

(b) → (d), (c) → (e): If $x \in T$, then $x = A^{12}_{ST}Ax$.

(d) → (b): The left part of the implication in (b) may be written as

$$AA^{12}_{ST}Ax \in AA^{12}_{ST}Q.$$

Since $A^{12}_{ST}Ax \in T$, the right part of the implication in (b) follows from (d).

(e) → (c) is proved similarly. ■

7 EXERCISES

(7.1) Let T be a complementary subspace of $N(A)$. Verify the logical implications between the following types of monotonicity.

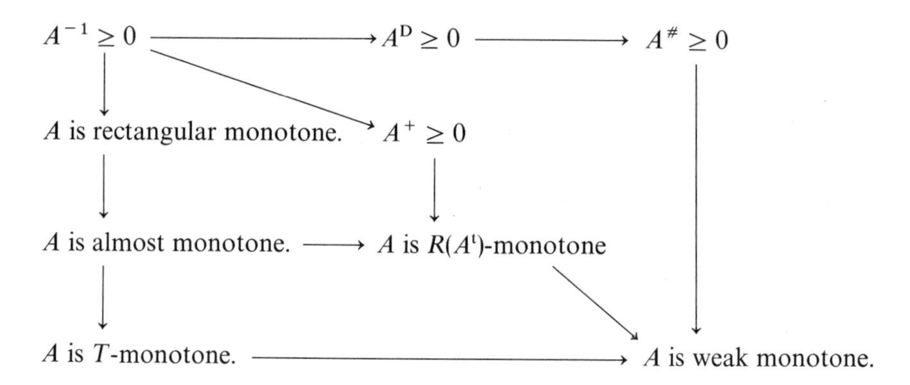

(See Berman and Plemmons [1976].)

(7.2) Let $A = \lambda I - B$ be an irreducible singular M-matrix; i.e., $B \geq 0$ is irreducible and $\lambda = \rho(B)$. Show that A^+ cannot be nonnegative. (Berman and Plemmons [1972]). Extend this result by showing that if A is an M-matrix then $A^+ \geq 0 \leftrightarrow A = 0$ or there is a permutation matrix P such that

$$PAP^t = \begin{bmatrix} M & 0 \\ 0 & 0 \end{bmatrix},$$

M a nonsingular M-matrix (Kuo[1977]).

(7.3) Let $A \geq 0$ be irreducible. Show that if A^+ is nonnegative, then it is irreducible.

(7.4) Verify that

$$A = \begin{bmatrix} 1 & 0 \\ 0 & 1 \\ 1 & 1 \end{bmatrix}$$

has a nonnegative left inverse and is $R(A^t)$-monotone but A^+ is not nonnegative.

(7.5) Let

$$A_1 = \begin{bmatrix} 1 & 0 & 1 \\ 1 & 0 & 1 \\ 0 & 1 & 0 \end{bmatrix} \quad \text{and} \quad A_2 = \begin{bmatrix} 1 & 0 & 0 \\ 0 & 1 & 0 \\ 1 & 1 & 0 \end{bmatrix}.$$

Verify that A_1^+ and $A_2^{\#}$ are nonnegative but $A_1^{\#}$ and A_2^+ are not.

(7.6) Let A be a doubly stochastic matrix. Show that A^+ is doubly stochastic if and only if $A = AXA$ has a double stochastic solution, in which case $A^+ = A^t$ (Plemmons and Cline [1972]).

(7.7) Let A be nonnegative. Show that $A = A^{\#}$ if and only if it has a nonegative rank factorization $A = BG$ such that $(GB)^{-1} = GB$ and that if $A = A^{\#} \geq 0$, then every nonnegative rank factorization of A has this property.

In the last three problems, A is an $m \times n$ real matrix, S is a complementary subspace of $R(A)$ and T a complementary subspace of $N(A)$, P is a nonempty set in R^m and Q a nonempty set in R^n.

(7.8) Show that $A_{ST}^{12}(Q \cap R(A)) \subseteq P$ if and only if $Ax \in Q$, $x \in T \rightarrow x \in P$.

(7.9) Show that $A_{ST}^{12}Q \subseteq P + N(A)$ if and only if $Q \subseteq AP + S$.

(7.10) Show that the following statements are equivalent:

(a) $A_{ST}^{12}(Q \cap R(A)) \subseteq P + N(A)$,
(b) $Ax \in Q \to x \in P + N(A)$,
(c) $Ax \in Q, x \in T \to x \in P + N(A)$,
(d) $Q \cap R(A) \subseteq AP$,
(e) $Q \cap R(A) \subseteq AP + S$.

8 NOTES

(8.1) The definition of monotononicity is due to Collatz [1952], who pointed out the equivalence of (1.1) and (1.2). Schröder, e.g., [1961, 1970, 1972], uses the term inverse-positive. Mangasarian [1968] spoke of "matrices of monotone kind" when he studies what we called rectangular monotonicity.

(8.2) Result (2.11) is due to Fan [1958]. Theorem 2.14 and its corollaries are from Price [1968].

(8.3) Section 3 is based on Berman and Gaiha [1972] which extends to general partial orders, some of the results of Fiedler and Ptak [1966].

(8.4) The terminology on generalized inverses in Section 4 is based on Ben-Israel and Greville [1973]. The reader is referred to this book for proofs of basic properties of generalized inverses described here without a proof. Theorem 4.24 is due to Neumann and Plemmons [1979].

(8.5) A splitting of a nonsingular matrix A to $A = B - C$ where B is monotone and C is nonnegative, is called *regular*. Regular splittings were introduced by Varga [1960] and Schröder [1961]. Schneider [1965] applies the concept to inertia theory. In Chapter 7 we shall define regular splittings of singular and rectangular matrices and describe the applications of regular splittings of nonsingular and singular matrices to iterative methods.

(8.6) The second part of Section 4 is based on works of Berman and Plemmons [1972, 1974b, 1976].

(8.7) Theorem 5.1 is from Barker [1972].

(8.8) Theorem 5.2 is due to Plemmons and Cline [1972]. A geometric interpretation and proof of the theorem was given by Smith [1974].

(8.9) Theorem 5.5 is from Berman and Plemmons [1974b]. Characterizations of λ-monotonicity for all λ with $1 \in \lambda$ were given by Berman and Plemmons [1974c].

(8.10) Theorem 5.6 was proved by Berman [1974b] and extended by Haynsworth and Wall [1979] and Jain, *et al.* [1979a, 1979b] for the polynomial case.

(8.11) Section 6 is based on Carlson [1976].

CHAPTER 6

M-MATRICES

1 INTRODUCTION

Very often problems in the biological, physical, and social sciences can be reduced to problems involving matrices which, due to certain constraints, have some special structure. One of the most common situations is where the matrix A in question has nonpositive off-diagonal and nonnegative diagonal entries, that is, A is a finite matrix of the type

$$A = \begin{bmatrix} a_{11} & -a_{12} & -a_{13} & \cdots \\ -a_{21} & a_{22} & -a_{23} & \cdots \\ -a_{31} & -a_{32} & a_{33} & \cdots \\ \vdots & \vdots & \vdots & \ddots \end{bmatrix},$$

where the a_{ij} are nonnegative. Since A can then be expressed in the form

$$A = sI - B, \qquad s > 0, \qquad B \geq 0, \tag{1.1}$$

it should come as no surprise that the theory of nonnegative matrices plays a dominant role in the study of certain of these matrices.

Matrices of the form (1.1) very often occur in relation to systems of linear or nonlinear equations or eigenvalue problems in a wide variety of areas including finite difference methods for partial differential equations, input–output production and growth models in economics, linear complementarity problems in operations research, and Markov processes in probability and statistics.

We adopt here the traditional notation by letting

$$Z^{n \times n} = \{A = (a_{ij}) \in R^{n \times n} : a_{ij} \leq 0, i \neq j\}$$

Our purpose in this chapter is to give a systematic treatment of a certain subclass of matrices in $Z^{n \times n}$ called M-matrices.

(1.2) Definition Any matrix A of the form (1.1) for which $s \geq \rho(B)$, the spectral radius of B, is called an M-*matrix*.

In Section 2 we consider nonsingular M-matrices, that is, those of the form (1.1) for which $s > \rho(B)$. Characterization theorems are given for $A \in Z^{n \times n}$ and $A \in R^{n \times n}$ to be a nonsingular M-matrix and the symmetric and irreducible cases are considered. In Section 3, the theory of completely monotonic functions is used to study further the inverse-positive property of nonsingular M-matrices. Section 4 is concerned with the total class of M-matrices. Characterization theorems are given here, too, and M-matrices leading to semiconvergent splittings are investigated. These M-matrices arise quite often in solving sparse singular systems of linear equations.

2 NONSINGULAR M-MATRICES

Before proceeding to the main characterization theorem for nonsingular M-matrices, it will be convenient to have available the following lemma, which is the matrix version of the Neumann lemma for convergent series. (Also see Exercise 1.3.11.)

(2.1) Lemma The nonnegative matrix $T \in R^{n \times n}$ is *convergent*; that is, $\rho(T) < 1$, if and only if $(I - T)^{-1}$ exists and

$$(I - T)^{-1} = \sum_{k=0}^{\infty} T^k \geq 0. \tag{2.2}$$

Proof If T is convergent then (2.2) follows from the identity

$$(I - T)(I + T + \cdots + T^k) = I - T^{k+1}, \qquad k \geq 0,$$

by letting k approach infinity.

For the converse let $Tx = \rho(T)x$ for some $x > 0$. Such an x exists by Theorem 2.1.1 or the Perron-Frobenius theorem. Then $\rho(T) \neq 1$ since $(I - T)^{-1}$ exists and thus

$$(I - T)x = [1 - \rho(T)]x$$

implies that

$$(I - T)^{-1}x = \frac{1}{1 - \rho(T)}x.$$

Then since $x > 0$ and $(I - T)^{-1} > 0$, it follows that $\rho(T) < 1$. ∎

For practical purposes in characterizing nonsingular M-matrices, it is evident that we can often begin by assuming that $A \in Z^{n \times n}$. However, many of the statements of these characterizations are equivalent without this assumption. We have attempted here to group together all such statements into certain categories. Moreover, certain other implications follow without assuming that $A \in Z^{n \times n}$ and we point out such implications in the following inclusive theorem. All matrices and vectors considered in this theorem are real.

(2.3) Theorem Let $A \in R^{n \times n}$. Then for each fixed letter $\mathscr{C}$ representing one of the following conditions, conditions $\mathscr{C}_i$ are equivalent for each i. Moreover, letting $\mathscr{C}$ then represent any of the equivalent conditions $\mathscr{C}_i$, the following implication tree holds:

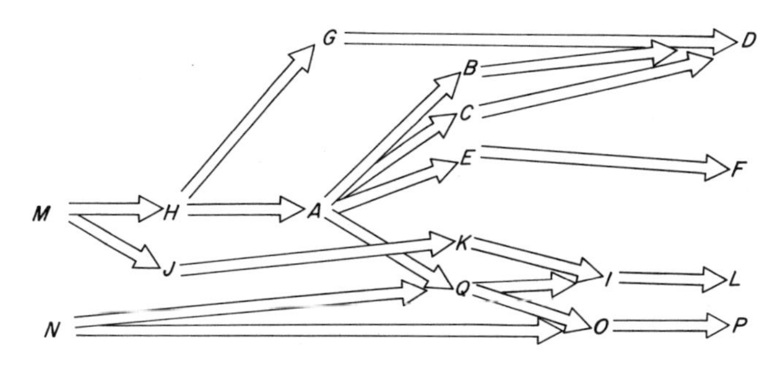

Finally, if $A \in Z^{n \times n}$ then each of the following 50 conditions is equivalent to the statement: "A is a nonsingular M-matrix".

Positivity of Principal Minors

(A_1) All of the principal minors of A are positive.
(A_2) Every real eigenvalue of each principal submatrix of A is positive.
(A_3) For each $x \neq 0$ there exists a positive diagonal matrix D such that

$$x^t A D x > 0.$$

(A_4) For each $x \neq 0$ there exists a nonnegative diagonal matrix D such that

$$x^t A D x > 0.$$

(A_5) A does not reverse the sign of any vector; that is, if $x \neq 0$ and $y = Ax$, then for some subscript i,

$$x_i y_i > 0.$$

(A_6) For each *signature matrix* S (here S is diagonal with diagonal entries ± 1), there exists an $x \gg 0$ such that

$$SASx \gg 0.$$

(B_7) The sum of all the $k \times k$ principal minors of A is positive for $k = 1, \ldots, n$.

(C_8) A is nonsingular and all the principal minors of A are nonnegative.

(C_9) A is nonsingular and every real eigenvalue of each principal submatrix of A is nonnegative.

(C_{10}) A is nonsingular and $A + D$ is nonsingular for each positive diagonal matrix D.

(C_{11}) $A + D$ is nonsingular for each nonnegative diagonal matrix D.

(C_{12}) A is nonsingular and for each $x \neq 0$ there exists a nonnegative diagonal matrix D such that

$$x^t Dx \neq 0 \quad \text{and} \quad x^t ADx > 0.$$

(C_{13}) A is nonsingular and if $x \neq 0$ and $y = Ax$, then for some subscript i, $x_i \neq 0$, and

$$x_i y_i \geq 0.$$

(C_{14}) A is nonsingular and for each signature matrix S there exists a vector $x > 0$ such that

$$SASx \geq 0.$$

(D_{15}) $A + \alpha I$ is nonsingular for each $\alpha \geq 0$.

(D_{16}) Every real eigenvalue of A is positive.

(E_{17}) All the leading principal minors of A are positive.

(E_{18}) There exist lower and upper triangular matrices L and U, respectively, with positive diagonals such that

$$A = LU.$$

(F_{19}) There exists a permutation matrix P such that PAP^t satisfies condition (E_{17}) or (E_{18}).

Positive Stability

(G_{20}) A is *positive stable*; that is, the real part of each eigenvalue of A is positive.

(G_{21}) There exists a symmetric positive definite matrix W such that

$$AW + WA^t$$

is positive definite.

(G_{22}) $A + I$ is nonsingular and

$$G = (A + I)^{-1}(A - I)$$

is convergent,

(G_{23}) $A + I$ is nonsingular and for

$$G = (A + I)^{-1}(A - I)$$

there exists a positive definite matrix W such that

$$W - G^{t}WG$$

is positive definite.

(H_{24}) There exists a positive diagonal matrix D such that

$$AD + DA^{t}$$

is positive definite.

(H_{25}) There exists a positive diagonal matrix E such that for $B = E^{-1}AE$, the matrix

$$(B + B^{t})/2$$

is positive definite.

(H_{26}) For each positive semidefinite matrix Q, the matrix QA has a positive diagonal element.

Semipositivity and Diagonal Dominance

(I_{27}) A is *semipositive*; that is, there exists $x \gg 0$ with $Ax \gg 0$.

(I_{28}) There exists $x > 0$ with $Ax \gg 0$.

(I_{29}) There exists a positive diagonal matrix D such that AD has all positive row sums.

(J_{30}) There exists $x \gg 0$ with $Ax > 0$ and

$$\sum_{j=1}^{i} a_{ij}x_j > 0, \qquad i = 1, \ldots, n.$$

(K_{31}) There exists a permutation matrix P such that PAP^{t} satisfies J_{30}.

(L_{32}) There exists $x \gg 0$ with $y = Ax > 0$ such that if $y_{i_0} = 0$, then there exists a sequence of indices $i_1, \ldots, i_r$ with $a_{i_{j-1}i_j} \neq 0$, $j = 1, \ldots, r$, and with $y_{i_r} \neq 0$.

(L_{33}) There exists $x \gg 0$ with $y = Ax > 0$ such that the matrix $\hat{A} = (\hat{a}_{ij})$ defined by

$$\hat{a}_{ij} = \begin{cases} 1 & \text{if} \quad a_{ij} \neq 0 \quad \text{or} \quad y_i \neq 0, \\ 0 & \text{otherwise}, \end{cases}$$

is irreducible.

(M_{34}) There exists $x \gg 0$ such that for each signature matrix S,

$$SASx \gg 0.$$

(M_{35}) A has all positive diagonal elements and there exists a positive diagonal matrix D such that AD is *strictly diagonally dominant*; that is,

$$a_{ii}d_i > \sum_{j \neq i} |a_{ij}|d_j, \qquad i = 1, \ldots, n.$$

(M_{36}) A has all positive diagonal elements and there exists a positive diagonal matrix E such that $E^{-1}AE$ is strictly diagonally dominant.

(M_{37}) A has all positive diagonal elements and there exists a positive diagonal matrix D such that AD is *lower semistrictly diagonally dominant*; that is,

$$a_{ii}d_i \geq \sum_{j \neq i} |a_{ij}|d_j, \qquad i = 1, \ldots, n,$$

and

$$a_{ii}d_i > \sum_{j=1}^{i-1} |a_{ij}|d_j, \qquad i = 2, \ldots, n.$$

Inverse-Positivity and Splittings

(N_{38}) A is *inverse-positive*; that is, A^{-1} exists and

$$A^{-1} \geq 0.$$

(N_{39}) A is *monotone*; that is,

$$Ax \geq 0 \Rightarrow x \geq 0 \qquad \text{for all} \qquad x \in \mathscr{R}^n.$$

(N_{40}) There exist inverse-positive matrices B_1 and B_2 such that

$$B_1 \leq A \leq B_2.$$

(N_{41}) There exists an inverse-positive matrix $B \geq A$ such that $I - B^{-1}A$ is convergent.

(N_{42}) There exists an inverse-positive matrix $B \geq A$ and A satisfies I_{27}, I_{28}, or I_{29}.

(N_{43}) There exists an inverse-positive matrix $B \geq A$ and a nonsingular M-matrix C, such that

$$A = BC$$

(N_{44}) There exists an inverse-positive matrix B and a nonsingular M-matrix C such that

$$A = BC.$$

(N_{45}) A has a *convergent regular splitting*; that is, A has a representation

$$A = M - N, \qquad M^{-1} \geq 0, \qquad N \geq 0,$$

where $M^{-1}N$ is convergent.

(N_{46}) A has a *convergent weak regular splitting*; that is, A has a representation

$$A = M - N, \qquad M^{-1} \geq 0, \qquad M^{-1}N \geq 0,$$

where $M^{-1}N$ is convergent.

(O_{47}) Every weak regular splitting of A is convergent.

(P_{48}) Every regular splitting of A is convergent.

Linear Inequalities

(Q_{49}) For each $y \geq 0$ the set

$$S_y = \{x \geq 0 : A^t x \leq y\}$$

is bounded, and A is nonsingular.

(Q_{50}) $S_0 = \{0\}$; that is, the inequalities $A^t x \leq 0$ and $x \geq 0$ have only the trivial solution $x = 0$, and A is nonsingular.

Remarks We mention again that Theorem 2.3 identifies those conditions characterizing nonsingular M-matrices that are equivalent for an arbitrary matrix in $R^{n \times n}$, as well as certain implications that hold for various classes of conditions. For example, (A_1)–(A_6) are equivalent and $A \rightarrow B$ for an arbitrary matrix in $R^{n \times n}$. We remark also that if follows that a matrix $A \in Z^{n \times n}$ is a nonsingular M-matrix if and only if each principal submatrix of A is a nonsingular M-matrix and thus satisfies one of the equivalent conditions in Theorem 2.3. It should also be pointed out that some of the classes have left–right duals with A replaced by A^t. Next, we remark that further characterizations, related to mathematical programming, are given in Chapter 10.

Finally, it will be left to the reader to verify that conditions $\mathscr{C}_i$ are equivalent for each i and that the implication tree holds for the conditions (A)–(Q). These proofs are not directly related to the purpose of this book and they can be found in the references cited in Note 6.3.

Proof of Theorem 2.3 for $A \in Z^{n \times n}$ We show now that each of the 50 conditions, (A_1)–(Q_{50}), can be used to characterize a nonsingular M-matrix A, beginning with the assumption that $A \in Z^{n \times n}$.

Suppose first that A is a nonsingular M-matrix. Then in view of the implication tree in the statement of the theorem, we need only show that conditions (M) and (N) hold, for these conditions together imply each of the remaining conditions in the theorem for arbitrary $A \in R^{n \times n}$. Now by Definition 1.2, A has the representation $A = sI - B$, $B \geq 0$ with $s \geq \rho(B)$. Moreover, $s > \rho(B)$

since A is nonsingular. Letting $T = B/s$, it follows that $\rho(T) < 1$ so by Lemma 2.1,

$$A^{-1} = (I - T)^{-1}/s \geq 0.$$

Thus condition (N_{38}) holds. In addition, A has all positive diagonals since the inner product of the ith row of A with the ith column of A^{-1} is one for $i = 1, \ldots, n$. Let $x = A^{-1}e$ where $e = (1, \ldots, 1)^t$. Then for $D = \operatorname{diag}(x_1, \ldots, x_n)$, D is a positive diagonal matrix and

$$ADe = Ax = e \gg 0,$$

and thus AD has all positive row sums. But since $A \in Z^{n \times n}$, this means that A is strictly diagonally dominant and thus (M_{35}) holds.

We show next that if $A \in Z^{n \times n}$ satisfies any of the conditions (A)–(Q), then DA is a nonsingular M-matrix. Once again, in view of the implication tree, it suffices to consider only conditions (D), (F), (L), and (P).

Suppose condition (D_{16}) holds for A and let $A = sI - B$, $B \geq 0$, $s > 0$. Suppose that $s \leq \rho(B)$. Then if $Bx = \rho(B)x$, $x \neq 0$,

$$Ax = [s - \rho(B)]x,$$

so that $s - \rho(B)$ would be a nonpositive real eigenvalue of A, contradicting (D_{16}). Now suppose condition (F_{19}) holds for A. Thus suppose $PAP^t = LU$ where L is lower triangular with positive diagonals and U is upper triangular with positive diagonals. We first show that the off-diagonal elements of both L and U are nonpositive. Let $L = (r_{ij})$, $U = (s_{ij})$ so that

$$r_{ij} = 0 \text{ for } i < j \text{ and } s_{ij} = 0 \text{ for } i > j \text{ and } r_{ii} > 0,\ s_{ii} > 0 \text{ for } 1 \leq i,\ j \leq n.$$

We shall prove the inequalities $r_{ij} \leq 0$, $s_{ij} \leq 0$ for $i \neq j$ by induction on $i + j$. Let

$$A' = PAP^t = (a'_{ij}).$$

If $i + j = 3$ the inequalities $r_{21} \leq 0$ and $s_{12} \leq 0$ follow from $a'_{12} = r_{11}s_{12}$ and $a'_{21} = r_{21}s_{11}$. Let $i + j > 3$, $i \neq j$, and suppose the inequalities $r_{kl} \leq 0$ and $s_{kl} \leq 0$, $k \neq l$, are valid if $k + l < i + j$. Then if $i < j$ in the relation

$$a_{ij} = r_{ii}s_{ij} + \sum_{k<i} r_{ik}s_{kj},$$

we have

$$a_{ij} \leq 0, \qquad \sum_{k<i} r_{ik}s_{kj} \geq 0 \qquad \text{since} \quad r_{ik} \leq 0, \quad s_{kj} \leq 0$$

according to $i + k < i + j$, $k + j < i + j$. Thus $s_{ij} \leq 0$. Analogously, for $i > j$ the inequality $r_{ij} \leq 0$ can be proved. It is easy to see then that L^{-1} and U^{-1} exist and are nonnegative. Thus

$$A^{-1} = (P^tLUP)^{-1} = P^tU^{-1}L^{-1}P \geq 0.$$

Now letting $A = sI - B$, $s > 0$, $B \geq 0$ it follows that $(I - T)^{-1} \geq 0$ where $T = B/s$. Then $\rho(T) < 1$ by Lemma 2.1, and thus $s > \rho(B)$ and A is a nonsingular M-matrix. Next, assume that condition (L_{33}) holds for A. We write $A = sI - B$, $s > 0$, $B \geq 0$ and let $T = B/s$. Then since $y = Ax > 0$ for some $x \gg 0$, it follows that $Tx < x$. Now define $\hat{T} = (\hat{t}_{ij})$ by

$$\hat{t}_{ij} = \begin{cases} t_{ij} & \text{if} \quad t_{ij} \neq 0, \\ \varepsilon & \text{if} \quad t_{ij} = 0 \quad \text{and} \quad y_i \neq 0, \\ 0 & \text{otherwise.} \end{cases}$$

It follows then that $\hat{T}$ is irreducible since $\hat{A}$ defined in (L_{33}) is irreducible. Moreover, for sufficiently small $\varepsilon > 0$,

$$\hat{T}x < x,$$

so that $\rho(\hat{T}) < 1$ by Theorem 2.1.1 or by the Perron–Frobenius Theorem 2.1.4. Finally, since $T \leq \hat{T}$ it follows that

$$\rho(T) \leq \rho(\hat{T}) < 1,$$

by Corollary 2.1.5. Thus A is a nonsingular M-matrix. Next, assume that condition (P_{48}) holds for A. Then since A has a regular splitting of the form $A = sI - B$, $s > 0$, $B \geq 0$, it follows from (P_{48}) that $T = B/s$ is convergent, so that $s > \rho(B)$ and A is a nonsingular M-matrix. This completes the proof of Theorem 2.3 for $A \in Z^{n \times n}$. ■

Next, we consider the necessary and sufficient conditions for an arbitrary $A \in R^{n \times n}$ to be a nonsingular M-matrix. In the following theorem it is not assumed in the hypothesis that A has all nonpositive off-diagonal entries.

(2.4) Theorem Let $A \in R^{n \times n}$, $n \geq 2$. Then each of the following conditions is equivalent to the statement: "A is a nonsingular M-matrix."

(i) $A + D$ is inverse-positive for each positive diagonal matrix D.
(ii) $A + \alpha I$ is inverse-positive for each scalar $\alpha \geq 0$.
(iii) Each principal submatrix of A is inverse-positive.
(iv) Each principal submatrix of A of orders 1,2, and n is inverse-positive.

Proof Suppose that A is a nonsingular M-matrix and let D be a positive diagonal matrix. Then since each principal minor of A is positive, each principal minor of $A + D$ is positive. Thus by condition (A_1) of Theorem 2.3, $A + D$ is a nonsingular M-matrix. Then by condition (N_{38}), $A + D$ is inverse-positive. Thus (i) and (ii) hold. Also from Theorem 2.3, each principal submatrix of a nonsingular M-matrix is also a nonsingular M-matrix. Thus (iii) and (iv) hold.

Now suppose that $A \in R^{n \times n}$ and that $A + \alpha I$ is inverse-positive for each $\alpha \geq 0$. It follows that $A^{-1} \geq 0$ by taking $\alpha = 0$. Thus by condition (N_{38})

of Theorem 2.3, to show that A is a nonsingular M-matrix we need only establish that $A \in Z^{n \times n}$. Assume that A has a positive off-diagonal entry, say, $a_{ij} > 0$, $i \neq j$. Then for sufficiently small $\alpha > 0$,

$$(I + \alpha A)^{-1} = I - \alpha A + (\alpha A)^2 - (\alpha A)^3 + \cdots$$

has its (i, j) entry negative, since the second term of the series dominates this entry. But this is a contradiction since by assumption

$$0 \leq (A + I/\alpha)^{-1} = \alpha(I + \alpha A)^{-1}.$$

Since (i) $\Rightarrow$ (ii) it also follows that if (i) holds for $A \in R^{n \times n}$ then A is a nonsingular M-matrix.

Next, let $A \in R^{n \times n}$ have the property that each of its principal submatrices of orders 1, 2, and n is inverse-positive. Note first then that A has all positive diagonal entries and $A^{-1} \geq 0$. Once again it remains to show that $A \in Z^{n \times n}$. Let

$$B = \begin{bmatrix} a & b \\ c & d \end{bmatrix}$$

be a principal submatrix of A of order 2. Then $a > 0$, $d > 0$, and

$$B^{-1} = (ad - bc)^{-1} \begin{pmatrix} d & -b \\ -c & a \end{pmatrix} \geq 0.$$

Then since the diagonal elements of B are positive it follows that $b \leq 0$ and $c \leq 0$. Thus we conclude that A has all nonpositive off-diagonal elements. Finally, since (iii) $\Rightarrow$ (iv) it also follows that if (iii) holds for $A \in R^{n \times n}$, then A is a nonsingular M-matrix. ■

We next consider two rather important special classes of M-matrices.

(2.5) Definition A symmetric nonsingular M-matrix is called a *Stieltjes matrix*.

(2.6) Exercise A symmetric matrix $A \in Z^{n \times n}$ is a Stieltjes matrix if and only if A is positive definite.

(2.7) Theorem Let $A \in Z^{n \times n}$ be irreducible. Then each of the following conditions is equivalent to the statement: "A is a nonsingular M-matrix."

(i) $A^{-1} \gg 0$.
(ii) $Ax > 0$ for some $x \gg 0$.

Proof Suppose A is a nonsingular M-matrix. Then by taking $K = R^n_+$ in Corollary 5.2.10, it follows that $A^{-1} \gg 0$ so that (i) holds. Condition (ii) follows immediately from condition (I) of Theorem 2.3. Conversely, if (i) holds then A is a nonsingular M-matrix by condition (N) of Theorem 2.3. Thus it remains to consider condition (ii). Suppose that $Ax > 0$ for some $x \gg 0$. Then since A is irreducible, it follows that $\hat{A}$ defined in condition (L_{33}) of Theorem 2.3 is irreducible, and thus A is a nonsingular M-matrix. ■

We conclude this section by indicating a relationship between nonsingular M-matrices and certain other matrices.

(2.8) Definitions For $A \in C^{n \times n}$, the $n \times n$ complex matrices, we define its *comparison matrix* $\mathcal{M}(A) = (m_{ij}) \in R^{n \times n}$ by

$$m_{ii} = |a_{ii}|, \qquad m_{ij} = -|a_{ij}|, \qquad i \neq j, \quad 1 \leq i, j \leq n,$$

and we define

$$\Omega(A) = \{B = (b_{ij}) \in C^{n \times n}: |b_{ij}| = |a_{ij}|, 1 \leq i, j \leq n\}$$

to be the set of *equimodular matrices* associated with A.

The complex case will be considered in Chapter 7, in conjunction with convergence criteria of iterative methods for solving systems of linear equations. The following exercise provides some characterizations for $\mathcal{M}(A)$ to be a nonsingular M-matrix, with $A \in R^{n \times n}$

(2.9) Exercise Let $A \in R^{n \times n}$ have all positive diagonal entries. Then $\mathcal{M}(A)$ is an M-matrix if and only if A satisfies one of the equivalent conditions (M_{34}), (M_{35}), (M_{36}), or (M_{37}) of Theorem 2.3 (see Plemmons [1977]).

3 M-MATRICES AND COMPLETELY MONOTONIC FUNCTIONS

Here the inverse-positivity property of nonsingular M-matrices is investigated further. The technique for developing these results is based in part on a connection between completely monotonic functions and nonnegative functions of nonnegative matrices. As this gives rise to perhaps more elegant proofs of results in Section 2 relating to inverse-positivity, this connection is of interest in itself.

We begin with the following definition.

(3.1) Definition Let $\mathcal{G}(x)$ be a function defined in the interval (a,b) where $-\infty \leq a < b \leq +\infty$. Then, $\mathcal{G}(x)$ is said to be *completely monotonic* in (a,b) if and only if

(3.2) $(-1)^j \mathcal{G}^{(j)}(x) \geq 0 \qquad$ for all $\quad a < x < b \quad$ and all $\quad j = 0, 1, 2, \ldots.$

It is known that if $\mathscr{G}(x)$ is completely monotonic in (a,b), then, for any $y \in (a,b)$, it can be extended to an analytic function in the open disk $|z - y| < y - a$ when b is finite and when $b = +\infty$, $\mathscr{G}$ is analytic in $\mathrm{Re}(z) > a$. Thus, for each y with $a < y < b$, $\mathscr{G}(z)$ is analytic in the open disk $|z - y| < \mathscr{R}(y)$, where $\mathscr{R}(y)$ denotes the radius of convergence of $\mathscr{G}(z)$ about the point $z = y$. It is clear that $\mathscr{R}(y) \geq y - a$ for $a < y < b$.

We now make the change of variables $z = y - \zeta$. Writing

$$\mathscr{G}(y - \zeta) = \sum_{j=0}^{\infty} b_j(y)\zeta^j, \qquad |\zeta| < \mathscr{R}(y),$$

it follows that the coefficients $b_j(y)$ are given by

$$b_j(y) = \frac{(-1)^j \mathscr{G}^{(j)}(y)}{j!}, \qquad j = 0,1,2. \ldots$$

Thus, if $\mathscr{G}(x)$ is completely monotonic in (a,b) and y satisfies $a < y < b$, then the coefficients $b_j(y)$ are, from (3.2), all nonnegative; i.e.,

$$b_j(y) \geq 0 \qquad \text{for} \quad j = 0,1,2, \ldots.$$

We now make use of some matrix notation. If $B \geq 0$, let $j(B)$ denote the order of the largest Jordan block for the eigenvalue $\rho(B)$ in the Jordan normal form for the matrix B.[See Definition 2.3.16 where $j(B)$ is denoted by $v(B)$.] If $B \geq 0$ is irreducible, then we know that $j(B) = 1$. With this notation, we now prove the following.

(3.3) Theorem Let $\mathscr{G}(x)$ be completely monotonic in (a,b), let $B \in R^{n \times n}$ with $B \geq 0$, and let y be any number with $a < y < b$. Then,

$$\mathscr{G}(yI - B) \equiv \sum_{j=0}^{\infty} b_j(y)B^j \tag{3.4}$$

is convergent as a matrix series and defines a matrix with nonnegative entries if and only if $\rho(B) \leq \mathscr{R}(y)$, with $\rho(B) = \mathscr{R}(y)$ only if the series

$$(-1)^m \mathscr{G}^{(m)}(y - \mathscr{R}(y)) = \sum_{j=m}^{\infty} b_j(y) \frac{j!(\mathscr{R}(y))^{j-m}}{(j-m)!} \tag{3.5}$$

and convergent for all $0 \leq m \leq j(B) - 1$.

Proof If $r > 0$ is the radius of convergence of the power series $f(z) = \sum_{j=0}^{\infty} \alpha_j z^j$, then we make use of the well-known fact that the matrix series $f(A) = \sum_{j=0}^{\infty} \alpha_j A^j$ for an $n \times n$ matrix A is convergent if and only if $\rho(A) < r$, with $\rho(A) = r$ only if the series for $f(\lambda_i), \ldots, f^{(m_i - 1)}(\lambda_i)$ are all convergent for any λ_i with $|\lambda_i| = \rho(A) = r$, where m_i is the largest order of the Jordan blocks for the eigenvalue λ_i for the matrix A. If the coefficients α_j of the power series are all nonnegative numbers and if A is itself a nonnegative matrix, it is clear that the preceding result can be simplified to state that $f(A) = \sum_{j=0}^{\infty} \alpha_j A^j$ is convergent if and only if $\rho(A) < r$, with $\rho(A) = r$ only if the series for $f^{(m)}(r)$ are all convergent for $0 \le m \le j(A) - 1$. Now, by the hypotheses of the theorem, it is evident that the coefficients $b_j(y)$ of (3.4) are all nonnegative, and that $B \ge 0$. Thus, to complete the proof, we simply apply the preceding result, noting that the series of (3.5), when convergent, defines a nonnegative matrix.

To extend Theorem 3.3, it is convenient to state the following definition.

(3.6) Definition Let $\mathscr{G}(x)$ be defined in the interval (a,b) where $-\infty \le a < b \le +\infty$. Then, $\mathscr{G}(x)$ is said to be s-*completely monotonic* in (a,b) if and only if

$$(3.7)\quad (-1)^j \mathscr{G}^{(j)}(x) > 0 \qquad \text{for all} \quad a < x < b \quad \text{and all} \quad j = 0,1,2,\ldots.$$

Here the symbol "s" means "strictly."

(3.8) Theorem Let $\mathscr{G}(x)$ be s-completely monotonic in (a,b), let $B \in R^{n \times n}$ with $B \ge 0$, and let y be any number with $a < y < b$. Then, $\mathscr{G}(yI - B) = \sum_{j=0}^{\infty} b_j(y)B^j$ is convergent as a matrix series and defines a matrix with positive entries if and only if B is irreducible and $\rho(B) \le \mathscr{R}(y)$, with $\rho(B) = \mathscr{R}(y)$ only if the series of (3.5) is convergent for $m = 0$.

Proof First, assuming that $\rho(B) \le \mathscr{R}(y)$, with $\rho(B) = \mathscr{R}(y)$ only if the series of (3.5) are convergent for all $0 \le m \le j(B) - 1$, we know from Theorem 3.3 that the matrix $\mathscr{G}(yI - B)$, defined by the convergent power series of (3.4) is a nonnegative matrix. But as $B \ge 0$ and $\mathscr{G}(y)$ is s-completely monotonic, there exists a positive constant k such that

$$\mathscr{G}(yI - B) = \sum_{j=0}^{\infty} b_j(y)B^j \ge k(I + B)^{n-1}.$$

If B is irreducible, it follows that $j(B) = 1$ and that $(I + B)^{n-1} \gg 0$, whence $\mathscr{G}(yI - B) \gg 0$. Conversely, assume that the matrix series of (3.4) is convergent and defines a positive matrix. Using the result of Theorem 3.3, it is only necessary to show that B is irreducible. Assume the contrary. Then, there exists a pair of integers i and j, with $i \ne j$ and $1 \le i, j \le n$, such that $(B^m)_{i,j} = 0$ for all $m = 0,1,2,\ldots$. It is clear that this implies that $(\mathscr{G}(yI - B))_{i,j} = 0$ also, which contradicts the assumption that $\mathscr{G}(yI - B) > 0$. ■

Perhaps the simplest way to show that a function is completely monotonic in $(0,\infty)$ is to use the famous Bernstein result that $\mathscr{G}(x)$ is completely monotonic in $(0,\infty)$ if and only if $\mathscr{G}(x)$ is the Laplace–Stieltjes transform of $\alpha(t)$:

$$(3.9)\qquad \mathscr{G}(x) = \int_0^{\infty} e^{-xt}\, d\alpha(t),$$

where $\alpha(t)$ is nondecreasing and the integral of (3.9) converges for all $0 < x < \infty$. In this case, $\mathscr{G}(z)$ is analytic in $\text{Re}(z) > 0$, and $\mathscr{R}(x) \geq x$. Next, if $\mathscr{G}(x)$ is completely monotonic on $(0,\infty)$, then $\mathscr{G}(x)$ is s-completely monotonic if and only if the nondecreasing function $\alpha(t)$ of (3.9) has at least one positive point of increase; i.e., there exists a $t_0 > 0$ such that

$$\alpha(t_0 + \delta) - \alpha(t_0) > 0 \qquad \text{for any} \quad \delta > 0.$$

This follows from the inequalities of

$$(3.10) \qquad (-1)^j \mathscr{G}^{(j)}(x) = \int_0^\infty e^{-xt} t^j \, d\alpha(t) \geq \int_{t_0}^{t_0+\delta} e^{-xt} t^j \, d\alpha(t)$$
$$\geq \exp[-x(t_0 + \delta)] t_0^j (\alpha(t_0 + \delta) - \alpha(t_0)) > 0$$

for all $0 < x < \infty$ and all $j = 0,1,2, \ldots$. More simply stated, this shows that if $\mathscr{G}(x)$ is completely monotonic in $(0,\infty)$, then $\mathscr{G}(x)$ is s-completely monotonic there if and only if $\mathscr{G}(x)$ does not identically reduce to a constant. Finally, if $\mathscr{G}(x)$ is completely monotonic on $(0,\infty)$, suppose that the nondecreasing function $\alpha(t)$ of (3.9) is such that for some $t_1 > 0$, $\alpha(t) = \alpha(t_1)$ for all $t \geq t_1$, where $\alpha(t_1)$ is finite. It then follows from (3.10) that

$$|\mathscr{G}^{(j)}(x)| = \int_0^{t_1} e^{-xt} t^j \, d\alpha(t) \leq t_1^j [\alpha(t_1) - \alpha(0)]$$
$$\text{for} \quad 0 \leq x < \infty, \quad j = 0,1,2, \ldots.$$

Thus since

$$\frac{|\mathscr{G}^{(j)}(0)|}{j!} \leq \frac{t_1^j [\alpha(t_1) - \alpha(0)]}{j!} \qquad \text{for all} \quad j = 0,1,2, \ldots,$$

it follows that $\mathscr{G}(z)$ in this case is an entire function; i.e., $\mathscr{G}(z)$ is analytic for all complex numbers z. Consequently, for any s with $0 \leq s < \infty$, we have that $\mathscr{R}(s) = +\infty$.

The preceding observations can be used to obtain some results (the first of which is given is Section 2) on nonsingular M-matrices as simple cases of Theorems 3.3 and 3.8. As the first example, we have the following.

(3.11) Theorem Let $B \in R^{n \times n}$ with $B \geq 0$. If $A \equiv sI - B$ where $s > 0$, then A is nonsingular and $A^{-1} \geq 0$ if and only if $s > \rho(B)$. Moreover, $A^{-1} \gg 0$ if and only if $\rho(B) < s$ and B is irreducible.

Proof If we write $\mathscr{G}_1(x) = (1/x) = \int_0^\infty e^{-xt} \, d\alpha_1(t)$ for $0 < x < \infty$, where $\alpha_1(t) = t$ for $t \geq 0$, then $\mathscr{G}_1(x)$ is s-completely monotonic on $(0,\infty)$, and $\mathscr{R}(s) = s$ for $s > 0$. Since $\mathscr{G}_1(x)$ is unbounded for $x = 0$, the series (3.5) for $\mathscr{G}_1(0) = \mathscr{G}_1(s - \mathscr{R}(s))$ is divergent. Then, apply Theorems 3.3 and 3.8. ■

(3.12) Theorem Let C be any *essentially nonnegative matrix*; that is, $C + sI \geq 0$ for all real s sufficiently large. Then, for all $t \geq 0$,

$$\exp(tC) = \sum_{j=0}^{\infty} \frac{(tC)^j}{j!} \geq 0.$$

Moreover, $\exp(tC) \gg 0$ for some (and hence all) $t > 0$ if and only if C is irreducible.

Proof Writing $\mathscr{G}_2(x) = e^{-x} = \int_0^\infty e^{-xt}\, d\alpha_2(t)$ for $0 < x < \infty$, where $\alpha_2(t) = 0$ for $0 \leq t < 1$, and $\alpha_2(t) = 1$ for $t \geq 1$, then $\mathscr{G}_2(x)$ is s-completely monotonic on $(0,\infty)$ and $\mathscr{G}_2(z)$ is an entire function. Thus, $\mathscr{R}(y) = +\infty$ for any $0 < y < \infty$. By hypothesis, for any $t \geq 0$, $B \equiv tC + sI$ is a nonnegative matrix for all positive s sufficiently large, and thus $\mathscr{G}_2(sI - B) = \exp(tC) \geq 0$ from Theorem 3.3. The remainder follows from Theorem 3.8. ■

While it is true that not all results on functions of nonnegative matrices fall out as consequences of Theorems 3.3 and 3.8, we nevertheless can generate some interesting results on M-matrices such as the following.

(3.13) Theorem Let C be any essentially nonnegative matrix. Then $\{I - \exp(tC)\}(-C)^{-1} \geq 0$ for all $t \geq 0$. Moreover, $\{I - \exp(tC)\}(-C)^{-1} \gg 0$ for all $t > 0$ if and only if C is irreducible.

Proof Writing $\mathscr{G}_3(x) = (1 - e^{-x})/x = \int_0^\infty e^{-xt}\, d\alpha_3(t)$ for $0 < x < \infty$, where $\alpha_3(t) = t$ for $0 \leq t \leq 1$ and $\alpha_3(t) = 1$ for $t \geq 1$, then $\mathscr{G}_3(x)$ is s-completely monotonic on $(0,\infty)$ and $\mathscr{G}_3(z)$ is an entire function. By hypothesis, for any $t \geq 0$, $B \equiv tC + sI$ is a nonnegative matrix for all positive s sufficiently large, and the conclusions follow from Theorems 3.3 and 3.8. ■

This section is concluded by listing some important corollaries of Theorem 3.13. The second corollary provides another necessary and sufficient condition for $A \in R^{n \times n}$ to be a nonsingular M-matrix, thus extending Theorem 2.4.

(3.14) Corollary Let A be a nonsingular M-matrix. Then

$$\{I - \exp(-tA)\}A^{-1} > 0 \quad \text{for all } t \geq 0$$

and

$$\{I - \exp(-tA)\}A^{-1} \gg 0 \quad \text{for all } t > 0$$

if and only if A is irreducible.

(3.15) Corollary Let $A = (a_{i,j})$ be an inverse-positive matrix; that is, A is nonsingular and $A^{-1} \geq 0$. Then $\{I - \exp(-tA)\}A^{-1} > 0$ for all $t \geq 0$ if and only if A is a nonsingular M-matrix. Similarly, $\{I - \exp(-tA)\}A^{-1} \gg 0$ for all $t > 0$ if and only if A is an irreducible nonsingular M-matrix.

4 GENERAL M-MATRICES

In this section we investigate some of the properties of the total class of M-matrices. Singular M-matrices, that is, matrices A of the form $A = \rho(B)I - B$, $B \geq 0$, are perhaps almost as prevalent in the application areas as the nonsingular M-matrices. However, the theory here is not yet so fully developed—perhaps because the concepts are considerably more difficult to study.

We begin with a lemma that shows that the total class of M-matrices can be thought of as the closure of the class of nonsingular M-matrices.

(4.1) Lemma Let $A \in Z^{n \times n}$. Then A is an M-matrix if and only if

$$A + \varepsilon I$$

is a nonsingular M-matrix for all scalars $\varepsilon > 0$.

Proof Let A be an M-matrix of the form $A = sI - B$, $s > 0$, $B \geq 0$. Then for any $\varepsilon > 0$

$$A + \varepsilon I = sI - B + \varepsilon I = (s + \varepsilon)I - B = s'I - B, \tag{4.2}$$

where $s' = s + \varepsilon > \rho(B)$ since $s \geq \rho(B)$. Thus $A + \varepsilon I$ is nonsingular.

Conversely if $A + \varepsilon I$ is a nonsingular M-matrix for all $\varepsilon > 0$, then it follows that A is an M-matrix by considering Eq. (4.2) and letting ε approach zero. ■

The primary objective of this section is to investigate those conditions in Theorem 2.3 that have some corresponding form that can be used to characterize the total class of M-matrices. Before extending parts of Theorem 2.3 to the general case, the following exercises are in order.

(4.3) Exercises Let $A \in R^{n \times n}$ and let A^{D} denote the *Drazin inverse* of A; that is, A^{D} is the operator inverse of A where the domain of A is restricted to $R(A^k)$, $k = \text{index } A$, as defined in Definition 5.4.10.

(a) Show that if $A = sI - B$, $T = B/s$, then $A^{\mathrm{D}} = (1/s)(I - T)^{\mathrm{D}}$.

(b) Show that if $A = I - T$, $T \geq 0$ with $\rho(T) = 1$ then for any $0 < \alpha < 1$

$$(1 - \alpha)I + \alpha T$$

is also nonnegative and has only the eigenvalue one on the unit circle.

The following is a natural extension of Lemma 2.1 to the singular cases. Recall that by Definition 5.4.17, a matrix A is nonnegative on a set $S \subseteq R^n$ if and only if

$$Ax \geq 0 \qquad \text{whenever} \quad x \in S \quad \text{and} \quad x \geq 0.$$

In particular, $A \geq 0$ if and only if A is nonnegative on R^n_+.

(4.4) Lemma If $T \in R^{n \times n}$ and $T \geq 0$ then $\rho(T) \leq 1$ if and only if $(I - T)^{\mathrm{D}}$ is nonnegative on $R[(I - T)^k]$, $k = \operatorname{index}(I - T)$. Moreover in this case if $0 < \alpha < 1$ and $T_\alpha = (1 - a)I + \alpha T$, then

$$(I - T)^{\mathrm{D}} = \alpha \sum_{j=0}^{\infty} T_\alpha^j E, \tag{4.5}$$

where E is the projector $(I - T)(I - T)^{\mathrm{D}}$.

Proof Suppose that $\rho(T) \leq 1$. If $\rho(T) < 1$ we can apply Lemma 2.1; thus suppose $\rho(T) = 1$. For $0 < \alpha < 1$ let

$$T_\alpha = (1 - \alpha)I + \alpha T.$$

Then $\lambda \in \sigma(T_\alpha)$, $|\lambda| = 1$ implies $\lambda = 1$, by Exercises 4.3. Moreover

$$I - T_\alpha = I - (1 - \alpha)I - \alpha T = \alpha(I - T).$$

Thus $(I - T)^{\mathrm{D}} = \alpha(I - T_\alpha)^{\mathrm{D}}$ and

$$E = (I - T)(I - T)^{\mathrm{D}} = (I - T_\alpha)(I - T_\alpha)^{\mathrm{D}}.$$

By considering the Jordan normal forms for T_α and E it follows that $\rho(T_\alpha E) < 1$ and that

$$(I - T_\alpha)^{\mathrm{D}} = (I - T_\alpha E)^{-1} + E - I.$$

But by Lemma 2.1

$$(I - T_\alpha E)^{-1} = \sum_{j=0}^{\infty} (T_\alpha E)^j = I + \sum_{j=1}^{\infty} T_\alpha^j E.$$

Thus

$$(I - T)^{\mathrm{D}} = \alpha \sum_{j=0}^{\infty} T_\alpha^j E$$

and (4.5) holds.

Now suppose $x \in R[(I - T)^k]$, $k = \operatorname{index}(I - T)$, and let $x \geq 0$. Then since E is a projector onto $R[(I - T)^k]$ it follows that $x = Ex$. Then since $T_\alpha \geq 0$ by Exercise 4.3,

$$(I - T)^{\mathrm{D}} x = \alpha \sum_{j=0}^{\infty} T_\alpha^j E x = \alpha \sum_{j=0}^{\infty} T_\alpha^j x \geq 0.$$

Thus $(I - T)^{\mathrm{D}}$ is nonnegative on $R[(I - T)^k]$.

For the converse let

$$Tx = \rho(T)x, \qquad x > 0,$$

and suppose $\rho(T) \neq 1$. Then

$$(I - T)x = [1 - \rho(T)]x$$

and moreover

$$(I - T)^j x = [1 - \rho(T)]^j x$$

for all $j \geq 0$. Thus if $k = \text{index}(I - T)$,

$$x \in R[(I - T)^k].$$

Then

$$x = (I - T)^{\mathrm{D}}(I - T)x = [1 - \rho(T)](I - T)^{\mathrm{D}}x$$

and since $(I - T)^{\mathrm{D}}x > 0$, it follows that $\rho(T) < 1$ and thus $\rho(T) \leq 1$ and the lemma is proved. ■

The following theorem extends Theorem 2.3 somewhat by providing certain characterizations of an arbitrary M-matrix.

(4.6) Theorem Let $A \in R^{n \times n}$. Then for each fixed letter $\mathscr{C}$ representing one of the following conditions, conditions $\mathscr{C}_i$ are equivalent for each i. Moreover letting $\mathscr{C}$ represent any of the equivalent conditions $\mathscr{C}_i$, the following implications hold:

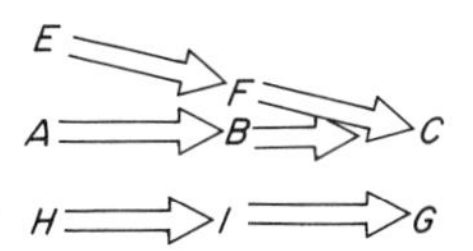

Finally, if $A \in Z^{n \times n}$ then each of the following conditions is equivalent to the statement: "A is an M-matrix."

Nonnegativity of Principal Minors

(A_1) All the principal minors of A are nonnegative.
(A_2) Every real eigenvalue of each principal submatrix of A is nonnegative.
(A_3) $A + D$ is nonsingular for each positive diagonal matrix D.
(A_4) For each $x \neq 0$ there exists a nonnegative diagonal matrix D such that

$$x^{\mathrm{t}}Dx \neq 0 \qquad \text{and} \qquad x^{\mathrm{t}}ADx \geq 0.$$

(A_5) If $x \neq 0$ and $y = Ax$, then there is a subscript i such that $x_i \neq 0$ and

$$x_i y_i \geq 0.$$

(A_6) For each signature matrix S there exists a vector $x > 0$ such that

$$SASx \geq 0.$$

(B_7) The sum of all the $k \times k$ principal minors of A is nonnegative for $k = 1, \ldots, n$.

(C_8) Every real eigenvalue of A is nonnegative.
(C_9) $A + \alpha I$ is nonsingular for each $\alpha > 0$.
(D_{10}) There exists a permutation matrix P and lower and upper triangular matrices L and U, respectively, with nonnegative diagonal entries, such that

$$PAP^{t} = LU.$$

Nonnegative Stability

(E_{11}) The real part of each nonzero eigenvalue of A is positive.
(F_{12}) A is *nonnegative stable*; that is, the real part of each eigenvalue of A is nonnegative.
(F_{13}) $A + I$ is nonsingular and for

$$G = (A + I)^{-1}(A - I),$$

it follows that $\rho(G) \leq 1$.

Generalized Inverse-Positivity and Splittings

(G_{14}) A is *generalized left inverse-positive*; that is, there exists a matrix Y satisfying

$$Y \geq 0 \qquad \text{and} \qquad YA^{k+1} = A^{k} \qquad \text{for some} \quad k \geq 1.$$

(G_{15}) A has a generalized left inverse Y such that Y is *nonnegative on*

$$V_A = \bigcap_{m=0}^{\infty} R(A^m);$$

that is,

$$x \geq 0 \qquad \text{and} \qquad x \in V_A \to Yx \geq 0.$$

(G_{16}) Every generalized left inverse of A is nonnegative on V_A.
(G_{17}) A is *monotone on* V_A; that is,

$$Ax \geq 0 \qquad \text{and} \qquad x \in V_A \to x \geq 0.$$

(H_{18}) A has a *regular splitting* whose iteration matrix has spectral radius at most one; that is, A has a splitting

$$A = M - N, \quad M^{-1} \geq 0, \quad N \geq 0, \quad \text{with} \quad V_{M^{-1}A} = V_A \quad \text{and} \quad \rho(M^{-1}N) \leq 1.$$

(H_{19}) There exists an inverse-positive matrix $B \geq A$ and an M-matrix $C = I - T$, $T \geq 0$, $V_C = V_A$, such that

$$A = BC.$$

(I_{20}) A has a *weak regular splitting* whose iteration matrix has spectral radius at most one, that is, A has a splitting

$$A = M - N, \quad M^{-1} \geq 0, \quad M^{-1}N \geq 0, \quad \text{with } V_{M^{-1}A} = V_A, \quad \text{and } \rho(M^{-1}N) \leq 1.$$

(I_{21}) There exists an inverse-positive matrix B and an M-matrix C, $V_C = V_A$, such that

$$A = BC.$$

Proof As was the case in the proof of Theorem 2.3, we shall only prove the part of the theorem here where $A \in Z^{n \times n}$, with the remainder of the proof being left to the reader.

Let $A \in Z^{n \times n}$ be an M-matrix. Then in view of the implication tree in the statement of the theorem, we need only establish that conditions (A), (D), (E), and (H) hold. That condition (A) is true for A follows immediately from condition (A_1) of Theorem 2.3 and Lemma 4.1. For if all the principal minors of $A + \epsilon I$ are positive for each $\epsilon > 0$, then all the principal minors of A must be nonnegative. Now condition (D_{10}) is just a special case of Theorem 4.18, to be established later. That condition (E_{11}) holds follows immediately; for if λ is an eigenvalue of $A = sI - B$, $s > 0$, $B \geq 0$, then $s - \lambda$ is an eigenvalue of B. Thus either $\lambda = 0$ or $\operatorname{Re} \lambda > 0$ since $s \leq \rho(B)$. To establish condition (H), we shall verify condition (H_{18}). Let $A = sI - B$, $s > 0$, $B \geq 0$. Then $s \geq \rho(B)$ so that

$$\rho(T) \leq 1$$

for $T = B/s$. Thus we have a regular splitting of A satisfying condition (H_{18}), with $M = sI$ and $N = B$.

For the converse we establish for $A \in Z^{n \times n}$, that each of conditions (C), (D), and (G) imply that A is an M-matrix. Suppose that condition (C_8) holds for $A \in Z^{n \times n}$ and let $A = sI - B$, $s > 0$, $B \geq 0$. Then since $s - \rho(B)$ is a real eigenvalue of A, $s \geq \rho(B)$, and A is an M-matrix. Now let $A \in Z^{n \times n}$ satisfy condition (D_{10}). Then since $PAP^t \in Z^{n \times n}$, it follows in a proof similar to the nonsingular case of condition (F_{19}) of Theorem 2.3, that L and U both have all nonpositive off-diagonal entries. But then L and U are M-matrices and thus LU is an M-matrix since $LU \in Z^{n \times n}$ (see Exercise 5.2). This means that $A = P^t LUP$ is an M-matrix. Finally, suppose (G_{16}) holds and let $A = sI - B$, $s > 0$, $B \geq 0$. Then for $T = B/s$, $(I - T)^D$ is nonnegative on $R(A^k) = R[(I - T)^k]$, $k = \text{index } A$. Thus $\rho(T) \leq 1$ by Lemma 4.4, and consequently $s \geq \rho(B)$, completing the proof of the theorem for $A \in Z^{n \times n}$. ∎

As was the case with nonsingular M-matrices, it is also possible to give necessary and sufficient conditions for $A \in R^{n \times n}$ to be an M-matrix. The proofs of these characterizations closely parallel part of the proof of Theorem 2.4 and are left as an exercise.

(4.7) Exercise Let $A \in R^{n \times n}$. Then each of the following conditions is equivalent to the statement: "A is an M-matrix."

(i) $A + D$ is inverse-positive for each positive diagonal matrix D.
(ii) $A + \alpha I$ is inverse-positive for each scalar $\alpha > 0$.

We next investigate an important proper subclass of the class of M-matrices. This class properly contains the nonsingular M-matrices but shares many of their important properties. These matrices will be used in Chapter 7 to establish convergence criteria for iterative methods for solving singular systems of linear equations.

(4.8) Definition A matrix $T \in R^{n \times n}$ is said to be *semiconvergent* whenever

$$\lim_{j \to \infty} T^j \text{ exists.}$$

Of course T is convergent if and only if $\rho(T) < 1$, so that by considering the Jordan form for T this limit is then zero. The following exercise is also established by use of the Jordan form for T.

(4.9) Exercise Let $T \in R^{n \times n}$. Then T is semiconvergent if and only if each of the following conditions hold.

(1) $\rho(T) \leq 1$ and
(2) if $\rho(T) = 1$ then all the elementary divisors associated with the eigenvalue 1 of T are linear; that is, $\operatorname{rank}(I - T)^2 = \operatorname{rank}(I - T)$, and
(3) if $\rho(T) = 1$ then $\lambda \in \sigma(T)$ with $|\lambda| = 1$ implies $\lambda = 1$.

Recall that if $A = sI - B$, $s > 0$, $B \geq 0$ and A is a nonsingular M-matrix then $T = B/s$ is convergent. We next extend this important property to certain singular M-matrices.

(4.10) Definition An M-matrix A is said to have "*property c*" if it can be split into $A = sI - B$, $s > 0$, $B \geq 0$, where the matrix $T = B/s$ is semiconvergent.

Notice that all nonsingular M-matrices have "property c," but that not all M-matrices share the property. For example consider the matrix

$$A = \begin{pmatrix} 0 & -1 \\ 0 & 0 \end{pmatrix}.$$

If A is represented as $A = sI - B$, $s > 0$, $B \geq 0$ then $T = B/s$ must have the form

$$T = \begin{pmatrix} 1 & 1/s \\ 0 & 1 \end{pmatrix}$$

so that T cannot be semiconvergent.

Notice also that from the Perron–Frobenius theorem, if A is an M-matrix and $A = sI - B$, $B \geq 0$ with $s > \max_i a_{ii}$, then condition (3) of Exercise 4.9 is automatically satisfied by $T = B/s$. For example let

$$A = \begin{pmatrix} 1 & -1 \\ -1 & 1 \end{pmatrix}.$$

Then for $s = 1$, $T = B/s$ is not semiconvergent. However for any $s > 1 = \max_i a_{ii}$, $T = B/s$ is semiconvergent. Thus this matrix A has "property c."

We turn now to the characterization of M-matrices with "property c." First, we give the following lemma.

(4.11) Lemma An M-matrix A has "property c" if and only if index $A \leq 1$.

Proof Let A be an M-matrix and split A into $A = sI - B$, $s > \max_i a_{ii}$, and $B \geq 0$. Let $T = B/s$ and suppose that S is a nonsingular matrix such that

$$S^{-1}TS = \begin{bmatrix} J & 0 \\ 0 & K \end{bmatrix}$$

is a Jordan form for T where $\rho(K) < 1$. Then

$$S^{-1}(I - T)S = \begin{bmatrix} I - J & 0 \\ 0 & I - K \end{bmatrix}$$

so that $I - K$ is nonsingular. Then by Exercise 4.9, T is semiconvergent if and only if $I - J = 0$. But this is true if and only if $\operatorname{rank}(I - T)^2 = \operatorname{rank}(I - T)$, thus if and only if $\operatorname{rank} A^2 = \operatorname{rank} A$, since $A = sI - B$. ■

It turns out that many of the conditions in Theorem 4.6 can be used to characterize matrices $A \in Z^{n \times n}$ that are M-matrices with "property c." This is provided, of course, that the condition index $A \leq 1$ is carried along. Some of these characterizations are given in the following theorem and exercises. The theorem is based upon the conditions in Theorem 4.6 related to generalized inverse-positivity.

(4.12) Theorem Let $A \in R^{n,n}$. Then for any fixed letter $\mathscr{C}$ representing one of the following conditions, conditions $\mathscr{C}_i$ are equivalent for each i. Moreover, letting $\mathscr{C}$ represent any of the equivalent conditions $\mathscr{C}_i$, the following implications hold.

$$B \Rightarrow C \Rightarrow D \begin{matrix} \nearrow E \\ \searrow A \Rightarrow F \end{matrix}$$

Finally, if $A \in Z^{n,n}$, then each of the conditions is equivalent to the statement: "A is an M-matrix with "property c.""

For some fixed complement S of null space A:

(A_1) A has a nonnegative $\{1\}$-*inverse* B with range $BA = S$, that is, there exists $B \geq 0$ such that

$$A = ABA, \qquad \text{range}\, BA = S.$$

(A_2) A has a $\{1\}$-inverse B with range $BA = S$, such that B is nonnegative on range A.

(A_3) Every $\{1\}$-inverse B of A with range $BA = S$ is nonnegative on range A.

(A_4) A has a $\{1,2\}$-inverse C with range $C = S$, such that C is nonnegative on range A; that is, there exists a matrix C, nonnegative on range A, such that

$$A = ACA, \qquad C = CAC, \qquad \text{range}\, C = S.$$

(A_5) Every $\{1,2\}$-inverse C of A with range $C = S$ is nonnegative on range A.

(A_6) A is monotone on S.

(B_7) A has a regular splitting $A = M - N$ such that range $M^{-1}A =$ range A and such that the powers of $M^{-1}N$ converge.

(B_8) There exist an inverse-positive matrix $M \geq A$ and an M-matrix $B = I - T$, $T \geq 0$, where the powers of T converge and with range $B =$ range A, such that

$$A = MB.$$

(C_9) A has a weak regular splitting $A = M - N$ such that range $M^{-1}A =$ range A and such that the powers of $M^{-1}N$ converge.

(C_{10}) There exist an inverse-positive matrix M and an M-matrix B with "property c" and with range $B =$ range A, such that

$$A = MB.$$

(D_{11}) Condition (A) holds for the case where $S =$ range A.

(E_{12}) Index $A \leq 1$ and A is *weak monotone*; that is,

$$Ax \geq 0 \rightarrow x = u + v, \qquad u \geq 0, \quad Av = 0.$$

(F_{13}) For every regular splitting of A into $A = M - N$, it follows that $\rho(M^{-1}N) \leq 1$ and index$(I - M^{-1}N) \leq 1$.

Proof As was the case in the proofs of Theorems 2.3 and 4.6, we shall only prove the part of the theorem where $A \in Z^{n \times n}$, with the remainder of the proof being left to the reader, as before.

Let $A \in Z^{n \times n}$ be an M-matrix with "property c." Then in view of the implication tree in the statement of the theorem, we need only establish that condition (B) holds. But since A has a regular splitting $A = sI - B$, $s > 0$,

$B \geq 0$ with B/s semiconvergent, condition (B_7) clearly holds with $M = sI$ and $N = B$.

For the converse we establish that if $A \in Z^{n \times n}$, then each of conditions (E) and (F) imply that A is an M-matrix with "property c." Suppose (E_{12}) holds. Then since index $A \leq 1$, it suffices to show that A is an M-matrix. For that purpose let A have the representation

$$A = sI - B, \qquad B \geq 0, \quad s > 0$$

and suppose for the purpose of contradiction that $s < \rho(B)$. By the Perron–Frobenius theorem there exists a vector $y > 0$, such that

$$By = \rho(B)y.$$

Then

$$Ay = [s - \rho(B)]y \leq 0,$$

so that $A(-y) \geq 0$ since $(s - \rho(B)) < 0$ and $y > 0$. Then

$$-y = u + v, \qquad u \geq 0, \quad Av = 0$$

so that $u + y \neq 0$ and

$$A(u + y) = A(-v) = 0$$

and thus

$$B(u + y) = s(u + y).$$

But then s is an eigenvalue of B, contradicting the assumption that $s < \rho(B)$. Then $s \geq \rho(B)$ and thus A is an M-matrix. Finally, if (F_{13}) holds for $A \in Z^{n \times n}$, then since any representation $A = sI - B$, $s > 0$, $B \geq 0$, is a regular splitting of A, it follows that A is an M-matrix with "property c." This completes the proof of the theorem for $A \in Z^{n \times n}$. ■

Some other characterizations of M-matrices with "property c" can be obtained from Theorems 4.6 and 4.12. As an example we give the following exercise.

(4.13) Exercise Show that $A \in Z^{n \times n}$ is an M-matrix with "property c" if and only if there exists a symmetric positive definite matrix W such that

$$AW + WA^t$$

is positive semidefinite (Berman *et al.* [1978]).

In addition, we state

(4.14) Exercise Show that if $A \in Z^{n \times n}$ and there exists $x \gg 0$ such that $Ax \geq 0$, then A is an M-matrix with "property c."

The following exercise is the extension of Exercise 2.6 to general M-matrices.

(4.15) Exercise Let $A \in Z^{n \times n}$ be symmetric. Then A is an M-matrix if and only if A is positive semidefinite. Moreover, in this case A has "property c."

Some important properties of singular, irreducible M-matrices are given next. Such matrices arise quite often as coefficient matrices for systems of linear equations resulting from finite difference methods for certain elliptic type partial differential equations, such as the Neumann boundary value problem or Possion's equation on a sphere (see Chapter 7).

(4.16) Theorem Let A be a singular, irreducible M-matrix of order n. Then

(1) A has rank $n - 1$ (see also Theorem 5.3.5).
(2) There exists a vector $x \gg 0$ such that $Ax = 0$.
(3) A has "property c."
(4) Each principal submatrix of A other than A itself is a nonsingular M-matrix.
(5) A is *almost monotone*. That is, $Ax \geq 0 \Rightarrow Ax = 0$.

Proof Let $A = sI - B$, $s > 0$, $B \geq 0$. Then B is also irreducible and thus $\rho(B)$ is a simple eigenvalue of B by the Perron–Frobenius theorem. Hence $0 = s - \rho(B)$ is a simple eigenvalue of A and thus A has rank $n - 1$, so that (1) holds for A. Also, by the Perron–Frobenius theorem, there exists a positive vector x such that $Bx = \rho(B)x$. Then $Ax = 0$ and (2) holds. That (3) holds for A then follows from (2) and Exercise 4.14. To establish (4) it suffices to consider the case where $n > 1$. By (2) and the dual for A^t there exist vectors $x \gg 0$ with $Ax = 0$ and $y \gg 0$ with $A^t y = 0$. The adjoint matrix B of A, whose elements are the cofactors A_{ij} of the elements a_{ij} of A, is known to have the form $\delta y^t x$. Here $\delta \neq 0$ since A has rank $n - 1 \geq 1$ by (1). Now if $\delta > 0$ there would exist an $\varepsilon > 0$ such that the adjoint B_ε of $A + \varepsilon I$ would satisfy $B_\varepsilon \ll 0$, which contradicts Lemma 4.1 since $\varepsilon > 0$. Thus $\delta > 0$ and so $B \gg 0$. It follows directly then that all principal minors of order $n - 1$ of the matrix A are positive. Thus from condition (A_1) of Theorem 2.3, all the principal minors of order $k \leq n - 1$ are then positive. Then each proper principal submatrix of A is a nonsingular M-matrix; that is, (4) holds. Finally suppose that $Ax \geq 0$ for some $x \in R^n$. By the dual of (2) for A^t, we can find a vector $y \gg 0$ such that $y^t A = 0$. But if $Ax \neq 0$ then $y^t Ax \neq 0$, a contradiction. Thus (5) holds. ∎

In view of Theorem 4.16, it is not too surprising that a singular, irreducible M-matrix might satisfy many of the conditions listed in Theorem 2.3 for a nonsingular M-matrix. As an illustration of this fact we prove the following important result which corresponds to condition (E_{18}) of Theorem 2.3.

(4.17) Corollary Let A be a singular, irreducible M-matrix of order n. Then there exists a lower triangular nonsingular M-matrix L and an upper triangular M-matrix U such that

$$A = LU$$

Proof By Theorem 4.16, each proper principal submatrix of A is a nonsingular M-matrix. Thus we can partition A into

$$A = \begin{bmatrix} A_1 & a \\ b^t & a_{nn} \end{bmatrix},$$

where A_1 is a nonsingular M-matrix of order $n-1$, $a \leq 0$, $b \leq 0$ are in R^{n-1} and $a_{nn} = b^t A_1^{-1} a$. By condition (E_{18}) of Theorem 2.3, there exist lower and upper triangular matrices L_1 and U_1, respectively, such that

$$A_1 = L_1 U_1$$

Moreover, by the proof of Theorem 2.3, L_1 and U_1 are nonsingular M-matrices. Set

$$L = \begin{bmatrix} L_1 & 0 \\ b^t U_1^{-1} & 1 \end{bmatrix} \quad \text{and} \quad U = \begin{bmatrix} U_1 & L_1^{-1} a \\ 0 & 0 \end{bmatrix}.$$

Then $b^t U_1^{-1} \leq 0$ since $U_1^{-1} \geq 0$ and $b \leq 0$ and $L_1^{-1} a \leq 0$ since $L_1^{-1} \geq 0$ and $a \leq 0$. Thus L is a nonsingular M-matrix, U is an M-matrix and $A = LU$. ■

In general, a reducible M-matrix need not have the LU decomposition described in Corollary 4.17. For example

$$A = \begin{bmatrix} 0 & -1 & 0 \\ 0 & 0 & 0 \\ -1 & 0 & 0 \end{bmatrix}$$

has no such decomposition. However, we can prove the following generalization of condition (F_{19}) of Theorem 2.3. It is useful in the investigation of iterative methods for singular systems of linear equations.

(4.18) Theorem Let A be an M-matrix. Then there exists a permutation matrix P, a lower triangular nonsingular M-matrix L and an upper triangular M-matrix U such that

$$PAP^t = LU.$$

Proof It is sufficient to consider the case where $A \neq 0$ is singular and reducible. Let P be a permutation matrix such that

$$PAP^t = \begin{bmatrix} A_1 & B \\ 0 & A_2 \end{bmatrix},$$

where A_1 is irreducible. We consider the case first where A_2 is also irreducible. Then by Corollary 4.17 there exist lower triangular nonsingular M-matrices L_1 and L_2 and upper triangular M-matrices U_1 and U_2 such that $A_1 = L_1U_1$ and $A_2 = L_2U_2$. Then

$$L = \begin{bmatrix} L_1 & 0 \\ 0 & L_2 \end{bmatrix} \quad \text{and} \quad U = \begin{bmatrix} U_1 & A_1^{-1}B \\ 0 & U_2 \end{bmatrix}$$

satisfy the conditions of the theorem since $A_1^{-1}B \leq 0$. If A_2 is reducible, then the proof is completed by using induction on the number of irreducible blocks in the reduced normal form for A. ■

Finally, we remark that Theorem 4.18 is a stronger form of condition (D_{10}) in Theorem 4.6, since L, or dually U, can always be chosen to be nonsingular.

5 EXERCISES

(5.1) Let $A,B \in Z^{n \times n}$. Prove or give a counterexample.

(a) $A + B$ is an M-matrix.

(b) AB is a nonsingular M-matrix if A and B are nonsingular and $AB \in Z^{n \times n}$.

(c) If A is an M-matrix and $B \geq A$ then B is an M-matrix.

(d) If A is a nonsingular M-matrix and $B \geq A$ then

(i) $\det B > \det A$.
(ii) B is a nonsingular M-matrix.
(iii) $B^{-1} \geq A^{-1}$.
(iv) $A^{-1}B \geq I$, $BA^{-1} \geq I$, $B^{-1}A \geq I$, and $AB^{-1} \geq I$.
(v) $B^{-1}A$ and AB^{-1} are nonsingular M-matrices.

(5.2) Show that if A and B are $n \times n$ M-matrices and if $AB \in Z^{n \times n}$, then AB is an M-matrix. (Hint: Use Lemma 4.1.)

(5.3) Show that if A is an M-matrix then there exists $x > 0$ such that $Ax \geq 0$, but that x cannot always be chosen positive.

(5.4) Let $A = (a_{ij})$ be an M-matrix. Show that there exists a real nonnegative eigenvalue $l(A)$ of A such that $l(A) \leq \text{Re}(\lambda)$ for all $\lambda \in \sigma(A)$ and that $l(A) \geq a_{ii}$ for each i, where $\sigma(A)$ is the set of eigenvalues of A.

(5.5) Let A be a nonsingular M-matrix. Show that $A - l(A)I$ is an M-matrix and that every singular M-matrix is obtained in this way. (See Exercise 5.4.)

(5.6) Let $A = (a_{ij})$ be an M-matrix and denote by $c(A)$ the circular region in the complex plane with center at $a = \max_i a_{ii}$ and radius $\rho(aI - A)$. Show that $c(A)$ contains the entire spectrum of A.

(5.7) Let A be an M-matrix. Show that $A^+ \geq 0$ if and only if either $A = 0$ or else there exists a permutation matrix P such that

$$PAP^{\mathrm{t}} = \begin{pmatrix} M & 0 \\ 0 & 0 \end{pmatrix},$$

where M is a nonsingular M-matrix (Kuo [1977]).

(5.8) Show that the Schur complement of a nonsingular M-matrix is also a nonsingular M-matrix (Lynn [1964]).

(5.9) Let A and B be M-matrices and let $\cdot$ denote the Hadamard product operation for matrices. Show that

(a) $\mathcal{M}(A \cdot B)$ is an M-matrix (Fan [1964]), and
(b) if B is nonsingular then $\mathcal{M}(A \cdot B^{-1})$ is an M-matrix (Johnson [1977]).

(5.10) Use condition (I_{27}) of Theorem 2.3 to show that if $B \in R^{n \times n}$ is nonnegative then

$$\rho(B) = \max \min[(PB)_{ii}/P_{ii}],$$

where the minimum is computed over all indices i such that $P_{ii} > 0$ and the maximum is taken over all nonzero positive semidefinite matrices P (Berman [1978]).

(5.11) Let $A \in R^{n \times n}$ and let $A[\alpha]$ denote the principal submatrix obtained from A by choosing rows and columns whose indices come from the increasing sequence $\alpha \subseteq \{1, \ldots, n\}$. Then A is called an *ω-matrix* if

(a) $\sigma(A[\alpha])$ contains a real eigenvalue for each $\phi \neq \alpha \subseteq \{1, \ldots, n\}$ and
(b) $l(A[\beta]) \leq l(A[\alpha])$ for each $\phi \neq \alpha \subseteq \beta \subseteq \{1, \ldots, n\}$, where $l(\cdot)$ is the smallest real eigenvalue.

If in addition A satisfies

(c) $l(A) \geq 0$

then each principal minor of A is nonnegative and A is called a *τ-matrix*.
Show that each matrix in $Z^{n \times n}$ is an ω-matrix and that the τ-matrices in $Z^{n \times n}$ are precisely the M-matrices.

(5.12) Show that a τ-matrix (defined in Exercise 5.11) satisfies the Hadamard–Fischer inequality for determinants, namely

$$0 \leq \det A \leq \det A[\alpha] \det A(\alpha),$$

where $A(\alpha) = A[\{1, \ldots, n\} \backslash \alpha]$ (Engel and Schneider [1977]). (Compare with Exercise 2.6.27.)

(5.13) If α and β are increasing sequences on $\{1, \ldots, n\}$ let $A[\alpha|\beta]$ denote the submatrix obtained from $A \in R^{n \times n}$ by choosing row indices from α and column indices from β. If α and β have the same length and differ in at most one entry then $\det A[\alpha|\beta]$ is called an almost principal minor of A. Show that if A is an M-matrix then the product of any two symmetrically placed almost principal minors of A is nonnegative.

(5.14) Show that if A is an irreducible, singular M-matrix, then index $A = 1$ and $I - AA^{\mathrm{D}} \gg 0$ (see Meyer and Stadelmaier [1978]).

(5.15) Show that if $A \in Z^{n \times n}$, then A is an M-matrix if and only if there exists $b > 0$ such that $A + t(I - AA^{\mathrm{D}})$ is inverse-positive for all $0 < t < b$ (see Exercise 5.14, Meyer and Stadelmaier [1978], and Section 3).

(5.16) Show that a matrix $A \in Z^{n \times n}$ has a unique factorization $A = LU$, where L is a nonsingular lower triangular M-matrix with unit diagonal and U is an upper triangular M-matrix, if and only if A is an M-matrix and each proper leading principal minor of A is positive.

6 NOTES

(6.1) No attempt will be made here to systematically trace the history of the development of the theory of M-matrices. It appears however, that the term M-matrix was first used by Ostrowski [1937] in reference to the work of Minkowski [1900, 1907] who proved that if $A \in Z^{n \times n}$ has all of its row sums positive, then $\det A > 0$. Papers following the early work of Ostrowski have primarily been produced by two groups of researchers, one in mathematics, the other in economics. The mathematicians have mainly had in mind the applications of M-matrices to the establishment of bounds on eigenvalues of nonnegative matrices and on the establishment of convergence criteria for iterative methods for the solution of large sparse systems of linear equations (see Chapter 7). Meanwhile, the economists have studied M-matrices in connection with gross substitutability, stability of a general equilibrium, and Leontief's input–output analysis of economic systems (see Chapters 9 and 10).

(6.2) The first systematic effort to characterize M-matrices was by Fiedler and Ptak [1962] and a partial survey of the theory of M-matrices was made by Poole and Boullion [1974]. Varga [1976] has surveyed the role of diagonal dominance in the theory of M-matrices. Plemmons [1977] has combined and extended these surveys. In addition, Schröder [1978] has listed several characterizations of nonsingular M-matrices using operator theory and partially ordered linear spaces. Finally, Kaneko [1978] has compiled a list of characterizations and applications of nonsingular M-matrices in terms of linear complementarity problems in operations research (see Chapter 10).

(6.3) The problem of giving proper credit to those originally responsible for the various characterizations listed in Theorem 2.3 is difficult, if not impossible. The situation is complicated by the fact that many of the characterizations are implicit in the work of Perron [1907] and Frobenius [1908, 1909, 1912] and in the work of Ostrowski [1937, 1956], but were not given there in their present form. Another complicating factor is that the diversification of the applications of M-matrices has led to certain conditions being derived independently. We attempt in this note only to give references to the literature where the various proofs can be found.

First of all condition (A_1), which is known as the Hawkins–Simon [1949] condition in the economics literature, was taken by Ostrowski [1937] as the definition for $A \in Z^{n \times n}$ to be a nonsingular M-matrix. He then proceeded to show the equivalence of his definition with ours; namely that A has a representation $A = sI - B$, $B \geq 0$, and $s > \rho(B)$. Condition (A_2) is also in

Ostrowski [1937]. Conditions (A_3)–(A_5) were listed in Fiedler and Ptak [1962]. Condition (A_5) is also in Gale and Nikaido [1965] and (A_6) was shown to be equivalent to (A_1) in Moylan [1977].

Next, condition (B_7) can be found in Johnson [1974a], and (C_8)–(C_{14}), (D_{15}), (D_{16}), (E_{17}), (E_{18}), and (F_{19}) are either immediate or can be found in Fiedler and Ptak [1962].

The stability condition (G_{20}) is in the work of Ostrowski [1937], while its equivalence with (G_{21}) is the Lyapunov [1892] theorem. The equivalence of (G_{22}) with (G_{23}) is the Stein [1952] theorem. Condition (H_{24}) is in the work of Tartar [1971] and of Araki [1975]. The equivalence of (H_{25}) and (H_{26}) is in Barker, *et al.* [1978].

Conditions (I_{27})–(I_{29}) are in Schneider [1953] and Fan [1958]. Condition (J_{30}) is in the work of Beauwens [1976], condition (K_{31}) is due to Neumann [1979] and condition (L_{32}) in a slightly weaker form is in Bramble and Hubbard [1964]. Condition (L_{33}) and its equivalence with (L_{32}) is believed to be new. Moreover, condition (M_{34}) is in Moylan [1977] while (M_{35}) and (M_{36}) are essentially in Fiedler and Ptak [1962]. Condition (M_{37}) and its equivalence to (M_{36}) is in Neumann [1979].

Next, condition (N_{38}) is in the pioneering paper by Ostrowski [1937] and condition (N_{39}) was shown to be equivalent to (N_{38}) by Collatz [1952]. Conditions (N_{40})–(N_{42}) are in Kuttler [1971] and conditions (N_{43}) and (N_{44}) are obvious. The equivalence of (N_{45}) with (I_{27}), (N_{38}), and (P_{48}), for $A \in Z^{n \times n}$, is essentially in Schneider [1965].

Condition (N_{45}) is in the work of Varga [1962] on regular splittings and condition (N_{46}) is due to Ortega and Rheinboldt [1967]. Conditions (O_{47}) and (P_{48}) are essentially in Price [1968]. Finally, conditions (Q_{49}) and (Q_{50}) are in Robert [1966, 1973].

Now let $\mathscr{C}$ represent any of the equivalent conditions (C_i) and let $A \in R^{n \times n}$. The implications $M \Rightarrow H \Rightarrow A$ are in Barker, *et al.* [1978], $I \Rightarrow Q$ follows from Exercise 1.3.7, $A \Rightarrow Q$ is in Gale and Nikaido [1965], $N \Rightarrow O$ is in Varga [1962], and the implications $M \Rightarrow J \Rightarrow K \Rightarrow L$ are in Neumann [1979]. The remaining implications are fairly immediate and can mostly be found in the standard texts on linear algebra.

(6.4) Conditions (i) and (ii) of Theorem 2.4 can be found in Willson [1971] and conditions (iii) and (iv) are to be found in Cottle and Veinott [1972].

(6.5) The material in Section 4 concerning completely monotonic functions is entirely due to Varga [1968].

(6.6) Theorem 2.7 and Lemma 4.1 are in Fiedler and Ptak [1962].

(6.7) Lemma 4.4 is believed to be new.

(6.8) Some of the conditions in Theorem 4.6 that characterize general M-matrices A are direct extensions of the characterizations in Theorem 2.3 for the nonsingular case. Conditions (B_7) and (D_{10}) are believed to be new. Conditions (G_{14})–(G_{17}), (H_{18}), (H_{19}), (I_{20}), and (I_{21}) are in Neumann and Plemmons [1979] or Rothblum [1979]. In addition, Meyer and Stadelmaier [1978] have studied the generalized inverse-positivity of M-matrices in terms of complementary perturbations $A + tE$, where $E = I - AA^{D}$. (See Exercises 5.14 and 5.15.)

(6.9) Most of the material in Section 4 on semiconvergent matrices and M-matrices with "property c" can be found in Meyer and Plemmons [1977], Neumann and Plemmons [1979], and Plemmons [1976b].

(6.10) Theorem 4.16 is essentially in Fiedler and Ptak [1962], except for part (3), which is in Plemmons [1976b].

(6.11) Corollary 4.17 and Theorem 4.18 can be found in Kuo [1977].

(6.12) There has been considerable activity in studying spectral properties of nonnegative matrices B in terms of the associated M-matrices

$$A = sI - B, \qquad s = \rho(B).$$

Schneider [1956] initiated much of the work in this area by investigating the eigenvectors and elementary divisors associated with the Perron root s of B in terms of the elementary divisors associated with the zero eigenvalue of the singular M-matrix A. Carlson [1963] extended Schneider's work to include the study of nonnegative solutions to linear systems of equations $Ax = b$, $b \geq 0$, where A is an M-matrix. Crabtree [1966a,b] and Cooper [1973] have also used this technique to study the spectral radius s of $B \geq 0$. Finally, Rothblum [1975] has investigated the generalized eigenvectors associated with s in terms of the combinatorial properties of M-matrices and further results can be found in Richman and Schneider [1978].

(6.13) Another approach to characterizing nonsingular M-matrices in a certain sense is that of characterizing nonsingular matrices $P \geq 0$ for which P^{-1} is a nonsingular M-matrix. This inverse M-matrix problem has important applications in operations research (see Mangasarian [1976a, 1976b, 1978]) and physics (see Leff [1971]). Markham [1972] has characterized those nonsingular $P \geq 0$ for which P^{-1} is an M-matrix where P is *totally nonnegative*; that is, all the minors of P are nonnegative (see Chapter 2). Also, Willoughby [1977] has studied the inverse M-matrix problem in terms of matrix scalings. In addition, Johnson *et al.* [1978], have studied sign patterns of inverse-positive matrices.

(6.14) Various degrees of nonnegative stability and semipositivity of singular M-matrices have been characterized and interrelated by Berman *et al* [1978]. These results are based upon Theorem 4.16, together with a reduced normal form for M-matrices.

(6.15) Finally, we mention that the τ-matrices defined in Exercise 5.11 contain, in addition to the M-matrices, the positive semidefinite as well as the totally nonnegative matrices defined in Note 6.13; so this concept offers a solution to the unification problem introduced by Taussky [1958].

CHAPTER 7

ITERATIVE METHODS FOR LINEAR SYSTEMS

1 INTRODUCTION

In this chapter we apply nonnegativity to the study of iterative methods for solving systems of linear equations of the form

$$Ax = b, \tag{1.1}$$

where A is a given $n \times n$ matrix and b is a given column vector of order n. It is desired to determine the unknown column vector x. Consider the case $n = 3$. The system (1.1) may then be written in the alternative forms

$$\begin{aligned} a_{11}x_1 + a_{12}x_2 + a_{13}x_3 &= b_1 \\ a_{21}x_1 + a_{22}x_2 + a_{23}x_3 &= b_2 \\ a_{31}x_1 + a_{32}x_2 + a_{33}x_3 &= b_3 \end{aligned}$$

or

$$\begin{bmatrix} a_{11} & a_{12} & a_{13} \\ a_{21} & a_{22} & a_{23} \\ a_{31} & a_{32} & a_{33} \end{bmatrix} \begin{bmatrix} x_1 \\ x_2 \\ x_3 \end{bmatrix} = \begin{bmatrix} b_1 \\ b_2 \\ b_3 \end{bmatrix}.$$

We shall mostly be concerned with cases where n is large, perhaps in the range 10^4 to 10^6, and where A is "sparse," that is, where A has a percentage of zero terms large enough so that it is advantageous to take into account their placements in operations on the matrix A. In addition, A will very often have additional important properties such as (1) having all nonpositive off-diagonal entries with the real part of each eigenvalue of A nonnegative; i.e., A may be an M-matrix or (2) being hermitian and having all nonnegative eigenvalues; i.e., A may be positive semidefinite. For the most part, we shall assume that A is nonsingular, so that (1.1) has the unique solution

$$x = A^{-1}b. \tag{1.2}$$

We shall study various iterative methods for approximating x. Such methods are usually ideally suited for problems involving large sparse matrices, much more so in most cases than direct methods such as Gaussian elimination. A typical iterative method involves the selection of an initial approximation x^0 to the solution x to (1.1) and the determination of a sequence $x^1, x^2, \ldots,$ according to some specified algorithm which, if the method is properly chosen, will converge to the exact solution x of (1.1). Since x is unknown, a typical method for terminating the iteration might be whenever

$$(1.3) \qquad |x_i^{k+1} - x_i^k| / |x_i^{k+1}| < \varepsilon, \quad i = 1, \ldots, n,$$

where ε is some pre-chosen small number, depending upon the precision of the computer being used. Essentially, the stopping criteria (1.3) for the iteration is that we will choose x^{k+1} as the approximation to the solution x whenever the relative difference between successive approximations x_i^k and x_i^{k+1} becomes sufficiently small for $i = 1, \ldots, n$.

The use of these iterative algorithms has the advantage that A is not altered during the computation. Hence, the problem of accumulation of roundoff errors is much less serious than for those methods, such as direct methods, where the matrix is decomposed during the iteration process.

The idea of solving large systems of linear equations by iterative methods is certainly not new, dating back at least to Gauss [1823]. For a discussion of the history of these iterative techniques the reader is referred to the excellent summary given by Varga [1962, pp. 1–2].

Our aim here is to show how, from this early beginning, the theory of nonnegative matrices and especially the theory of M-matrices have played a fundamental role in the development of efficient methods for establishing the convergence and in accelerating the convergence of these iterative methods. In particular, the Perron–Frobenius theory developed in Chapter 2 forms the foundation for the exposition of these results.

In Section 3, we develop the formulas for the classical Jacobi and Gauss–Seidel iterative methods. The fundamental convergence lemma for general splittings of the coefficient matrix A is also given and rates of convergence are discussed.

Section 4 is devoted to an extensive discussion of the widely used successive overrelaxation method. Techniques that can sometimes be used for choosing an optimum relaxation parameter are discussed and are illustrated by two rather different examples.

The basic convergence criteria for general iterative methods, and specifically for the Jacobi and SOR methods, are given in Section 5. These convergence results are related to two common classes of situations: (1) those in which A is a nonsingular M-matrix (or some generalization) and (2) those in which A is hermitian and/or positive definite. Here (1) is in conjunction with the topic of this book while (2) is added for completeness.

Section 6 is concerned with the often neglected singular case, the case where A is singular but the linear equations are still consistent. It is shown how these problems can often be handled by the methods developed for the nonsingular case in earlier sections. Some partial extensions of these methods to rectangular linear systems and to least-squares solutions to inconsistent linear systems are given in the exercises.

As usual, the last two sections are devoted to the exercises and the notes, respectively.

Next, in Section 2 we lay the framework for the material to be presented later by illustrating many of the important concepts with a particular example, that of solving a two-point boundary-value problem sometimes called the heat equation.

2 A SIMPLE EXAMPLE

There are two goals of this section. The first is to describe one way in which large sparse systems of linear equations can arise in practice. Here we explain a natural and important physical application that requires very little background. The second and primary goal here is to illustrate, by this application, the special properties related to nonnegativity that these coefficient matrices frequently have.

The example comes from modeling a continuous problem by a discrete one. The continuous problem will have infinitely many unknowns and of course cannot be solved exactly on a computer. Thus it has to be approximated by a discrete problem. The more unknowns we keep, the better the accuracy we may expect and, naturally, the greater the expense. As a simple but very typical continuous problem, we choose the following differential equation:

$$-\frac{d^2u}{dx^2} = g(x), \qquad 0 \le x \le 1. \tag{2.1}$$

This is then a linear second order ordinary differential equation with nonhomogeneous term $g(x)$. Since any combination $ax + b$ may be added to any solution $u(x)$ to (2.1) and produce another solution, we impose boundary conditions

$$u(0) = \alpha, \qquad u(1) = \beta \tag{2.2}$$

in order to remove the constants a and b Then (2.1) with (2.2) constitutes a *two-point boundary-value problem* with a unique solution $u = u(x)$. The problem describes a steady-state rather than a transient problem. For example, it may be used to describe the temperature in a rod of unit length, with fixed temperatures α at $x = 0$ and β at $x = 1$ and with a distributed heat source $g(x)$.

Since we seek a discrete model for this continuous problem, we cannot accept more than a finite amount of information about the function g, say, its values at the equally spaced partition points $x = h, x = 2h, \ldots, x = nh$, where

$$h = 1/(n + 1).$$

And what we compute will be approximations $u_1, \ldots, u_n$ to the true solution u at these same points; that is, u_j will be the approximation to $u(jh)$ for $j = 1, \ldots, n$. At the endpoints, we have $u_0 = u(0) = \alpha$ and $u_{n+1} = u(1) = \beta$ exactly.

In order to discretize the ordinary differential equation (2.1), we use the usual second central difference quotient

$$\frac{d^2u(x)}{dx^2} \approx \frac{u(x + h) - 2u(x) + u(x - h)}{h^2}. \tag{2.3}$$

The right-hand side of (2.3) approaches d^2u/dx^2 as $h \to 0$, but we must stop at a finite h. Then at a typical partition point $x = jh$ of (0,1), the differential equation (2.1) is replaced by this discrete analogue (2.3) and after multiplying through by h^2 we have

$$-u_{j+1} + 2u_j - u_{j-1} = h^2 g(jh), \qquad j = 1, \ldots, n. \tag{2.4}$$

Thus we obtain exactly n linear equations in the n unknowns $u_1, \ldots, u_n$. Note that the first and last equations include the expressions $u_0 = \alpha$ and $u_{n+1} = \beta$, respectively, and they will be shifted to the right-hand side of the equations. The structure of the Eqs. (2.4) for $h = \frac{1}{5}$, or $n = 4$, is given by

$$\begin{aligned}
2u_1 - u_2 \qquad\qquad &= h^2 g(h) + \alpha \\
-u_1 + 2u_2 - u_3 \qquad &= h^2 g(2h) \\
- u_2 + 2u_3 - u_4 &= h^2 g(3h) \\
- u_3 + 2u_4 &= h^2 g(4h) + \beta.
\end{aligned} \tag{2.5}$$

We now apply a very simple iterative method to the system (2.5). We simply solve the first equation for u_1, the second for u_2, etc., obtaining

$$\begin{aligned}
u_1 &= \tfrac{1}{2}u_2 + \tfrac{1}{2}[h^2 g(h) + \alpha] \\
u_2 &= \tfrac{1}{2}u_1 + \tfrac{1}{2}u_3 + \tfrac{1}{2}h^2 g(2h) \\
u_3 &= \tfrac{1}{2}u_2 + \tfrac{1}{2}u_4 + \tfrac{1}{2}h^2 g(3h) \\
u_4 &= \tfrac{1}{2}u_3 + \tfrac{1}{2}[h^2 g(4h) + \beta].
\end{aligned} \tag{2.6}$$

Then beginning with an initial approximating vector $u^0 = (u_1^0, u_2^0, u_3^0, u_4^0)^t$ to the solution u, at any stage, given $u^k = (u_1^k, u_2^k, u_3^k, u_4^k)^t$, we determine u^{k+1} by

$$\begin{aligned}
u_1^{k+1} &= \tfrac{1}{2}u_2^k + \tfrac{1}{2}[h^2 g(h) + \alpha] \\
u_2^{k+1} &= \tfrac{1}{2}u_1^k + \tfrac{1}{2}u_3^k + \tfrac{1}{2}[h^2 g(2h)] \\
u_3^{k+1} &= \tfrac{1}{2}u_2^k + \tfrac{1}{2}u_4^k + \tfrac{1}{2}[h^2 g(3h)] \\
u_4^{k+1} &= \tfrac{1}{2}u_3^k + \tfrac{1}{2}[h^2 g(4h) + \beta].
\end{aligned} \tag{2.7}$$

This method is known as the *Jacobi method*. It can be shown that the method converges to u for any u^0 for this simple example. However, a large number of iteration steps would be required if n is large; i.e., the partition width h is small. This number of iterations can be reduced by approximately 50% here by using the *Gauss–Seidel method*, which involves using values of u_j^{k+1} whenever available in place of u_j^k. Thus, for instance, the second equation of (2.7) would be

$$u_2^{k+1} = \tfrac{1}{2}u_1^{k+1} + \tfrac{1}{2}u_3^k + \tfrac{1}{2}[h^2 g(2h)].$$

Neither the Jacobi nor the Gauss–Seidel method is satisfactory for examples of this type for large n. A slight modification of the Gauss–Seidel method by the use of a relaxation parameter ω can be used to produce a method, known as the *SOR method*, which can dramatically reduce the number of iterations in certain cases.

Many of the tools that are used to establish the convergence and the convergence rates of these basic iterative methods are based upon the theory of nonnegative matrices and M-matrices. In matrix form $Au = b$, the linear system (2.5) becomes

$$\begin{bmatrix} 2 & -1 & 0 & 0 \\ -1 & 2 & -1 & 0 \\ 0 & -1 & 2 & -1 \\ 0 & 0 & -1 & 2 \end{bmatrix} \begin{bmatrix} u_1 \\ u_2 \\ u_3 \\ u_4 \end{bmatrix} = \begin{bmatrix} h^2 g(h) + \alpha \\ h^2 g(2h) \\ h^2 g(3h) \\ h^2 g(4h) + \beta \end{bmatrix}.$$

In general, for arbitrary $n \geq 2$, the coefficient matrix A for the resulting linear system is given by

$$A = \begin{bmatrix} 2 & -1 & 0 & \cdots & 0 \\ -1 & 2 & -1 & \cdots & 0 \\ 0 & \ddots & \ddots & \ddots & \vdots \\ \vdots & & \ddots & \ddots & -1 \\ 0 & \cdots & & -1 & 2 \end{bmatrix} \tag{2.8}$$

Moreover, A is a symmetric, irreducible, nonsingular M-matrix. Thus A is an irreducible Stieltjes matrix (see Definition 6.2.5), so that A is positive definite and $A^{-1} \gg 0$. This is valuable information as we shall see later. Some of these properties of A hold quite generally for systems of linear equations resulting from numerical methods of this type for approximating solutions to both ordinary and partial differential equations. For that reason we shall be able to extensively study iterative methods for the solution to systems associated with such problems.

3 BASIC ITERATIVE METHODS

We begin this section by describing two simple, but intrinsically important, iterative formulas for solving the linear system (1.1). Here we assume that the coefficient matrix $A = (a_{ij})$ for (1.1) is nonsingular and has all nonzero diagonal entries; i.e.,

$$a_{ii} \neq 0, \qquad i = 1, \ldots, n.$$

Assume that the kth approximating vector x^k to $x = A^{-1}b$ has been computed. Then the *Jacobi method*, mentioned earlier, for computing x^{k+1} is given by

$$x_i^{k+1} = \frac{1}{a_{ii}}\left(b_i - \sum_{j \neq i} a_{ij} x_j^k\right), \qquad i = 1, \ldots, n. \tag{3.1}$$

Now let $D = \operatorname{diag} A = \operatorname{diag}(a_{11}, \ldots, a_{nn})$ and $-L$ and $-U$ be the strictly lower and strictly upper triangular parts of A, respectively; that is,

$$L = -\begin{bmatrix} 0 & & & 0 \\ a_{21} & \ddots & & \\ \vdots & & & \\ a_{n1} & \cdots & a_{nn-1} & 0 \end{bmatrix}, \qquad U = -\begin{bmatrix} 0 & a_{12} & \cdots & a_{1n} \\ & & \ddots & \vdots \\ & & & a_{n-1n} \\ 0 & & & 0 \end{bmatrix}.$$

Then, clearly, (3.1) may be written in matrix form as

$$x^{k+1} = D^{-1}(L + U)x^k + D^{-1}b, \qquad k = 0, 1, \ldots. \tag{3.2}$$

A closely related iteration may be derived from the following observation. If we assume that the computations of (3.1) are done sequentially for $i = 1, \ldots, n$, then at the time we are ready to compute x_i^{k+1} the new components $x_1^{k+1}, \ldots, x_{i-1}^{k+1}$ are already available, and it would seem reasonable to use them instead of the old components; that is, we compute

$$x_i^{k+1} = \frac{1}{a_{ii}}\left(b_i - \sum_{j=1}^{i-1} a_{ij} x_j^{k+1} - \sum_{j=i+1}^{n} a_{ij} x_j^k\right), \qquad i = 1, \ldots, n. \tag{3.3}$$

This is the *Gauss–Seidel method* mentioned in Section 2. It is easy to see that (3.3) may be written in the form

$$Dx^{k+1} = b + Lx^{k+1} + Ux^k,$$

so that the matrix form of the Gauss–Seidel method (3.3) is given by

$$x^{k+1} = (D - L)^{-1}Ux^k + (D - L)^{-1}b, \qquad k = 0,1,\ldots. \tag{3.4}$$

Of course the Eqs. (3.1) and (3.3) should be used rather than their matrix formulations (3.2) and (3.4) in programming these methods.

We shall return to these two fundamental procedures later, but first we consider the more general iterative formula

$$x^{k+1} = Hx^k + c, \qquad k = 0,1,\ldots. \tag{3.5}$$

The matrix H is called the iteration matrix for (3.5) and it is easy to see that if we split A into

$$A = M - N, \qquad M \text{ nonsingular},$$

then for $H = M^{-1}N$ and $c = M^{-1}b$, $x = Hx + c$ if and only if $Ax = b$. Clearly, the Jacobi method is based upon the choice

$$M = D, \qquad N = L + U,$$

while the Gauss–Seidel method is based upon the choice

$$M = D - L, \qquad N = U.$$

Then for the Jacobi method $H = M^{-1}N = D^{-1}(L + U)$ and for the Gauss–Seidel method $H = M^{-1}N = (D - L)^{-1}U$. We next prove the basic convergence lemma for (3.5).

(3.6) Lemma Let $A = M - N \in C^{n \times n}$ with A and M nonsingular. Then for $H = M^{-1}N$ and $c = M^{-1}b$, the iterative method (3.5) converges to the solution $x = A^{-1}b$ to (1.1) for each x^0 if and only if $\rho(H) < 1$.

Proof If we subtract $x = Hx + c$ from (3.5), we obtain the error equation

$$x^{k+1} - x = H(x^k - x) = \cdots = H^{k+1}(x^0 - x).$$

Hence the sequence $x^0,x^1,x^2,\ldots,$ converges to x for each x^0 if and only if

$$\lim_{k\to\infty} H^k = 0;$$

that is, if and only if $\rho(H) < 1$, by considering the Jordan form for H. ■

In short we shall say that a given iterative method *converges* if the iteration (3.5) associated with that method converges to the solution to the given linear system for every x^0.

Notice that Lemma 3.6 reduces the convergence analysis of (3.5) to that of showing that $\rho(H) < 1$. We shall return to this subject in Section 5, where it will be shown how nonnegativity plays a fundamental role in the analysis for many practical problems. Next, we discuss the general topic of rates of convergence.

(3.7) Definition For $H \in C^{n \times n}$ assume that $\rho(H) < 1$ and let $x = Hx + c$. Then for

(3.8)
$$\alpha = \sup\left\{ \lim_{k \to \infty} \|x^k - x\|^{1/k} : x^0 \in C^n \right\}$$

the number

(3.9)
$$R_\infty(H) = -\ln \alpha$$

is called the *asymptotic rate of convergence* of the iteration (3.5).

The supremum is taken in (3.8) so as to reflect the worst possible rate of convergence for any x^0. Clearly, the larger the $R_\infty(H)$, the smaller the α and thus the faster the convergence of the process. Also note that α is independent of the particular vector norm used in (3.8). In the next section, the Gauss–Seidel method is modified using a relaxation parameter ω and it is shown how, in certain instances, to choose ω in order to maximize the asymptotic rate of convergence for the resulting process. This section is concluded with the following important exercise.

(3.10) Exercise Let $H \in R^{n \times n}$ and assume that $\rho(H) < 1$. Show that α defined by (3.8) satisfies $\alpha = \rho(H)$. Thus the asymptotic rate of convergence of (3.5) is

$$R_\infty(H) = -\ln \rho(H)$$

(see Ortega [1972, p. 126]).

4 THE SOR METHOD

Here we investigate in some detail a procedure that can sometimes be used to accelerate the convergence of the Gauss–Seidel method. We first note that the Gauss–Seidel method can be expressed in the following way. Let $\bar{x}_i^{k+1}$ be the ith component of x^{k+1} computed by the formula (3.3) and set

$$\Delta x_i = \bar{x}_i^{k+1} - x_i^k.$$

Then for $\omega = 1$, the Gauss–Seidel method can trivially be restated as

(4.1)
$$x_i^{k+1} = x_i^k + \omega x_i, \qquad i = 1, \ldots, n, \quad \omega > 0.$$

It was discovered during the years of hand computation (probably by accident) that the convergence is often faster if we go beyond the Gauss–Seidel correction Δx_i. If $\omega > 1$ we are "overcorrecting" while if $\omega < 1$ we are "undercorrecting." As just indicated, if $\omega = 1$ we recover the Gauss–Seidel method (3.3). In general the method (4.1) is called the *successive over-relaxation* (*SOR*) *method*. Of course the problem here is to choose the relaxation parameter ω so as to maximize the asymptotic rate of convergence of (4.1). This is the primary topic of the present section.

In order to write this procedure in matrix form, we replace $\bar{x}_i^{k+1}$ in (4.1) by by the expression in (3.3) and rewrite (4.1) as

$$x_i^{k+1} = (1-\omega)x_i^k + \frac{\omega}{a_{ii}}\left(b_i - \sum_{j=1}^{i-1} a_{ij}x_j^{k+1} - \sum_{j=i+1}^{n} a_{ij}x_j^k\right). \tag{4.2}$$

Then (4.2) can be rearranged into the form

$$a_{ii}x_i^{k+1} + \omega \sum_{j=1}^{i-1} a_{ij}x_j^{k+1} = (1-\omega)a_{ii}x_i^k - \omega \sum_{j=i+1}^{n} a_{ij}x_j^k + \omega b_i.$$

This relation of the new iterates x_i^{k+1} to the old x_i^k holds for $i = 1, \ldots, n$, and, by means of the decomposition $A = D - L - U$, developed in Section 3, may be written as

$$Dx^{k+1} - \omega Lx^{k+1} = (1-\omega)Dx^k + \omega Ux^k + \omega b$$

or, under the assumption that $a_{ii} \neq 0$, $i = 1, \ldots, n$,

$$x^{k+1} = H_\omega x^k + \omega(D - \omega L)^{-1}b, \qquad k = 0,1,\ldots, \tag{4.3}$$

where

$$H_\omega = (D - \omega L)^{-1}[(1-\omega)D + \omega U]. \tag{4.4}$$

We first prove a result that gives the maximum range of values of $\omega > 0$ for which the SOR iteration can possibly converge.

(4.5) Theorem Let $A \in C^{n \times n}$ have all nonzero diagonal elements. Then the SOR method (4.1) converges only if

$$0 < \omega < 2. \tag{4.6}$$

Proof Let H_ω be given by (4.4). In order to establish (4.6) under the assumption that $\rho(H_\omega) < 1$, it suffices to prove that

$$|\omega - 1| \leq \rho(H_\omega). \tag{4.7}$$

Because L is strictly lower triangular,

$$\det D^{-1} = \det(D - \omega L)^{-1}.$$

Thus

$$\begin{aligned}\det \mathrm{H}_\omega &= \det(D - \omega L)^{-1} \det[(1-\omega)D + \omega U] \\ &= \det[(1-\omega)I + \omega D^{-1}U] \\ &= \det[(1-\omega)I] \\ &= (1-\omega)^n,\end{aligned}$$

because $D^{-1}U$ is strictly upper triangular. Then since det H_ω is the product of its eigenvalues, (4.7) must hold and the theorem is proved. ■

It will be shown in Section 5 that for certain important classes of matrices, (4.6) is also a sufficient condition for convergence of the SOR method. For other classes we shall show that the method converges if and only if $0 < \omega < c$ for some fixed $c > 0$ which depends upon A.

The introduction of the parameter ω into the Gauss–Seidel iteration is not done so as to force convergence but, rather, to enhance the rate of convergence. We next show how, in certain instances, to determine an ω_b, $0 < \omega_b < 2$ such that

$$R_\infty(H_{\omega_b}) \geq R_\infty(H_\omega), \qquad 0 < \omega < 2. \tag{4.8}$$

The scalar ω_b may be called an *optimum* SOR *relaxation parameter* since it maximizes the asymptotic convergence rate of the SOR method.

Thus far we have discussed only point iterative methods. The methods (3.1), (3.3), and (4.1) are called the *point Jacobi, Gauss–Seidel, and SOR methods*, respectively. However, a similar analysis holds if we replace the elements of A by matrix blocks, so long as the diagonal blocks are square and nonsingular. This leads, then, to the *block Jacobi, Gauss–Seidel, and SOR methods.* We now suggest a theory of cyclically ordered block matrices in order to study some important methods for determining the optimum SOR parameter ω_b.

Let $A \in C^{n \times n}$, $n \geq 2$, be nonsingular and be partitioned into the block form

$$A = \begin{bmatrix} A_{11} & A_{12} & \cdots & A_{1k} \\ A_{21} & A_{22} & \cdots & A_{2k} \\ \vdots & \vdots & & \vdots \\ A_{k1} & A_{k2} & \cdots & A_{kk} \end{bmatrix}, \tag{4.9}$$

where the diagonal blocks A_{ii} are square and nonsingular, so that the block diagonal matrix $D = \text{diag}(A_{11}, \ldots, A_{kk})$ is nonsingular. In the usual notation, let $-L$ and $-U$ denote the block strictly lower and upper parts of A given in (4.9), respectively. Then the block Jacobi matrix J corresponding to (4.9) is given by

$$J = D^{-1}(L + U) = \begin{bmatrix} 0 & J_{12} & \cdots & J_{1k} \\ J_{21} & 0 & \cdots & J_{2k} \\ \vdots & \vdots & & \vdots \\ J_{k1} & J_{k2} & \cdots & 0 \end{bmatrix}, \tag{4.10}$$

where $J_{ij} = A_{ii}^{-1}A_{ij}$ for all $i \neq j$.

(4.11) Definition The matrix J given in (4.10) is *weakly cyclic of index p* if there exists a permutation matrix Q such that QJQ^t has the block partitioned form (this is the transposed form of (2.2.21)) given by

$$QJQ^t = \begin{bmatrix} 0 & & \cdots & & B_{1p} \\ B_{21} & 0 & \cdots & & 0 \\ 0 & B_{32} & 0 & \cdots & 0 \\ \vdots & & & & \vdots \\ 0 & \cdots & & B_{p\,p-1} & 0 \end{bmatrix}, \tag{4.12}$$

where the null diagonal blocks are square.

(4.13) Definition A matrix A with the block partitioned form (4.9) is *p-cyclic* relative to the given partitioning if the block Jacobi matrix J of (4.10) is weakly cyclic of index p.

(4.14) Definition A matrix A with the block partitioned form (4.9) is *consistently order p-cyclic* if A is p-cyclic and if all the eigenvalues of the matrix

$$J(\alpha) = \alpha L + \alpha^{1-p}U, \tag{4.15}$$

where $-L$ and $-U$ are the lower and upper block partitioned parts of A, respectively, are independent of α for all $\alpha \neq 0$.

With these definitions in hand, we proceed to study the SOR method for matrices A satisfying Definition 4.14. Many of the results along these lines are quite technical and their proofs will be omitted (see Varga [1959]).

For a system of linear equations $Ax = b$, where A is nonsingular and satisfies Definition 4.14, one can give an expression for the optimum relaxation parameter ω_b for the block SOR method. This expression is based upon the following:

(4.16) Theorem Let the matrix A of (4.9) be a consistently ordered p-cyclic matrix. If $\omega \neq 0$ and λ is a nonzero eigenvalue of the block SOR iteration matrix H_ω of (4.4) and if δ satisfies

$$(\lambda + \omega - 1)^p = \lambda^{p-1}\omega^p\delta^p, \tag{4.17}$$

then δ is an eigenvalue of the block Jacobi matrix J of (4.10). Conversely, if δ is an eigenvalue of J and λ satisfies (4.17), then λ is an eigenvalue of H_ω (see Varga [1959 or 1962, Chapter 4]).

The choice of $\omega = 1$ is such that H_1 reduces to the block Gauss–Seidel iteration matrix; thus as an immediate consequence of Theorem 4.16, we have the following result:

(4.18) Corollary Let the matrix A of (4.9) be a consistently ordered p-cyclic matrix. Then the block Jacobi method converges if and only if the block Gauss–Seidel method converges, and if both converge, then

$$\rho(H_1) = [\rho(J)]^p < 1$$

and consequently

$$R_\infty(H_1) = pR_\infty(J). \tag{4.19}$$

Notice that (4.19) gives the exact ratio value for the asymptotic rates of convergence of the Jacobi and Gauss–Seidel methods in this case. We now consider the general case where $0 < \omega < 2$. The main result concerning the optimum SOR relaxation parameter for consistently ordered p-cyclic matrices is contained in the following fundamental theorem:

(4.20) Theorem Let the matrix A of (4.9) be a consistently ordered p-cyclic matrix and suppose that all the eigenvalues of J^p are nonnegative (nonpositive). If $\rho(J) < 1$ then

(1) $\rho(H_\omega) < 1$ for all $0 < \omega < p/(p-1)$,
(2) the optimum SOR relaxation parameter ω_b defined by (4.8) is unique and satisfies:

$$(s\rho(J)\omega_b)^p = [P^p(p-1)^{1-p}](\omega_b - 1), \tag{4.21}$$

where $s = 1$ $[-1]$ if the sign of the eigenvalues of J^p is positive [negative] and
(3) $\rho(H_{\omega b}) = (\omega_b - 1)(p-1)$ (see Varga [1959 or 1962, Chapter 4]).

The proof of this theorem, as well as the proof of Theorem 4.16, is very technical and can also be found in Young [1972]. We shall concentrate upon applying these results. With this in mind, this section is concluded with two examples, each of which illustrates very different, but interesting, situations in which Theorem 4.20 can be applied.

(4.22) Example We will show how Theorem 4.20 can be used to solve the linear system associated with the *two-point boundary problem* (the *heat* *equation), discussed in Section 2,* by the point SOR method. It turns out that the SOR method is dramatically superior to the Jacobi and Gauss–Seidel methods here, so long as the number of unknowns is not exceedingly large.

Recall that the second order central difference quotient discretization of the two-point boundary value problem $-d^2u/dx^2 = g(x)$, $0 \leq x \leq 1$, $u(0) = \alpha$, $u(1) = \beta$ leads to the linear system of equations $Au = b$ where A is the $n \times n$ matrix given by (2.8). Then $D = \operatorname{diag} A = 2I$ so that the point Jacobi iteration matrix associated with A is

$$J = \frac{1}{2}\begin{bmatrix} 0 & 1 & & \cdots & 0 \\ 1 & 0 & 1 & \cdots & 0 \\ 0 & 1 & & \ddots\cdots & 0 \\ \vdots & \vdots & \ddots & \ddots & \vdots \\ & & & \ddots & 1 \\ 0 & 0 & \cdots & 1 & 0 \end{bmatrix}. \tag{4.23}$$

From Exercise 7.7 it follows that every block tri-diagonal matrix with nonsingular diagonal blocks is two-cyclic and is consistently ordered. In particular then, A, given by (2.8), is a point consistently ordered two-cyclic matrix. Moreover, since J in (4.23) is symmetric its eigenvalues are real, so that the eigenvalues of J^2 are all nonnegative. In particular, the eigenvalues of J are given (see Exercise 7.5) by

$$\lambda_k = \cos[k\pi/(n+1)], \qquad k = 1, \ldots, n, \tag{4.24}$$

so that

$$\rho(J) = \cos[\pi/(n+1)], \tag{4.25}$$

where n is the order of A, that is, the number of unknowns in (2.4). Thus A satisfies all the hypotheses of Theorem 4.20 with $p = 2$. Consequently,

$$\rho(H_\omega) < 1 \qquad \text{for} \quad 0 < \omega < 2, \tag{4.26}$$

so that the point SOR method converges for the entire permissible range of ω. Now for $p = 2$, Eq. (4.21) takes the form

$$\rho(J)^2\omega_b^2 - 4\omega_b + 4 = 0$$

and $\rho(H_\omega)$ is minimized at the smallest positive root

$$\omega_b = \frac{2}{1 + [1 - \rho(J)^2]^{1/2}}. \tag{4.27}$$

Thus for our case, the optimum SOR parameter is given by (4.27) with $\rho(J)$ given by (4.25). Moreover from Theorem 4.20, part (3),

$$\rho(H_{\omega_b}) = (\omega_b - 1) = \frac{1 - [1 - \rho(J)^2]^{1/2}}{1 + [1 - \rho(J)^2]^{1/2}}.$$

We remark that even the SOR method with the optimum parameter ω_b may not be effective for this problem if n is chosen very large, since $\rho(J) = \cos[\pi/(n+1)]$ approaches one as n increases.

(4.28) Example Here we consider a block SOR iterative method that can sometimes be used to solve the important *large sparse linear least-squares problem.*

Let $A \in R^{m \times n}$, $m \geq n$, have rank n and let b be a real m-vector. Then *the linear least-squares problem* can be stated as follows. One wishes to minimize

$$\|b - Ax\|_2, \tag{4.29}$$

where $\|\cdot\|_2$ denotes the Euclidean norm. That is, one wishes to find a vector y such that

$$\|b - Ay\|_2 = \min\|b - Ax\|_2 \qquad \text{for all} \quad x \in R^n.$$

Note that since A has full column rank, the vector y is unique in this case. An equivalent formulation of the problem is the following. One wishes to determine y and r so that

$$r + Ay = b \qquad \text{and} \qquad A^t r = 0. \tag{4.30}$$

Now suppose it is possible to permute the rows of A to obtain the block partitioned form

$$A = \begin{bmatrix} A_1 \\ A_2 \end{bmatrix},$$

where $A_1 \in R^{n \times n}$ and is nonsingular. Suppose that r and b are partitioned according to the given partition of A. Under these conditions we can write (4.30) in the form

$$\begin{bmatrix} A_1 \\ A_2 \end{bmatrix} y + \begin{bmatrix} r_1 \\ r_2 \end{bmatrix} = \begin{bmatrix} b_1 \\ b_2 \end{bmatrix}, \qquad [A_1^t, A_2^t]\begin{bmatrix} r_1 \\ r_2 \end{bmatrix} = 0. \tag{4.31}$$

Then we have the block system of linear equations

$$\begin{aligned} A_1 y \qquad\quad &+ \quad r_1 = b_1 \\ A_2 y + \quad r_2 \quad &\qquad\quad = b_2 \\ A_2^t r_2 + A_1^t r_1 &= 0 \end{aligned} \tag{4.32}$$

and this system can be written in block matrix form $Cz = d$, where

$$C = \begin{bmatrix} A_1 & 0 & I \\ A_2 & I & 0 \\ 0 & A_2^t & A_1^t \end{bmatrix}, \qquad z = \begin{bmatrix} y \\ r_2 \\ r_1 \end{bmatrix}, \qquad d = \begin{bmatrix} b_1 \\ b_2 \\ 0 \end{bmatrix}. \tag{4.33}$$

The matrix C given in (4.33) is nonsingular since A has rank n. Thus the linear least-squares problem (4.29) has been reduced to the problem of solving the linear system of (4.32) of $m + n$ linear equations in the $m + n$ unknown components of y and r. We are primarily interested in the case here where n is large and A is sparse, so that (4.32) represents a large sparse system of linear equations.

Let $D = \operatorname{diag}(A_1, I, A_1^t)$. Then the block Jacobi matrix for C is given by

$$J = \begin{bmatrix} 0 & 0 & -A_1^{-1} \\ -A_2 & 0 & 0 \\ 0 & -A_1^{-t}A_2^t & 0 \end{bmatrix}. \tag{4.34}$$

Then J is weakly cyclic of index 3 since it already has the block canonical form (4.12). Thus the block coefficient matrix C of the system (4.32) given in (4.33) is three-cyclic. Moreover, C is consistently ordered three-cyclic since by (4.15)

$$J(\alpha) = \begin{bmatrix} 0 & 0 & -\alpha^{-2}A_1^{-1} \\ -\alpha A_2 & 0 & 0 \\ 0 & -\alpha A_1^{-t}A_2^t & 0 \end{bmatrix},$$

so that $J(\alpha)^3$ is independent of α and, consequently, the eigenvalues of $J(\alpha)$ are independent of α.

Moreover, for $P = A_2 A_1^{-1}$,

$$\begin{aligned} J^3 &= \operatorname{diag}(-A_1^{-1}A_1^{-t}A_2^tA_2, -A_2A_1^{-1}A_1^{-t}A_2^t, -A_1^{-t}A_2^tA_2A_1^{-1}) \\ &= \operatorname{diag}(-A_1^{-1}P^tPA_1, -PP^t, -P^tP) \end{aligned}$$

so that the eigenvalues of J^3 are the eigenvalues of $-P^tP$ and are thus nonpositive real numbers, since P^tP is positive semidefinite and has all nonnegative eigenvalues. Thus the hypotheses of Theorem 4.20 are satisfied whenever $\rho(J) < 1$. In general, one would not normally have $\rho(J) < 1$ since

$$\rho(J) < 1 \leftrightarrow \rho(J)^3 < 1 \leftrightarrow \rho(P^tP) < 1 \leftrightarrow \rho((A_1^tA_1)^{-1}(A_2^tA_2)) < 1 \leftrightarrow \|A_2A_1^{-1}\|_2 < 1. \tag{4.35}$$

However, situations where (4.35) is satisfied by the matrix A in (4.29) often arise, for example, in certain methods for solving the linear least-squares problem with equality constraints. Various sufficient conditions on A in order that (4.35) be satisfied can also be given in terms of diagonal dominance-type conditions on A_1.

Suppose $\rho(J) < 1$. Then since $p = 3$, the block SOR method for (4.32) converges for all $0 < \omega < \frac{3}{2}$ by Theorem 4.20. Since all the eigenvalues of J^3 are nonpositive, Eq. (4.21) takes the form

$$-4\omega_b^3\rho(J)^3 - 27\omega_b + 27 = 0. \tag{4.36}$$

It can be shown, then, that the optimum SOR relaxation parameter ω_b for the block SOR method for solving the linear system (4.32) is the unique positive root of (4.36).

Some elementary analysis shows that the optimum SOR parameter here in fact satisfies

$$0.9 < \omega_b \leq 1. \tag{4.37}$$

This is in contrast to Example (4.22), where it is seen by (4.27) that in the two-cyclic case, $\omega_b \geq 1$. From (4.28) it is evident that the Gauss–Seidel method may sometimes be an appropriate choice for solving the linear system (4.32). In this case the asymptotic rate of convergence is about three times that of the Jacobi method, since from (4.19), $R_\infty(H_1) = 3R_\infty(J)$. In conclusion then, we have used the SOR analysis in this section to show that the Gauss–Seidel method, $\omega = 1$, is often a good choice for solving the three-block system (4.32) associated with the large sparse linear least-squares problem (4.29), whenever

$$A = \begin{pmatrix} A_1 \\ A_2 \end{pmatrix}$$

with A_1 nonsingular and with $\|A_2 A_1^{-1}\|_2 < 1$.

5 NONNEGATIVITY AND CONVERGENCE

In this section we investigate several important convergence criteria for the iterative methods for solving systems of linear equations $Ax = b$, derived in Sections 3 and 4. The primary tools in these investigations include the Perron–Frobenius theory of nonnegative matrices discussed in Chapter 2, the theory of M-matrices and generalizations discussed in Chapter 6, and the theory of quadratic forms and positive definite matrices.

It should be mentioned that much of the material in this section could be presented in terms of partial orderings induced by cones in R^n other than the nonnegative orthant. In particular, we could use the theory of proper cones, developed in Chapter 1, and the resulting generalized Perron–Frobenius theory to establish more general convergence criteria (see Exercise 7.17). Rather we shall be concerned here with the usual nonnegativity assumptions and leave the extension of our results to the more general partial orderings induced by cones to the exercises or the reader. All matrices will be considered real unless specifically stated otherwise.

For $A \in R^{n \times n}$, nonsingular, we shall be concerned first with general splittings

$$A = M - N, \qquad M \text{ nonsingular}, \tag{5.1}$$

and conditions under which $\rho(H) < 1$ where $H = M^{-1}N$. Thus we are concerned with convergence criteria for the general iterative formula (3.5). We begin with the following application of the Perron–Frobenius theory.

(5.2) Theorem Let $A = M - N$ with A and M nonsingular, and suppose that $H \geq 0$, where $H = M^{-1}N$. Then

$$\rho(H) < 1$$

if and only if

$$A^{-1}N \geq 0$$

in which case

$$\rho(H) = \frac{\rho(A^{-1}N)}{1 + \rho(A^{-1}N)}. \tag{5.3}$$

Proof Assume that $\rho(H) < 1$. Then clearly

$$A^{-1}N = [M(I - M^{-1}N)]^{-1}N = (I - H)^{-1}H$$

$$= \sum_{k=1}^{\infty} H^k \geq 0,$$

since $H \geq 0$. Conversely assume that $A^{-1}N \geq 0$. Then since $H \geq 0$,

$$Hx = \rho(H)x$$

for some $x > 0$ by Theorem 2.1.1 or the Perron–Frobenius theorem, Theorem 2.1.4. In this case

$$A^{-1}Nx = (I - H)^{-1}Hx = \frac{\rho(H)x}{1 - \rho(H)}.$$

This already shows that $\rho(H) < 1$ since $A^{-1}Nx \geq 0$ and $x > 0$. Moreover since

$$\rho(A^{-1}N) \geq \frac{\rho(H)}{1 - \rho(H)},$$

it follows that

$$\rho(H) \leq \frac{\rho(A^{-1}N)}{1 + \rho(A^{-1}N)}.$$

Similarly since $A^{-1}N \geq 0$,

$$Hy = (I + A^{-1}N)^{-1}A^{-1}Ny = \frac{\rho(A^{-1}N)y}{1 + \rho(A^{-1}N)}$$

for some $y > 0$, once again by the Perron–Frobenius theorem. Thus

$$\rho(H) \geq \frac{\rho(A^{-1}N)}{1 + \rho(A^{-1}N)}$$

and (5.3) follows. ■

The following corollary provides further conditions under which the iteration (3.5) converges for each x^0.

(5.4) Corollary Let $A = M - N$ with A and M nonsingular. If A, M, and N satisfy

$$A^t y \geq 0 \rightarrow N^t y \geq 0 \qquad \text{and} \qquad M^t y \geq 0 \rightarrow N^t y \geq 0,$$

then $\rho(H) < 1$.

Proof Note that

$$A^{-1} \geq 0 \leftrightarrow N^t A^{-t} \geq 0 \leftrightarrow x \geq 0 \rightarrow N^t A^{-t} x \geq 0.$$

Thus if $A^t y \geq 0 \rightarrow N^t y \geq 0$, then for $x = A^t y$, $x \geq 0 \rightarrow N^t A^{-t} x \geq 0$ so that $A^{-1}N \geq 0$. Conversely, if $A^{-1}N \geq 0$ then for $x = A^t y \geq 0$, it follows that $N^t y = N^t A^{-t} x \geq 0$. Thus the implication $A^t y \geq 0 \rightarrow N^t y \geq 0$ is equivalent to the condition $A^{-1}N \geq 0$. Similarly $M^t y \geq 0 \rightarrow N^t y \geq 0$ is equivalent to $H = M^{-1}N \geq 0$. The result then follows from Theorem 5.2. ■

We now investigate certain types of general splittings, discussed first in Theorem 6.2.3, in terms of characterizations of nonsingular M-matrices.

(5.5) Definitions The splitting $A = M - N$ with A and M nonsingular is called a *regular splitting* if $M^{-1} \geq 0$ and $N \geq 0$. It is called a *weak regular splitting* if $M^{-1} \geq 0$ and $M^{-1}N \geq 0$.

Clearly, a regular splitting is a weak regular splitting. The next result relates the convergence of (3.5) to the inverse-positivity of A. We shall call it the (weak) *regular splitting theorem.*

(5.6) Theorem Let $A = M - N$ be a weak regular splitting of A. Then the following statements are equivalent:

(1) $A^{-1} \geq 0$; that is, A is inverse-positive, so that each of conditions (N_{38})–(N_{46}) of Theorem 6.2.3 holds for A.
(2) $A^{-1}N \geq 0$.
(3) $\rho(H) = \rho(A^{-1}N)/(1 + \rho(A^{-1}N))$ so that $\rho(H) < 1$.

Proof The equivalence of (2) and (3) follows from Theorem 5.2. That (1) → (2) follows from the implication N → O of Theorem 6.2.3. Finally, if

(3) holds then (1) follows since

$$A^{-1} = (I - H)^{-1}M^{-1} = \sum_{k=0}^{\infty} H^k M^{-1} \geq 0. \blacksquare$$

Notice that from Theorem 6.2.3, part (O), weak regular splittings of a nonsingular M-matrix satisfy Theorem 5.6. It is clear that parts (1), (2), and (3) of Theorem 5.6 are also equivalent whenever $A = M - N$ is a regular splitting of A. A comparison of convergence rates for regular splittings is given next.

(5.7) Corollary Let A be inverse-positive and let $A = M_1 - N_1$ and $A = M_2 - N_2$ be two regular splittings of A where $N_2 \leq N_1$. Then for $H = M_1^{-1}N_1$ and $K = M_2^{-1}N_2$,

$$\rho(K) \leq \rho(H) < 1$$

so that

$$R_\infty(K) \geq R_\infty(H).$$

Proof The proof follows from Theorem 5.6, part (3), together with the fact that $\alpha(1 + \alpha)^{-1}$ is an increasing function of α for $\alpha \geq 0$. $\blacksquare$

We now turn to iterative procedures for two sequences of vectors which not only approximate the solution to the given system of linear equations but also yield bounds for this solution. The starting vectors of the iterative procedure are to satisfy conditions which are closely related to matrix monotonicity (see Chapter 5, Section 4).

(5.8) Theorem Let $A = M - N$ with A and M nonsingular and suppose that $H \geq 0$ where $H = M^{-1}N$. Consider the system $Ax = b$ of linear equations and the iterative formula (3.5).

(i) If there exist vectors v^0 and w^0 such that $v^0 \leq v^1$, $v^0 \leq w^0$, and $w^1 \leq w^0$, where v^1 and w^1 are computed from the iterative formulas $v^{k+1} = Hv^k + M^{-1}b$ and $w^{k+1} = Hw^k + M^{-1}b$, $i = 0,1, \ldots$, respectively, then

$$v^0 \leq v^1 \leq \cdots \leq v^i \leq \cdots \leq A^{-1}b \leq \cdots \leq w^i \leq \cdots \leq w^1 \leq w^0 \tag{5.9}$$

and for any scalar λ

$$A^{-1}b = \lambda \lim_{i \to \infty} v^i + (1 - \lambda) \lim_{i \to \infty} w^i. \tag{5.10}$$

(ii) If $\rho(H) < 1$ then the existence of such v^0 and w^0 is assured.

Proof It follows by induction and the assumption that $H \geq 0$, that

$$v^i \leq v^{i+1},\ v^i \leq w^i \qquad \text{and} \qquad w^{i+1} \leq w^i \qquad \text{for each } i.$$

Thus the bounded sequences $\{v^i\}$ and $\{w^i\}$ converge and thus they converge to $A^{-1}b$ by Lemma 3.6. This establishes (5.9) and (5.10) and shows, moreover, that $\rho(H) < 1$.

For part (ii), assume that $\rho(H) < 1$. Then there exists $z > 0$ such that $Hz = \rho(H)z < z$ by Theorem 2.1.1 or the Perron–Frobenius theorem, Theorem 2.1.4. Let

$$v^0 = A^{-1}b - z.$$

Then

$$\begin{aligned} v^1 &= Hv^0 + M^{-1}b = HA^{-1}b + M^{-1}b - Hz \\ &= A^{-1}b - \rho(H)z \geq A^{-1}b - z = v^0. \end{aligned}$$

Letting

$$w^0 = A^{-1}b + z,$$

we see that it follows in a similar fashion that $w^1 \leq w^0$. Moreover $w^0 - v^0 = 2z \geq 0$ and the proof is complete. ■

As indicated earlier, every [weak] regular splitting of a nonsingular M-matrix is convergent by Theorem 6.2.3, part [(O)] (P). Clearly for such matrices, the Jacobi and Gauss–Seidel methods defined by (3.2) and (3.4) are based upon regular splittings. Moreover, if $0 < \omega \leq 1$, then the SOR method defined by (4.3) is based upon a regular splitting. These concepts will now be extended to an important class of complex matrices.

Let $A \in C^{n \times n}$ have all nonzero diagonal elements and let $A = D - L - U$, where as usual $D = \operatorname{diag} A$ and where $-L$ and $-U$ represent the lower and the upper parts of A, respectively. Then the *comparison matrix* for A, $\mathcal{M}(A)$, as given by Definition 6.2.8, satisfies

$$\mathcal{M}(A) = |D| - |L| - |U|. \tag{5.11}$$

In addition, the set of *equimodular matrices* associated with A, $\Omega(A)$, as given by Definition 6.2.8, satisfies

$$\Omega(A) = \{B \in C^{n \times n} | \mathcal{M}(B) = \mathcal{M}(A)\}. \tag{5.12}$$

We are interested in conditions under which the Jacobi, Gauss–Seidel, and more generally the SOR methods converge for the complex matrix A. With this in mind we give the following definition.

(5.13) Definition The matrix $A \in C^{n \times n}$ is called an H-*matrix* if $\mathcal{M}(A)$ is an M-matrix.

Our next theorem provides some characterizations of nonsingular H-matrices in terms of the convergence of the Jacobi and SOR methods. Its proof is rather technical and will be omitted. Rather, we shall prove some of the results later in the special case where A is a nonsingular M-matrix.

(5.14) Theorem Let $A \in C^{n \times n}$ have all nonzero diagonal elements. Then the following statements are equivalent.

(1) A is a nonsingular H-matrix, that is, $\mathcal{M}(A)$ satisfies any one of the 50 equivalent conditions of Theorem 6.2.3.
(2) For each $B \in C^{n \times n}$, $\mathcal{M}(B) \geq \mathcal{M}(A)$ implies that B is nonsingular.
(3) The Jacobi method (3.2) converges for each $B \in \Omega(A)$.
(4) The SOR method (4.3) converges for each $B \in \Omega(A)$ whenever

$$0 < \omega < \frac{2}{1 + \rho(|J|)}, \tag{5.15}$$

where J is the Jacobi iteration matrix for A.

(See Varga [1976].)

It is interesting to interpret parts of Theorem 5.14 in terms of diagonal dominance and generalizations of diagonal dominance.

(5.16) Definition The matrix $A \in C^{n \times n}$ is *generalized column diagonally dominant* if there exists $x \in R^n$, $x = (x_i) \gg 0$ such that

$$|a_{ii}|x_i \geq \sum_{j \neq i} |a_{ij}|x_j, \qquad i = 1, \ldots, n. \tag{5.17}$$

Then A is called *strictly generalized column diagonally dominant* if strict inequality holds in (5.17), for $i = 1, \ldots, n$, and *irreducibly generalized column diagonally dominant* if A is irreducible and generalized column diagonally dominant and strict inequality of (5.17) holds for at least one i.

Clearly, the concept of generalized column diagonal dominance reduces to the usual one whenever $x = e$, the vector of all ones. We now state an important convergence result based upon these principles.

(5.18) Theorem If $A \in C^{n \times n}$ is either strictly or irreducibly generalized column diagonally dominant then the Jacobi method given by (3.2) converges for A and the SOR method given by (4.1) converges for A whenever

$$0 < \omega < \frac{2}{1 + \rho(|J|)},$$

where J denotes the Jacobi iteration matrix for A.

Proof Since $A \in \Omega(A)$, it suffices to show that A is a nonsingular H-matrix, so that Theorem 5.14, parts (3) and (4), may be applied. But (5.17) holds if and only if $\mathcal{M}(A)x > 0$. Thus if A is strictly generalized column diagonally dominant then $\mathcal{M}(A)$ is a nonsingular M-matrix by Theorem 6.2.3, part(I). Similarly, if A is irreducibly generalized column diagonally dominant then $\mathcal{M}(A)$ is a nonsingular M-matrix by Theorem 6.2.7, part (ii). ■

We now return to the special case where $A \in R^{n \times n}$ and A is a nonsingular M-matrix. In the following analysis we may assume, without loss of generality, that $D = \operatorname{diag} A = I$. In this case the Jacobi iteration matrix for A is given by

$$J = L + U, \tag{5.19}$$

while the SOR iteration matrix for A is given by

$$H_\omega = (I - \omega L)^{-1}[(1 - \omega)I + \omega U]. \tag{5.20}$$

We first prove an extension of the Stein and Rosenberg theorem [1948].

(5.21) Theorem Let $A = I - L - U \in R^{n \times n}$ where $L \geq 0$ and $U \geq 0$ are strictly lower and upper triangular, respectively. Then for $0 < \omega \leq 1$,

(1) $\rho(J) < 1$ if and only if $\rho(H_\omega) < 1$.

(2) $\rho(J) < 1$ (and $\rho(H_\omega) < 1$) if and only if A is a nonsingular M-matrix, in which case

$$\rho(H_\omega) \leq 1 - \omega + \omega\rho(J).$$

(3) if $\rho(J) \geq 1$ then $\rho(H_\omega) \geq 1 - \omega + \omega\rho(J) \geq 1$.

Proof Let $\lambda = \rho(J)$ and for fixed $0 < \omega \leq 1$ let $\delta = \rho(H_\omega)$.

Now L is strictly lower triangular so that $L^n = 0$. Then since $L \geq 0$ and $U \geq 0$ and $0 < \omega \leq 1$,

$$(I - \omega L)^{-1} = I + \omega L + \cdots + \omega^{n-1}L^{n-1} \geq 0,$$

and thus

$$H_\omega = (I - \omega L)^{-1}[(1 - \omega)I + \omega U] \geq 0.$$

By the Perron–Frobenius theorem, $\lambda = \rho(J)$ and $\delta = \rho(H_\omega)$ are eigenvalues of J and H_ω, respectively. For some $x \neq 0$, $H_\omega x = \delta x$ and so

$$(\delta L + U)x = [(\delta + \omega - 1)/\omega]x.$$

Now since $(\delta + \omega - 1)/\omega$ is an eigenvalue of $\delta L + U$, we have

$$\delta + \omega - 1 \leq \omega\rho(\delta L + U).$$

If $\delta \leq 1$, then $\rho(\delta L + U) \leq \rho(L + U) = \rho(J) = \lambda$, since $0 \leq \delta L + U \leq L + U$. Thus

$$\delta \leq \omega\lambda + 1 - \omega.$$

On the other hand, if $\delta \geq 1$, then

$$(\delta + \omega - 1)/\omega \leq \rho(\delta L + U) \leq \rho(\delta L + \delta U) = \delta\lambda$$

and

$$\lambda = (\delta + \omega - 1)/\omega\delta = 1 + [(1 - \omega)(\delta - 1)/\omega\delta] \geq 1.$$

This establishes the following inequalities.

(a) If $\delta \leq 1$, then $\delta \leq \omega\lambda + 1 - \omega$.
(b) If $\delta \geq 1$, then $\lambda \geq 1$, which implies
(c) if $\lambda < 1$, then $\delta < 1$.

Similarly, there exists $y \neq 0$ such that $Jy = \lambda y$ and

$$Sy = (1 - \omega + \omega\lambda)y,$$

where

$$S = (I - \alpha L)^{-1}[(1 - \omega)I + \omega U],$$

$$\alpha = \omega/(1 - \omega + \omega\lambda).$$

Thus

$$1 - \omega + \omega\lambda \leq \rho(Q).$$

But if $\lambda \geq 1$, then since $\alpha \leq \omega$, we have

$$(I - \alpha L)^{-1} = I + \alpha L + \cdots + \alpha^{n-1}L^{n-1} \leq I + \omega L + \cdots + \omega^{n-1}L^{n-1} = (I - \omega L)^{-1}.$$

Moreover $S \leq H_\omega$ and hence

$$1 - \omega + \omega\lambda \leq \rho(S) \leq \rho(H_\omega) = \delta.$$

Thus we have the following inequalities.

(d) If $\lambda \geq 1$, then $\delta \geq 1 - \omega + \omega\lambda \geq 1$, which implies

(e) if $\delta < 1$ then $\lambda < 1$.

Now by (c) and (e) we have statement (1). Statement (2) follows from Theorem 6.2.3, part (P), since the Jacobi and the SOR splittings (with $0 < \omega \leq 1$) constitute regular splittings of A. Finally, statement (3) follows from (d) and the proof is complete. ■

It is perhaps worthwhile to restate Theorem 5.21 in terms of the Jacobi and Gauss–Seidel methods, thus recovering the Stein–Rosenberg theorem.

(5.22) Corollary Let A be as in Theorem 2.1. Then

(1) $\rho(J) < 1$ if and only if $\rho(H_1) < 1$.
(2) $\rho(J) < 1$ and $\rho(H_1) < 1$ if and only if A is a nonsingular M-matrix; moreover, if $\rho(J) < 1$ then

$$\rho(H_1) \leq \rho(J).$$

(3) If $\rho(J) \geq 1$ then

$$\rho(H_1) \geq \rho(J) \geq 1.$$

We next give a comparison theorem of the convergence rates of the SOR method for nonsingular M-matrices.

(5.23) Theorem Let A be a nonsingular M-matrix and let $0 < \omega_1 \leq \omega_2 \leq 1$. Then

$$\rho(H_{\omega_2}) \leq \rho(H_{\omega_1}) < 1$$

so that

$$R_\infty(H_{\omega_2}) \geq R(H_{\omega_1}).$$

Proof Let $A = D - L - U$. Then

$$H_\omega = M_\omega^{-1} N_\omega,$$

where

$$M_\omega = \omega^{-1} D - L, \qquad N_\omega = (\omega^{-1} - 1)D + U.$$

But $M_\omega^{-1} \geq 0$ and $N_\omega \geq 0$ for $0 < \omega \leq 1$ as before, and $A = M_\omega - N_\omega$ is a regular splitting of A. Now since ω^{-1} is a decreasing function of ω for $0 < \omega \leq 1$, it follows that if $0 < \omega_1 \leq \omega_2 \leq 1$, then

$$N_{\omega_2} \leq N_{\omega_1}.$$

The result then follows from Corollary 5.7. ∎

Now if A is an M-matrix, then $\rho(J) < 1$ by Theorem 5.21, and by Theorem 5.23, $\rho(H_\omega)$ is a nonincreasing function of ω in the range $0 < \omega \leq 1$. Moreover, $\rho(H_\omega) < 1$. By the continuity of $\rho(H_\omega)$ as a function of ω, it must follow that $\rho(H_\omega) < 1$ for $0 < \omega \leq \alpha$ with some $\alpha > 1$. The following simplified version of part (4) of Theorem 5.14 provides such a bound.

(5.24) Theorem If A is an M-matrix then

$$\rho(H_\omega) < 1$$

for all ω satisfying

$$0 < \omega < \frac{2}{1 + \rho(J)}. \tag{5.25}$$

Proof The convergence follows from Theorem 5.21, whenever $0 < \omega \leq 1$, so assume that $\omega \geq 1$. Assume that $D = \operatorname{diag} A = I$, as usual, and define the matrix

$$T_\omega = (I - \omega L)^{-1}[\omega U + (\omega - 1)I].$$

Then clearly

$$T_\omega \geq 0 \qquad \text{and} \qquad |H_\omega| \leq T_\omega.$$

Let $\lambda = \rho(T_\omega)$. Then for some $x > 0$, we have $Tx = \lambda x$ by the Perron–Frobenius theorem, and so

$$(\omega U + \omega\lambda L)x = (\lambda + 1 - \omega)x.$$

Hence

$$\lambda + 1 - \omega \leq \rho(\omega U + \omega\lambda L)$$

and if $\lambda \geq 1$, then

$$\lambda + 1 - \omega \leq \omega\lambda\rho(J)$$

since $\lambda L \geq L$. In this case then

$$\omega \geq \frac{1 + \lambda}{1 + \lambda\rho(J)} \geq \frac{2}{1 + \rho(J)}.$$

Hence if (5.25) hold then we must have $\lambda < 1$. Then from $|H_\omega| \leq T_\omega$ it follows that $\rho(H_\omega) \leq \lambda < 1$, and the theorem is proved. ■

This concludes our applications of the Perron–Frobenius theory of nonnegative matrices and the theory of M-matrices. We now digress from the main topic of the book in order to give, for completeness, another important class of tools in such investigations which involves the use of quadratic forms and in particular the nonnegativity of certain quadratic forms. Here we are interested in the cases where the coefficient matrix A is hermitian and/or positive definite. These concepts will be investigated next. Included in these presentations will be the important Ostrowski–Reich theorem, as well as some of its generalizations. We begin with some useful terminology and notation.

Let $A \in C^{n \times n}$. Then A^* denotes the conjugate transpose of A. Recall that if $A = A^*$, that is, A is hermitian, then A is positive definite if

(5.26) $$x^*Ax > 0 \qquad \text{for all} \quad x \in C^n, \quad x \neq 0.$$

With this in mind we state the following definition.

(5.27) Definition The matrix $A \in C^{n \times n}$ will be called *positive definite* if

(5.28) $$\text{Re}(x^*Ax) > 0, \qquad \text{for all} \quad x \in C^n, \quad x \neq 0.$$

We note that (5.28) reduces to (5.26) if A is hermitian and that (5.28) is equivalent to $A + A^*$ satisfying (5.26). Although matrices satisfying (5.28) appear in the literature in a variety of contexts, there is no generally accepted terminology for them.

We shall also need the concept of an eigenset of an iteration matrix.

(5.29) Definition Let $H \in C^{n \times n}$. Then a subset E of C^n is called an *eigenset* for H if E consists of a set of nonzero eigenvectors of H with at least one eigenvector associated with each distinct eigenvalue of H.

The following technical lemma provides a useful identity.

(5.30) Lemma Let $A = M - N \in C^{n \times n}$ with A and M nonsingular and let $H = M^{-1}N$. Then

$$A - H^*AH = (I - H^*)(M^*A^{-*}A + N)(I - H). \tag{5.31}$$

Proof The proof requires some algebraic manipulation. Since $AH = N(I - H)$, it follows that

$$\begin{aligned} H^*AH &= H^*N(I - H) - N(I - H) + AH \\ &= -(I - H^*)N(I - H) + AH \\ &= -S + AH + (I - H^*)M^*(I - H), \end{aligned}$$

where we have set

$$S = (I - H^*)(M^* + N)(I - H).$$

Then since $(I - H^*)M^* = A^*$, it follows that

$$\begin{aligned} A - H^*AH &= A + S - AH - A^*(I - H) \\ &= S + (A - A^*)(I - H) \\ &= (I - H^*)[M^* + N + M^*A^{-*}(A - A^*)](I - H) \\ &= (I - H^*)(M^*A^{-*}A + N)(I - H). \quad \blacksquare \end{aligned}$$

We are now ready to state a very general convergence result involving certain quadratic forms.

(5.32) Theorem Let $A = M - N \in C^{n \times n}$ with A and M nonsingular and let $H = M^{-1}N$. If A and M satisfy the conditions

$$x^*Ax \neq 0 \quad \text{and} \quad \frac{x^*(M^*A^{-*}A + N)x}{x^*Ax} > 0 \tag{5.33}$$

for every x in some eigenset E of H, then $\rho(H) < 1$. Conversely, if $\rho(H) < 1$, then either (5.33) holds or

$$x^*Ax = x^*(M^*A^{-*}A + N)x = 0 \tag{5.34}$$

holds for every eigenvector x of H.

Proof Let λ be any eigenvalue of H and let x be a corresponding eigenvector in E. Note that $\lambda \neq 1$; otherwise, $Mx = Nx$ which contradicts the nonsingularity of A. Then by Lemma 5.30,

$$x^*Ax - (Hx)^*A(Hx) = [(I - H)x]^*(M^*A^{-*}A + N)(I - H)x$$

or

$$(1 - |\lambda|^2)x^*Ax = |1 - \lambda|^2 x^*(M^*A^{-*}A + N)x. \tag{5.35}$$

Thus, by the assumption (5.33),

$$1 - |\lambda|^2 = |1 - \lambda|^2 \frac{x^*(M^*A^{-*}A + N)x}{x^*Ax} > 0$$

so that $|\lambda| < 1$. Thus $\rho(H) < 1$.

Conversely, if $\rho(H) < 1$ and λ, x are any eigenpair of H, then (5.35) holds and since $1 - |\lambda|^2 > 0$ and $|1 - \lambda|^2 > 0$, then either (5.33) or (5.34) must be true and the proof is complete. ■

The condition that $x^*Ax \neq 0$ for all $x \in E$ appears to be a drawback to Theorem 5.32. However, note that $x^*Ax \neq 0$ for an eigenvector x of H implies that $x^*Mx \neq 0$; conversely, $x^*Mx \neq 0$ implies that either $x^*Ax \neq 0$ or $\lambda = 1$. Note also, that if $Hx = \lambda x$ then $Nx = \lambda Mx$ and so $Ax = (M - N)x = (1 - \lambda)Mx$. Thus x is an eigenvector of the generalized eigenvalue problem

$$Ax = \delta Mx.$$

In general, the quantities x^*Ax and $x^*(M^*A^{-*}A + N)x$, as well as their product, are complex although, as shown in the proof, the quotient in (5.33) is always real for an eigenvector x of H. However, if A is hermitian, then $M^*A^{-*}A + N = M^* + N$ which is also hermitian. Thus the condition (5.33) is that the quantities x^*Ax and $x^*(M^* + N)x$, which are necessarily real, have the same sign. It is useful to isolate this as in the following.

(5.36) Definition Let S be a subset of C^n. Then two hermitian matrices $P, Q \in C^{n \times n}$ are *quadratic form sign equivalent* (QFSE) on S if either

$$(x^*Px)(x^*Qx) > 0 \tag{5.37}$$

or

$$x^*Px = x^*Qx = 0 \tag{5.38}$$

for all $x \in S$, and *strongly* QFSE (SQFSE) on S if (5.37) holds for $x \neq 0$ in S.

In terms of this definition, Theorem 5.32 reduces for hermitian A to the following useful result.

(5.39) Corollary If A is hermitian and A and $M^* + N$ are SQFSE on some eigenset of H, then $\rho(H) < 1$. Conversely, if $\rho(H) < 1$, then A and $M^* + N$ are QFSE on every eigenset of H.

In the usual way, we say that a hermitian matrix B is *positive definite on a subset* S of C^n if $x^*Bx > 0$ for all $x \neq 0$ in S. In these terms, another result of Theorem 5.32 is the following.

(5.40) Corollary If A is hermitian and A and $M^* + N$ are positive definite on some eigenset of H, then $\rho(H) < 1$. Conversely if $x^*Mx > 0$ for all x in some eigenset E of H and $\rho(H) < 1$, then A and $M^* + N$ are positive definite on E.

Proof The converse follows immediately from the following identity

$$x^*Ax + x^*(M^* + N)x = 2x^*Mx > 0,$$

which holds for any eigenvector x of H. This shows that at least one of the terms on the left-hand side is positive and hence, by Corollary 5.39 and the definition of QFSE, both must be positive. ■

We next state a companion result to Theorem 5.32 under the assumption that $M^*A^{-*}A + N$ is positive definite. In this case a somewhat different form of the converse is given.

(5.41) Theorem Let $A = M - N \in C^{n \times n}$ with A and M nonsingular and let $H = M^{-1}N$. Assume that $M^*A^{-*}A + N$ is positive definite. Then $\rho(H) < 1$ if and only if A is positive definite.

Proof Assume that A is positive definite and for any $x \neq 0$, let

$$x^*Ax = \alpha + i\beta, \qquad x^*(M^*A^{-*}A + N)x = \gamma + i\delta.$$

Then by assumption α and γ are positive. If x is an eigenvector of H, then (5.35) shows that the quotient

$$\frac{\gamma + i\delta}{\alpha + i\beta} \tag{5.42}$$

is real. Thus $\beta\gamma = \alpha\delta$ so that

$$\alpha\gamma + \beta\delta = \frac{\alpha^2\gamma + \beta^2\gamma}{\alpha} > 0,$$

and hence the quotient (5.42) is positive. Then by applying Theorem 5.32, it follows that $\rho(H) < 1$.

Conversely, assume that $\rho(H) < 1$. Then the sequence generated from $x^{k+1} = Hx^k$, $k = 0,1,\ldots,$ converges to the zero vector for any x^0. By Theorem 5.32,

$$(x^k)^*Ax^k - (x^{k+1})^*Ax^{k+1} = (x^k - x^{k+1})^*(M^*A^{-*}A + N)(x^k - x^{k+1})$$

and by the assumption that $M^*A^{-*}A + N$ is positive definite,

$$\mathrm{Re}[(x^k)^*Ax^k - (x^{k+1})^*Ax^{k+1}] \geq 0,$$

so that

$$\mathrm{Re}[(x^{k+1})^*Ax^{k+1}] \leq \mathrm{Re}[(x^k)^*Ax^k], \qquad k = 0,1,\ldots. \tag{5.43}$$

Now if A were not positive definite, we could find an x^0 such that $\mathrm{Re}[(x^0)^*Ax^0] \leq 0$ and since $x^1 \neq x^0$ (otherwise $\rho(H) = 1$), the positive definiteness of $M^*A^{-*}A + N$ implies that

$$\mathrm{Re}[(x^1)^*Ax^1] < \mathrm{Re}[(x^0)^*Ax^0] \leq 0.$$

But this would preclude the convergence of this particular sequence $\{x^i\}$ to zero. Thus A must be positive definite. ■

As an immediate result of Theorem 5.42 we have the following useful corollary.

(5.44) Corollary Assume that A is hermitian and that $M^* + N$ is positive definite. Then $\rho(H) < 1$ if and only if A is positive definite.

These convergence results are now applied to the block Jacobi, Gauss–Seidel, and general SOR methods. For that purpose we let $A \in C^{n \times n}$ such that A has a block decomposition

$$A = D - L - U, \tag{5.45}$$

where D is a block diagonal matrix with square nonsingular blocks on the diagonal and where $-L$ and $-U$ are the block lower and upper parts of A, respectively, as usual.

The application of the general results, Theorems 5.32 and 5.41, is difficult in the non-hermitian case because of the presence of the matrix $A^{-*}A$. For example, with $A = D - L - U$ the application of Theorem 5.41 to the SOR method requires that we ascertain whether the matrix

$$(5.46)\quad M^*A^{-*}A + N = \omega^{-1}D^*A^{-*} - L^*A^{-*}A + \omega^{-1}[(1-\omega)D + \omega U]$$

is positive definite. One set of sufficient conditions for this is that D and DA^{-*} are positive definite, $L^*A^{-*} = UA^{-1}$, and $0 < \omega \le 1$, but these are rather unlikely to be satisfied in problems of interest. Another sufficient condition is that A is skew-hermitian so that $A^{-*}A = -I$ and (5.46) reduces to

$$\omega^{-1}D + L^* + \omega^{-1}[(1-\omega)D - \omega L^*] = \omega^{-1}(2-\omega)D.$$

Then the application of Theorem 5.32 requires that we determine when

$$x^*Ax \neq 0 \qquad \text{and} \qquad \frac{x^*Dx}{x^*Ax} > 0$$

for all x in some eigenset of the block SOR iteration matrix H. Similar statements apply to the block Jacobi method.

In contrast to this general situation, Theorem 5.41 provides some very powerful convergence results for the block Jacobi and SOR methods in case A is hermitian. For the block Jacobi method we give the following.

(5.47) Corollary Let $A \in C^{n\times n}$ have the block decomposition (5.45) and assume that A is hermitian and that D is positive definite. Then the block Jacobi method converges for A if and only if A and $2D - A$ are positive definite.

Proof Assume that A and $2D - A$ are positive definite. In order to apply Theorem 5.41, let $M = D$, $N = L + U$. Then

$$M^* + N = D + L + U = 2D - (D - L - U) = 2D - A$$

so that $\rho(D^{-1}(L + U)) < 1$ since A and $M^* + N$ are positive definite.

Conversely, suppose that the block Jacobi method converges for A. Then since $M = D$ is positive definite, it follows by the converse of Corollary 5.40 that A and $M^* + N = 2D - A$ are positive definite. ∎

Moreover, we recover the famous Ostrowski–Reich theorem for the block SOR method when A is hermitian.

(5.48) Corollary Let $A \in C^{n\times n}$ have the block decomposition (5.45) and assume that A is hermitian and that D is positive definite. Then the SOR method converges for all $0 < \omega < 2$ if and only if A is positive definite.

Proof Here we set

$$M = \omega^{-1}(D - \omega L), \qquad N = \omega^{-1}[(1-\omega)D + \omega L^*].$$

Then since

$$M^* + N = \omega^{-1}(2 - \omega)D$$

is positive definite for all $0 < \omega < 2$, the corollary follows immediately from Theorem 5.41. ∎

Finally, as an illustration of Corollaries 5.47 and 5.48, consider the finite difference formulation of the *two-point boundary-value problem*, the heat equation, discussed in Section 2. As is often the situation in such formulations, the coefficient matrix A given by (2.8) is positive definite. Moreover, for $D = 2I$, the matrix

$$2D - A = |A| = \begin{bmatrix} 2 & 1 & 0 & \cdots & 0 \\ 1 & 2 & 1 & & \\ 0 & \ddots & \ddots & \ddots & \vdots \\ \vdots & & & & 1 \\ 0 & \cdots & & 1 & 2 \end{bmatrix}$$

is also positive definite. Here then, as indicated in Example 4.22, the point Jacobi method converges and the point SOR method converges for all $0 < \omega < 2$.

6 SINGULAR LINEAR SYSTEMS

Under certain conditions that are discussed in this section, the iterative methods described in Sections 3 and 4 can, for the most part, be extended to the case where the matrix of coefficients is singular but the equations are consistent. In particular, let $A \in C^{n \times n}$ and consider the consistent system of linear equations

$$Ax = b, \qquad b \in R(A), \tag{6.1}$$

where $R(A)$ denotes the range or column space of A, as usual.

There are many practical problems, such as the Neumann problem and those for elastic bodies with free surfaces and Poisson's equation on a sphere and with periodic boundary conditions, whose finite difference formulations lead to singular systems of linear equations. Other classes of singular systems will be discussed in later chapters. In Chapter 8, we will see how the stationary distribution vector of a finite homogeneous Markov chain is a solution to a certain singular system involving the state transition matrix. Moreover, calculation of the production vector of a Leontief input–output economic model sometimes involves the solution of a singular system. This is discussed in Chapter 9. As usual, iterative methods for solving such problems are often quite useful whenever the coefficient matrix is large and sparse.

We now give a specific example of such situations. We discuss Poisson's equation on a rectangle with periodic boundary conditions, from the area of partial differential equations.

(6.2) Example Let R denote a rectangular region in the plane defined by the inequalities $a \leq x \leq b$ and $c \leq y \leq d$. Let $f(x,y)$ denote a continuous function defined on R. We seek a function $u(x,y)$, continuous in R which is twice continuously differentiable and satisfies *Poisson's equation*

$$\frac{\partial^2 u}{\partial x^2} + \frac{\partial^2 u}{\partial y^2} = -f(x,y), \qquad a < x < b, \quad c < y < d \tag{6.3}$$

and satisfies the periodic boundary conditions

$$\begin{aligned} u(a,y) &= u(b,y), \qquad c \leq y \leq d, \\ u(x,c) &= u(x,d), \qquad a \leq x \leq b. \end{aligned} \tag{6.4}$$

Then the problem described by (6.3) and (6.4) is called *Poisson's equation with periodic boundary condition* and it is known that the problem has a solution, but that the solution is not unique and that any two solutions differ by an additive constant (see Buzbee *et al.* [1970]).

Using the five-point central difference approximation to the left-hand side of (6.3), the finite difference equations can be represented by a matrix equation $Au = g$, where g is a vector determined by $f(x,y)$ and the boundary conditions and where A has the block form

$$A = \begin{bmatrix} D & -I & 0 & \cdots & 0 & 0 & -I \\ -I & D & -I & \cdots & 0 & 0 & 0 \\ 0 & -I & D & \cdots & 0 & 0 & 0 \\ \vdots & \vdots & \vdots & \ddots & \vdots & \vdots & \vdots \\ 0 & 0 & 0 & & D & -I & 0 \\ 0 & 0 & 0 & \cdots & -I & D & -I \\ -I & 0 & 0 & \cdots & 0 & -I & D \end{bmatrix} \tag{6.5}$$

and where the matrix D is given by

$$D = \begin{bmatrix} 4 & -1 & 0 & \cdots & 0 & 0 & -1 \\ -1 & 4 & -1 & \cdots & 0 & 0 & 0 \\ 0 & -1 & 4 & \cdots & 0 & 0 & 0 \\ \vdots & \vdots & \vdots & & \vdots & \vdots & \vdots \\ 0 & 0 & 0 & \cdots & 4 & -1 & 0 \\ 0 & 0 & 0 & \cdots & -1 & 4 & -1 \\ -1 & 0 & 0 & \cdots & 0 & -1 & 4 \end{bmatrix}. \tag{6.6}$$

It follows that the large sparse matrix A is singular of rank$(n-1)$, where n is the order of A. We shall see later how the iterative methods developed in this section for singular systems of linear equations may be used to approximate a solution to the problem.

Next suppose $A \in C^{n \times n}$ is split into $A = M - N$, as usual, with M nonsingular. Let $H = M^{-1}N$ and $c = M^{-1}b$ and consider the iteration

$$x^{k+1} = Hx^k + c. \tag{6.7}$$

In the case where the coefficient matrix A is nonsingular, we have seen by Lemma 3.6 that the iteration (6.7) converges to the solution $x = A^{-1}b$ to the system $Ax = b$ for every initial vector x^0, if and only if $\rho(H) < 1$, so that

$$\lim_{i \to \infty} H^i = 0. \tag{6.8}$$

However, this is *not* the case when A is singular. We proceed now to analyze this important situation.

Recall that a matrix H is called *semiconvergent* (see Definition 6.4.8) if the limit in (6.8) exists, although it need not be the zero matrix. Moreover, by Exercise (6.4.9), H is semiconvergent if and only if (1) $\rho(H) \le 1$ and (2) if $\rho(H) = 1$ then all the Jordan blocks associated with 1 for H are one-by-one and (3) if $\rho(H) = 1$ then $\lambda \in \sigma(H)$ and $|\lambda| = 1$ implies $\lambda = 1$. This leads to the following lemma.

(6.9) Lemma Let $H \in C^{n \times n}$. Then H is semiconvergent if and only if there exists a nonsingular matrix P such that

$$H = P\begin{bmatrix} I & 0 \\ 0 & K \end{bmatrix}P^{-1}, \tag{6.10}$$

where I is missing if $1 \notin \sigma(H)$ and where $\rho(K) < 1$.

Proof The proof follows from Exercise 6.4.9 by applying the Jordan canonical form theorem to H. ■

Recall also that the *index* of $A \in C^{n \times n}$, index A, is defined (see Definition 5.4.9) to be the least nonnegative integer k such that rank $A^k =$ rank A^{k+1}. From the definition of the Drazin inverse A^{D} of A (see Definition 5.4.10) it follows that A^{D} is the unique matrix given by

$$A^{\mathrm{D}}x = \begin{cases} y & \text{if } \; Ay = x, \quad x \in R(A^k), \\ 0 & \text{if } \; A^k x = 0. \end{cases}$$

With this in mind, we give

(6.11) Lemma Let H be semiconvergent. Then

$$\lim_{i \to \infty} H^i = I - E, \qquad E = (I - H)(I - H)^{\mathrm{D}}. \tag{6.12}$$

Proof The proof follows immediately from Lemma 6.9 and the characterization of the Drazin inverse just given. ■

We are now in a position to give a result which extends Lemma 3.6 to the singular case.

(6.13) Lemma Let $A = M - N \in C^{n \times n}$ with M nonsingular. Then for $H = M^{-1}N$ and $c = M^{-1}b$, the iterative method (6.7) converges to some solution $x(x^0)$ to $Ax = b$, for each x^0, if and only if H is semiconvergent. In this case

$$\lim_{i \to \infty} x^i = (I - H)^{\mathrm{D}}c + Ex^0, \qquad E = (I - T)(I - T)^{\mathrm{D}}. \tag{6.14}$$

Proof Suppose x is a fixed solution to $Ax = b$. Then $x = Hx + c$. Now subtracting this equation from (6.7) yields the error equation

$$x^{k+1} - x = H(x^k - x) = \cdots = H^{k+1}(x^0 - x).$$

Hence the sequence $\{x^i\}$ converges to some solution to $Ax = b$ if and only if the sequence $\{x^i - x\}$ converges to a vector in the null space of A. But that is the case if and only if

$$\lim_{i \to \infty} H^i(x^0 - x) \tag{6.15}$$

exists. It follows from Lemma 6.9 that (6.15) holds for all x^0 if and only H is semiconvergent. Moreover if H is semiconvergent then the identity (6.14) follows immediately from Lemma 6.11. ■

We now consider asymptotic rates of convergence. For $H \in C^{n \times n}$ with H semiconvergent, we adopt the notation

$$\delta(H) = \max\{|\lambda| : \lambda \in \sigma(H), \lambda \neq 1\}. \tag{6.16}$$

Then if $\rho(H) < 1$, $\delta(H) = \rho(H)$. Otherwise, $\delta(H)$ is the second largest of the moduli of the eigenvalues of H. With this in mind we see that $\delta(H) = \rho(K)$, where K is given by 6.10. This leads to the observation that if H is semiconvergent then the asymptotic rate of convergence of the iteration (6.7) is given by

$$R_\infty(H) = -\ln \delta(H), \tag{6.17}$$

where $\delta(H)$ is given by (6.16) (see Exercise 2.6.37).

It is clear that the Jacobi, Gauss–Seidel, and more generally the SOR methods can all be applied to the problem of iterating to a particular solution to $Ax = b$, $b \in R(A)$, with $a_{ii} \neq 0$, $i = 1, \ldots, n$, provided of course that the method in question converges. We next investigate the convergence of general and specific methods for singular linear systems. The approach taken here is similar to the approach taken in Section 5 for the nonsingular case; specifically, we use the tools of nonnegativity and M-matrices and of quadratic forms and positive definiteness to obtain convergence results for certain general splittings of A and then apply these results to the Jacobi and SOR methods. Not all the convergence theorems for the nonsingular case have been shown to have extensions to the singular case at the writing of this book. Ostensibly, this is because the concepts here are more difficult to work with.

The proofs of the convergence results in this section for the singular case are mostly more complicated versions of the proofs given in Section 5 for the nonsingular case. Accordingly, they will be mostly omitted.

We begin with a regular splitting theorem. But first, it will be convenient to state a result which combines many of the items in Theorems 5.4.24 and 6.4.12. Recall that $Z^{n\times n}$ consists of all $A \in R^{n\times n}$ with all nonpositive off-diagonal elements and that $A \in Z^{n\times n}$ is an M-matrix with "property c" if A is an M-matrix and index $A \leq 1$. With this in mind, we give

(6.18) Lemma Let $A \in R^{n\times n}$ and $S \subseteq R^n$. Then the six conditions that follow are equivalent for A and S. Moreover, if $A \in Z^{n\times n}$ and $S = R(A)$, then each of these conditions is equivalent to the statement: A is an M-matrix with "property c."

(i) S is a vector space complement of the null space of A, $N(A)$, and A is monotone on S. That is,

$$Ax \geq 0, \qquad x \in S \rightarrow x \geq 0.$$

(ii) A has a nonnegative $\{1\}$-inverse B where $R(BA) = S$. That is,

$$A = ABA, \qquad R(BA) = S, \qquad B \geq 0 \text{ is consistent.}$$

(iii) A has a $\{1\}$-inverse B where $R(BA) = S$ and where B is nonnegative on $R(A)$. That is,

$$x \in R(A),\ x \geq 0 \rightarrow Bx \geq 0.$$

(iv) Every $\{1\}$-inverse B of A, where $R(BA) = S$, is nonnegative on $R(A)$.

(v) A has a $\{1,2\}$-inverse C, where $R(C) = S$ and where C is nonnegative on $R(A)$.

(vi) Every $\{1,2\}$-inverse C of A, where $R(C) = S$, is nonnegative on $R(A)$.

(See Neumann and Plemmons [1979].) ■

The following result extends part of the regular splitting theorem, Theorem 5.6, to the singular case.

(6.19) Theorem Let $A = M - N$ be a regular splitting of $A \in R^{n \times n}$ and let $H = M^{-1}N$. Then the following statements are equivalent.

(1) A satisfies any one of the six equivalent conditions of Lemma 6.18 and $\delta(H) < 1$.

(2) H is semiconvergent; that is, the iteration (6.7) converges to a solution to (6.1) for every x^0.

(See Neumann and Plemmons [1978].) ■

As a consequence of these results we have another characterization of M-matrices with "property c."

(6.20) Theorem Let $A \in Z^{n \times n}$. Then A is an M-matrix with "property c" if and only if every regular splitting of A into $A = M - N$, $H = M^{-1}N$, satisfies $\rho(H) \leq 1$ and $\text{index}(I - H) \leq 1$.

Proof The proof follows immediately from Lemma 6.18, Theorem 6.19, and the definition of an M-matrix with "property c." ■

It turns out that if $A = M - N$ is only a weak regular splitting of A, then the conclusions of Theorem 6.20 do not necessarily hold. To illustrate this, consider

$$A = \begin{bmatrix} 0 & -1 \\ 0 & 1 \end{bmatrix} = \begin{bmatrix} 0 & 1 \\ 1 & -1 \end{bmatrix} - \begin{bmatrix} 0 & 2 \\ 1 & -2 \end{bmatrix} = M - N.$$

Here A is an M-matrix with "property c," but $\rho(M^{-1}N) = 2$.

We now conclude our M-matrix approach to convergence results with an extension of part of the Stein–Rosenberg result, Corollary 5.22, to the singular case.

(6.21) Theorem Let $A \in Z^{n \times n}$ have all positive diagonal entries. Let $A = D - L - U$ where, as usual, $D = \text{diag}\, A$ and $-L$ and $-U$ are the strictly lower and upper parts of A, respectively. Let $J = D^{-1}(L + U)$ and $H_1 = (D - L)^{-1}U$ denote the Jacobi and Gauss–Seidel iteration matrices, respectively. Then

(1) $\rho(J) \leq 1$ and index $(I - J) \leq 1$ if and only if $\rho(H_1) \leq 1$ and index $(I - H_1) \leq 1$.

(2) J and H_1 satisfy condition (1) if and only if A is an M-matrix with "property c," in which case

$$\delta(H_1) \leq \delta(J)$$

so that if $\delta(J) < 1$, then

$$R_\infty(H_1) \geq R_\infty(J). \quad ■$$

(See Neumann and Plemmons [1978].)

As an illustration of these ideas consider Example 6.2, concerning Poisson's equation with periodic boundary conditions. Note that the block partitioned coefficient matrix A, given by (6.5), for the finite difference formulation of the problem, belongs to $Z^{n \times n}$. Moreover the diagonal block matrix D of A given by (6.6) is a Stieltjes matrix and so $D^{-1} \gg 0$. Thus A is an M-matrix and moreover by Exercise 6.4.15, A has "property c" since A is symmetric and irreducible. Thus Theorem 6.21 applies to A in block form.

Another important tool in investigating the convergence of iterative methods for singular systems involves the use of quadratic forms and positive definiteness. At the writing of this book we are only able to consider the case where A is hermitian in such investigations. The following result extends Corollary 5.44 to the singular case.

(6.22) Theorem Let $A \in C^{n \times n}$ be hermitian. Let $A = M - N$ with M nonsingular and let $H = M^{-1}N$. Assume that $M^* + N$ is positive definite. Then H is semiconvergent if and only if A is positive semidefinite.

(See Keller [1965].)

Notice that the distinction between Corollary 5.44 and Theorem 6.22 is that in the former A is required to be positive definite. As a first corollary to Theorem 6.22 we consider that Jacobi iteration.

(6.23) Corollary Let $A \in C^{n \times n}$ have the block decomposition (5.45) and assume that A is hermitian and that D is positive definite. Then the block Jacobi iteration matrix J for A is semiconvergent if and only if $2D - A$ is positive definite and A is positive semidefinite.

For the SOR method we have the following extension of Corollary 5.48 to the singular case.

(6.24) Corollary Let $A \in C^{n \times n}$ have the block decomposition (5.45) and assume that A is hermitian and that D is positive definite. Then the block SOR iteration matrix H_ω for A is semiconvergent for all $0 < \omega < 2$ if and only if A is positive semidefinite.

The section is concluded with an application of these important concepts related to positive semidefiniteness to Example 6.2, Poisson's equation with periodic boundary conditions. Here we consider the block Jacobi and SOR methods applied to the solution of the finite difference formulation of the problem in the form $Au = g$, with A given by (6.5). We list some obvious properties of A as follows.

(1) A is symmetric.

(2) A is positive semidefinite by Exercise 6.4.15 since A is a symmetric M-matrix.

(3) A is irreducible so that A satisfies the five properties in Theorem 6.4.16.

(4) The block diagonal matrix

$$\tilde{D} = \text{block diag } A = \text{diag}(D, \ldots, D)$$

is positive definite since D is positive definite.

(5) The matrix $2\tilde{D} - A$ is positive definite.

Then from (1), (2), (4), and (5) the block Jacobi iteration matrix for A is semiconvergent, by Corollary 6.23. Finally, from (1), (2) and (4) it follows that the block SOR iteration matrix for A is semiconvergent for all $0 < \omega < 2$, by Corollary 6.24. Thus the block SOR method, applied to $Au = g$, will converge to an approximate solution to Poisson's equation with periodic boundary conditions for all $0 < \omega < 2$ and each initial approximation vector u^0.

7 EXERCISES

(7.1) For the linear system

$$\begin{bmatrix} 4 & -1 \\ -1 & 2 \end{bmatrix} \begin{bmatrix} x_1 \\ x_2 \end{bmatrix} = \begin{bmatrix} 3 \\ 1 \end{bmatrix}$$

use Eqs. (3.1) and (3.3) to carry out three iterations of the Jacobi and Gauss–Seidel methods, respectively, with the starting values $x_1^0 = 1, x_2^0 = 0$.

(7.2) Consider the two-point boundary-value problem (2.1) with $g(x) = 4\pi^2 \sin 2\pi x$ and with initial conditions $u(0) = u(1) = 0$. Show that for $n = 3$, the finite difference equations (2.4) become, in matrix form,

$$\begin{bmatrix} 2 & -1 & 0 \\ -1 & 2 & -1 \\ 0 & -1 & 2 \end{bmatrix} \begin{pmatrix} u_1 \\ u_2 \\ u_3 \end{pmatrix} = \frac{\pi^2}{4} \begin{pmatrix} 1 \\ 0 \\ -1 \end{pmatrix}.$$

Solve for the u_i in terms of π by a direct method and find the error in comparison to the true solution $u = \sin 2\pi x$, at $x = \frac{1}{4}$, $x = \frac{1}{2}$, and $x = \frac{3}{4}$.

(7.3) With $u^0 = (1,1,1)^t$, compute approximations u^1 and u^2 to the solution u to the linear system in Exercise 7.2 by (a) the Jacobi method and (b) the

Gauss-Seidel method. Compare the accuracy of your approximations with those obtained in Exercise 7.2.

(7.4) Show that the coefficient matrix A, given by (2.8), for the finite difference formulation of the heat equation is an irreducible Stieltjes matrix (see Definition 6.2.5).

(7.5) Consider the Jacobi matrix J given by (4.23) for the matrix A in (2.8). Show that the eigenvalues and associated eigenvectors of J can be expressed as

$$\lambda_k = \cos\frac{k\pi}{n+1}, \qquad x_k = \left(\sin\frac{k\pi}{n+1}, \sin\frac{2k\pi}{n+1}, \ldots, \sin\frac{nk\pi}{n+1}\right)^t \qquad k = 1, \ldots, n,$$

where n is the order of J. Conclude then that $\rho(J) = \cos[\pi/(n+1)] = -\cos[n\pi/(n+1)]$.

(7.6) By the techniques described in Example 4.22, compute the optimum SOR relaxation parameter, ω_b, for the example in Exercise 7.2. Use this ω_b rounded to three decimal places to compute u^1 and u^2 by the SOR method, with starting value $u^0 = (1,1,1)^t$. Compare your accuracy with that obtained by the Jacobi and Gauss–Seidel methods in Exercise 7.3. Compare $\rho(J)$, $\rho(H_1)$, and $\rho(H_{\omega_b})$ for this example.

(7.7) Show that every block tridiagonal matrix with nonsingular diagonal blocks is consistently ordered two-cyclic (see Varga [1962, Chapter 4]).

(7.8) Prove that a matrix A is point two-cyclic if and only if there exists a permutation matrix P such that

$$PAP^t = \begin{bmatrix} D_1 & C_1 \\ C_2 & D_2 \end{bmatrix},$$

where D_1 and D_2 are square diagonal matrices.

(7.9) Show that if a matrix $A \in C^{n\times n}$ is hermitian positive definite then all the eigenvalues of its Jacobi iteration matrix are real.

(7.10) Write a computer program to solve the two-point boundary-value problem given in Exercise 7.2 by the Jacobi method and by the SOR method using the optimum relaxation parameter, for $n = 3$, $n = 50$, and $n = 100$ (see Example 4.22). Interpret your results.

(7.11) Let A be a matrix with all nonzero diagonal elements and consider the symmetric SOR method (SSOR method) which involves two half-iterations using the SOR method. The first half-iteration is the ordinary SOR method and the second half-iteration is the SOR method using the reverse order of the equations. For $A = D - L - U$, as usual, express the SSOR method in equation form, such as is done for the SOR method in (4.2) and in the matrix form $x^{k+1} = Hx^k + c$, giving H and c in terms of D, L, and U.

(7.12) (a) Let

$$A = \begin{bmatrix} 1 & 2 & -2 \\ 1 & 1 & 1 \\ 2 & 2 & 1 \end{bmatrix}$$

and show that the Jacobi method converges but that the Gauss–Seidel method does not converge for A.

(b) Let

$$A = \begin{bmatrix} 2 & -1 & 1 \\ 2 & 2 & 2 \\ -1 & -1 & 2 \end{bmatrix}$$

and show that the Gauss–Seidel method converges but that the Jacobi method does not converge for A.

(c) Construct similar examples using 4×4 matrices.

(7.13) Show that the matrix

$$A = \begin{bmatrix} 1 & a & a \\ a & 1 & a \\ a & a & 1 \end{bmatrix}$$

is positive definite for $-\frac{1}{2} < a < 1$, but that the Jacobi method converges for A only for $-\frac{1}{2} < a < \frac{1}{2}$. Interpret this in view of Corollary 5.47.

(7.14) Compute $\rho(J)$ and $\rho(H_1)$ as a function of the scalar a for the matrix

$$A = \begin{bmatrix} 1 & -a \\ -4a & 1 \end{bmatrix}.$$

Verify Theorem 5.21 for the case $a \geq 0$. Also verify that $\rho(H_\omega) < 1$ for the case $a = \frac{1}{4}$, $\omega = \frac{1}{2}$, and show that

$$\rho(H_\omega) < 1 - \omega + \omega\rho(J).$$

In addition, verify Theorem 5.23 for the case $\omega_1 = \frac{1}{2}$, $\omega_2 = 1$.

(7.15) Let $A \in C^{n \times n}$ have all nonzero diagonal elements. Show that the SSOR method (see Exercise 7.11) converges for each $B \in \Omega(A)$ whenever

$$0 < \omega < \frac{2}{1 + \rho(|J|)},$$

if and only if A is an H-matrix (see Alefeld and Varga [1976]). This extends Theorem 5.14 to the SSOR method.

(7.16) Show that if $A \in C^{n \times n}$ is hermitian positive definite then the SSOR method (see Exercise 7.11) converges for all $0 < \omega < 2$ and moreover, all the eigenvalues of the SSOR iteration matrix are real and positive.

(7.17) State and prove extensions of Theorems 5.2, 5.6, and 5.8 to the case where the partial orderings induced by the nonnegative orthant R^n_+ are replaced by the partial orderings induced by a general proper cone in R^n (see Chapter 1).

(7.18) Let $A = M - N \in C^{n \times n}$ with A and M nonsingular. Show that $M^* + N$ may be replaced by $M + N$ in Corollaries 5.40 and 5.44.

(7.19) Let $A \in C^{n \times n}$ have a block decomposition $A = D - L - U$ where D is a block diagonal matrix with square nonsingular blocks on the diagonal. Assume that $x^*Dx \neq 0$ for all x in some eigenset E of the Jacobi iteration matrix J. Show that $\rho(J) < 1$ if and only if

$$|x^*Dx| > |x^*(L + U)x| \qquad \text{for all} \quad x \in E.$$

(7.20) Let A have the decomposition in Exercise 7.19. Show that $\rho(H_\omega) < 1$ if and only if

$$(x^*Dx)(x^*Ax) > 0 \qquad \text{for all} \quad x \in E.$$

(7.21) It is tempting to conjecture that the SOR method will converge for any $0 < \omega < 2$ if A is positive definite according to Definition 5.27. Show, however, that even the Gauss–Seidel method fails to converge for the simple matrix

$$A = \begin{bmatrix} 1 & 1 & 1 \\ -1 & 1 & 1 \\ -1 & -1 & 1 \end{bmatrix}$$

even though A satisfies Definition 5.27.

(7.22) Determine the finite difference equations and the resulting linear system $Au = b$, as in Section 2, that approximates

$$-\frac{d^2u}{dx^2} = g(x), \qquad \frac{du}{dx}(0) = \frac{du}{dx}(1) = 0,$$

where the boundary-value conditions are given by $u_0 = u_1$ and $u_{n+1} = u_n$. Show that the solution u to the continuous problem is not unique but that any two solutions differ by an additive constant. Verify this by showing that the finite difference approximation matrix A is singular of rank $n - 1$ and that $Ae = 0$, where $e = (1, \ldots, 1)^t$. Investigate when the techniques described in Section 6 can be applied to iterative methods to approximate the u_i in the finite difference formulation of the problem.

(7.23) Show that the matrix A given by (6.5) in Example 6.2 has a decomposition $A = LL^t$ where L is an M-matrix (see Corollary 6.4.17).

(7.24) Show that the matrix A given by (6.5) in Example 6.2 is block two-cyclic, but that A is not consistently ordered.

(7.25) Verify that the matrix $2\tilde{D} - A$ is positive definite, where A is given by (6.5) and $\tilde{D} = \operatorname{diag}(D, \ldots, D)$ with D given by (6.6).

(7.26) Let $A = M - N \in R^{n \times n}$ with M nonsingular and let $H = M^{-1}N$. Assume that $M^{-1} \geq 0$, that $\operatorname{index}(I - H) \leq 1$ and that $NE \geq 0$ where $E = (I - H)(I - H)^{\mathrm{D}}$. Prove that the following statements are equivalent:

(1) $B = (I - HE)^{-1}M^{-1}$ is a nonnegative $\{1\}$-inverse of A.
(2) $[A + N(I - E)]^{-1} \geq 0$.
(3) $BNE \geq 0$ where B is given in (1).
(4) H is semiconvergent (Meyer and Plemmons [1977]).

(7.27) Show that $A \in R^{n \times n}$ has a splitting that satisfies the hypotheses of Exercise 7.26 with H semiconvergent if and only if A has a nonnegative, nonsingular $\{1\}$-inverse, that is, A satisfies

$$A = AXA, \qquad X \geq 0, \quad X \text{ nonsingular.}$$

(7.28) Let $A \in C^{m \times n}$ and $b \in C^m$. Then the least squares solution to $Ax = b$ of minimum Euclidean norm is given by A^+b where A^+ denotes the Moore–Penrose generalized inverse of A (see Definition 5.4.8 and Ben-Israel and Greville [1974]). A splitting of the form

$$A = M - N, \qquad R(A) = R(M), \qquad N(A) = N(M)$$

is called a *proper splitting* of A. Show that for such a splitting

(1) $A = M(I - M^+N)$,
(2) $I - M^+N$ is nonsingular,
(3) $A^+ = (I - M^+N)^{-1}M^+$, and
(4) A^+b is the unique solution to the linear system $x = M^+Nx + M^+b$ (see Berman and Plemmons [1974a]).

(7.29) Let $A = M - N$ be a proper splitting of $A \in C^{m \times n}$. Show that the iteration

$$x^{k+1} = M^+Nx^k + M^+b$$

converges to A^+b for every x^0 if and only if $\rho(M^+N) < 1$ (see Berman and Plemmons [1974a]).

(7.30) Using the concepts developed in Exercises 7.28 and 7.29, state and prove theorems analogous to Theorem 5.2, 5.6, and 5.8 for establishing the convergence of iterative methods for approximating A^+b in the presence of a proper splitting of $A \in R^{m \times n}$ (see Berman and Plemmons [1974a]).

8 NOTES

(8.1) Probably the earliest mention of iterative methods for linear systems dates back to Gauss [1823] who, in a letter to his student, Gerling, suggested the use of iterative methods for solving certain systems of linear equations by iteration. For further informative historical remarks, see Bodewig [1959, p. 145], Forsythe and Wasow [1960], Ostrowski [1956], Varga [1962], and Young [1972].

(8.2) The two-point boundary problem, sometimes called the heat equation, discussed in Section 2, can be found in a variety of sources in the literature. Our treatment of the problem follows that in Strang [1976].

(8.3) For a historical discussion of the Jacobi and Gauss–Seidel methods studied in Section 3 see Forsythe [1953]. It has been noted (see Strang [1976, p. 284]) that the Gauss–Seidel method was apparently unknown to Gauss and not recommended by Seidel. For excellent bibliographical references to the older contributions of Gauss, Jacobi, and others on iterative methods see Bodewig [1959] and Ostrowski [1956]. The first paper to consider simultaneously the point Jacobi and Gauss–Seidel methods with sufficient conditions for their convergence was apparently the paper by Von Mises and Pollazek-Geiringer [1929]. However, much earlier Nekrasov

[1885] and Mehmke [1892] examined the point Gauss–Seidel iterative method and considered sufficient conditions for its convergence.

(8.4) The origin of Lemma 3.6 which gives necessary and sufficient conditions for the powers of a matrix to converge to the zero matrix goes back to Hensel [1926] and was independently rediscovered by Oldenburger [1940] (see Varga [1962, p. 25]). It is interesting to note that there are several proofs of this fundamental result which make no use of the Jordan form of a matrix. Householder [1958], for example, proves this result using only matrix norm concepts.

(8.5) The successive overrelaxation method, studied in Section 4, was essentially developed by Young [1950], who pointed out the advantage gained in certain cases by using a fixed relaxation parameter ω greater than unity. Varga [1962] states that the idea that "overrelaxing" could be beneficial occurred earlier. Explicit mention of the use of "overrelaxation" in certain iterative methods can be found, for example, in Southwell [1946].

(8.6) Our discussion of methods for choosing the optimum SOR parameter for consistently ordered p-cyclic methods is along the lines of that in Varga [1959, 1962, Chapter 4]. The terms consistent and inconsistent orderings are due to Young [1950], who used the term "property A" to denote the property that a matrix is point two-cyclic (see also Young [1972, p. 54]).

(8.7) Example 4.28 concerning an application of the SOR method to the large sparse linear least-squares problem is essentially in Melendez [1977] and is an extension of work begun earlier by Chen [1975]. For further work of this type see Plemmons [1979].

(8.8) The material in Section 5 concerning the application of the Perron–Frobenius theory of nonnegative matrices to the study of the convergence of iterative methods was motivated by the work on regular splittings by Varga [1960], who essentially established Theorem 5.2. Corollary 5.4 is due to Mangasarian [1970] while Theorem 5.6 concerning weak regular splittings is by Ortega and Rheinboldt [1967]. Corollary 5.7 is in Varga [1962, p. 90]. The monotone iteration theorem, Theorem 5.8, is attributed to work of Collatz and Schroder and can be found in Collatz [1966].

(8.9) The characterizations of nonsingular H-matrices in Theorem 5.14 constitute a collection of results and can be found in Varga [1976]. Alefeld and Varga [1976] have used the theory of H-matrices to establish a convergence result for the SSOR method (see Exercise 7.15). Theorem 5.18 on generalized diagonal dominance is in James and Riha [1974].

(8.10) The extension of the Stein–Rosenberg [1948] theorem given in Theorem 5.21 is due to Kahan [1958], who also proved Theorems 5.23 and 5.24.

(8.11) Our treatment of convergence criteria involving the nonnegativity of certain quadratic forms is along the lines of the material developed in Ortega and Plemmons [1979] where Lemma 5.30 and the general convergence theorems, 5.32 and 5.41, are given. A proof of Corollary 5.47 concerning convergence of the Jacobi method for hermitian positive definite matrices is in Faddeev and Faddeeva [1963, p. 191]. The important SOR convergence result for such matrices, Corollary 5.48, was proved by Reich [1949] for $\omega = 1$ and by Ostrowski [1954] for the more general case.

(8.12) Regarding Section 6, iterative methods for solving singular but consistent systems of linear equations have unfortunately been somewhat neglected in the literature. Perhaps this is due to some of the difficulties involved in establishing criteria for convergence. We hope that the material in Section 6 will motivate further work along these lines.

(8.13) Conditions under which the powers of a matrix converge to some matrix were given by Hensel [1926] (see Note 8.4). Our treatment of this subject in Lemmas 6.11 and 6.13 is along the lines in Meyer and Plemmons [1977]. Lemma 6.18 is in Neumann and Plemmons [1979] and the regular splitting theorem for the singular case, Theorem 6.19, is in Neumann and Plemmons [1978], as is the extension of the Stein–Rosenberg theorem to the singular case in Theorem 6.21. The material in Theorem 6.22 and Corollaries 6.23 and 6.24, concerning convergence results for hermitian positive semidefinite coefficient matrices, is due to Keller [1965].

(8.14) Further work on iterative methods for singular, consistent systems of linear equations can be found in Marchuk and Kuznetzov [1972] and Plemmons [1976a].

(8.15) Finally, we mention that iterative methods for solving the large sparse linear least-squares problem (see Example 4.28) by splittings of the coefficient matrix as described in Berman and Plemmons (see Exercises 7.28, 7.29, and 7.30), are also studied in Kammerer and Plemmons [1975], Berman and Neumann [1976a,b], Lawson [1975], and Marchuk and Kuznetzov [1972].

CHAPTER **8**

FINITE MARKOV CHAINS

1 INTRODUCTION

In the present chapter we consider n-state homogeneous Markov chains. The study of such chains provides some of the most beautiful and elegant applications of the theory of nonnegative matrices.

A Markov chain is a special type of stochastic process and a stochastic process is concerned with events that change in a random way with time. A typical stochastic process might predict the motion of an object that is constrained to be at any one time in exactly one of a number of possible states. It would then be a scheme for determining the probability of the object's being in a specific state at a specific time. The probability would generally depend upon a large number of factors, e.g., (1) the state, (2) the time, (3) some or all of the previous states the object has been in, and (4) the states other objects are in or have been in. For example, the object might be the reader, and the states might be the countries of the world. Clearly, aspects of all four factors just mentioned might influence the probability that the reader will be in a specific country at a specific time.

Informally, if the probability that an object will move to another state depends only on the two states involved (and not on earlier states, on time, or other factors) then the stochastic process is called a *homogeneous Markov process*, named after the Russian mathematician A. A. Markov (1856–1922), who introduced it in 1908. The theory of Markov processes comprises the largest and the most important part of the theory of stochastic processes; this importance is further enhanced by the many applications it has found in the physical, biological, and social sciences as well as in engineering and commerce. Some diverse and informative applications will be given in Section 2. But first, we give a slightly more formal mathematical treatment of these concepts. We shall not dwell here upon topics from probability theory, since such topics are beyond the scope and purpose of this chapter and book.

Markov chains serve as theoretical models for describing a system that can be in various fixed states. The system jumps at unit time intervals from one state to another according to the probabilistic law:

If the system is in the ith state s_i at time $k - 1$, $k \geq 1$, then *the next jump will take it to state s_j with probability $t_{ij}(k)$.*

The set of transition probabilities $t_{ij}(k)$ is prescribed for all i, j, and k and determines the probabilistic behavior of the system, once it is known how the system begins at time zero. If the set of states is finite, then the process is called *finite* and if the probabilities $t_{ij}(k)$ are independent of k, for each $k \geq 1$, then the process is called *homogenous.* In this case $T = (t_{ij})$ is a finite matrix, called the *transition matrix* for the process. More formally we have the following definition.

(1.1) Definition By a *finite homogeneous Markov chain* we will mean a system consisting of (1) a finite set of states $S = \{s_1, \ldots, s_n\}$, (2) an $n \times n$ matrix $T = (t_{ij})$, where t_{ij} is the probability that the system will move to state s_j, given only that the system is in state s_i, and (3) a vector $\pi^0 = (\pi_1^0, \ldots, \pi_n^0)$ where π_i^0 denotes the probability that the system is initially in state s_i, $i = 1, \ldots, n$.

We will use the term Markov chain to mean a finite homogeneous Markov chain and will use the notation $\mathcal{M} = (T, \pi^0)$ to denote such a chain.

Clearly, the transition matrix T for a Markov chain satisfies

$$t_{ij} \geq 0, \qquad 1 \leq i, j \leq n,$$

and

$$\sum_{j=1}^{n} t_{ij} = 1, \qquad 1 \leq i \leq n.$$

Thus, T is a (row) stochastic matrix. (Recall that idempotent stochastic matrices and some algebraic properties of multiplicative systems of stochastic matrices were studied in Chapters 2 and 3.)

The vector π^0 associated with a Markov chain is called the *initial (probability) distribution vector.* More generally we let $\pi^k = (\pi_i^k)$ denote the probability distribution vector defined by

$$\pi_i^k = \text{probability that the system is in state } s_i \text{ after } k \text{ steps.}$$

Then clearly $\pi_i^k \geq 0$ and $\sum_{i=1}^{n} \pi_i^k = 1$ for each k. For notationally convenient purposes π^k *will denote a row vector throughout this chapter.* The vector π^k will be called the *kth (probability) distribution vector.* As we shall see later, it is sometimes not convenient to require that the entries of π^0 sum to one. In this case π^0 is called simply the *initial distribution vector* associated with the chain and, accordingly, π^k is called the *kth distribution vector.*

The vectors π^k are then related by a simple rescursive formula in the following lemma.

(1.2) Lemma Let $\mathcal{M} = (T,\pi^0)$ denote a Markov chain with state transition matrix T and initial probability distribution vector π^0. Then for each positive integer k

$$\pi^k = \pi^{k-1}T = \pi^0 T^k, \tag{1.3}$$

where π^k denotes the row vector whose ith entry is the probability that the process is in state s_i after k steps.

Proof Let k be a positive integer. Then since the probability that the process moves to state s_i depends only upon the previous state, it follows that the probability that the process is in state s_m after $k-1$ steps and moves to state s_i is $\pi_m^{k-1}t_{mi}$. Thus, the probability that the process is in state s_i after k steps is the sum of these probabilities. That is,

$$\pi_i^k = \sum_{m=1}^{n} \pi_m^{k-1} t_{mi}.$$

Equation (1.3) then follows immediately. ■

A problem of major interest in the study of Markov chains is the tendency of π^k as $k \to \infty$. The study of this tendency requires the following notion.

(1.4) Definition An initial (probability) distribution vector π^0 for a Markov chain is said to be *stationary* if

$$\pi^k = \pi^0, \qquad k = 1,2,\ldots, \tag{1.5}$$

and is called a *stationary* (*probability*) *distribution vector.*

The Perron–Frobenius theory developed in Chapter 2 will be used in Section 3 to study stationary probability distribution vectors. It is imperative in Markov chain theory to classify states and chains of various kinds, and such classifications will also be developed in Section 3. One of the novel features of this book is the modern treatment of the analysis of Markov chains given in Section 4. Here, it will be shown how the use of the theory of singular M-matrices, developed and studied in Chapter 6, has both a theoretical and a computational advantage over the more traditional approaches. Certain parts of Chapters 5 and 7 also find application here. But first, we give some informative examples of Markov chains in Section 2.

2 EXAMPLES

The purpose of this section is threefold: (1) to give some interesting and important applications of the theory of Markov chains, (2) to acquaint the reader with certain notation and terminology conventions, and most importantly, (3) to provide material for future reference in later sections. We begin with an example of current significance.

(2.1) Example This example is an analysis of the population movement between cities and their surrounding suburbs in the United States. The numbers given are based upon statistics in *Statistical Abstracts of the U.S.A.*, 1969–1971.

The number of people, in thousands, of persons one year old or older who lived in cities in the U.S.A. during 1971 was 57,633. The number of people who lived in the surrounding suburbs was 71,549. This information will be represented by the initial population distribution vector

$$\pi^0 = [57{,}633 \quad 71{,}549].$$

Consider the population flow from cities to suburbs. During the period 1969–1971, the average probability per year of a person staying in the city was 0.96. Thus, the probability of moving to the suburbs was 0.04, assuming that all those moving went to the suburbs. Now consider the reverse population flow from suburbia to the city. The probability of a person moving to the city was 0.01, the probability of a person remaining in suburbia was 0.99.

Clearly then, this process can be modeled by a two-state Markov chain with state set $S = \{s_1, s_2\}$, where if s_1 represents the city and s_2 represents suburbia, then the state transition matrix is given by

$$T = \begin{array}{l} \text{City} \\ \text{Suburbia} \end{array} \overset{\begin{array}{cc} \text{City} & \text{Suburbia} \end{array}}{\begin{bmatrix} 0.96 & 0.04 \\ 0.01 & 0.99 \end{bmatrix}}.$$

Assuming that the population flow represented by the stochastic matrix T is unchanged over the years, the population distribution vector π^k after k years is given by Lemma 1.2 to be

$$\pi^k = \pi^0 T^k$$

For example, after five years, the relative populations of city centers and suburbia would be given by

$$\pi^5 = \pi^0 T^5 = [57{,}633 \quad 71{,}549] \begin{bmatrix} 0.96 & 0.04 \\ 0.01 & 0.94 \end{bmatrix}^5$$

$$= [50{,}440 \quad 78{,}742].$$

Notice here that we have not found it necessary nor convenient to convert π^0 to a probability vector since we are interested in population distributions.

A further modification of the model would allow for an overall population increase in the U.S.A. of 1% during the period 1969–1971. If we assume that the population will increase by the same amount annually during the years immediately following 1971, then the population distribution after k years would be given by

$$\pi^k = (101/100)^k \pi^0 T^k.$$

(2.2) Example As a hypothetical but perhaps more illustrative example, suppose the population of the lost continent Atlantis is 1,800. There are three cities in Atlantis, A, B, and C, and every year the entire population of each city moves, half each to each of the other two cities. If the populations of A, B, and C are currently 200, 600, and 1000, respectively, what will the population of each city be next year, the following year, and what are the long range population expectations for each city. Let a^k, b^k, and c^k be the populations of A, B, and C, respectively, after k years and

$$\pi^k = [a^k \quad b^k \quad c^k].$$

Clearly we have a three-state Markov chain with transition matrix and initial population vector given by

$$T = \begin{bmatrix} 0 & \frac{1}{2} & \frac{1}{2} \\ \frac{1}{2} & 0 & \frac{1}{2} \\ \frac{1}{2} & \frac{1}{2} & 0 \end{bmatrix}, \qquad \pi^0 = [200 \quad 600 \quad 1000],$$

respectively. Then

$$\pi^1 = \pi^0 T = [800 \quad 600 \quad 400],$$
$$\pi^2 = \pi^1 T = [500 \quad 600 \quad 700],$$

etc. Moreover, it follows by induction on k that

$$T^k = \begin{bmatrix} t_k & t_{k+1} & t_{k+1} \\ t_{k+1} & t_k & t_{k+1} \\ t_{k+1} & t_{k+1} & t_k \end{bmatrix}, \qquad t_i = \tfrac{1}{3}\left[1 + \frac{(-1)^i}{2^{i-1}}\right],$$

so that the powers of T converge to

$$\lim_{k\to\infty} T^k = \tfrac{1}{3}\begin{bmatrix} 1 & 1 & 1 \\ 1 & 1 & 1 \\ 1 & 1 & 1 \end{bmatrix}.$$

Consequently

$$\pi = \lim_{k \to \infty} \pi^k = [600 \quad 600 \quad 600]$$

is the unique stationary distribution vector associated with the Markov chain $\mathcal{M} = (T, \pi^0)$. Thus the population of city B remains at 600 while the populations of cities A and B approach 600 as a limit.

(2.3) Example Consider a particle which moves in a straight line in unit steps. Suppose it moves to the right with probability p and to the left with probability q. Suppose further that the particle can assume a finite number of positions s_i between two extreme points s_1 and s_n. Geometrically, we have

$$\left| \; s_1 \quad s_2 \quad \cdots \quad s_{n-1} \quad s_n \; \right|.$$

If the particle reaches an extreme point then it moves back one step with probability one.

This process can then be modeled by a Markov chain with transition matrix T of order n given by

$$T = \begin{bmatrix} 0 & 1 & & & & & \\ q & 0 & p & & & 0 & \\ & q & 0 & p & & & \\ & & \ddots & \ddots & \ddots & & \\ & & & & q & 0 & p \\ & & 0 & & & 1 & 0 \end{bmatrix}.$$

Clearly here, $0 < p < 1$ and $q = 1 - p$. This chain is called a *random walk* and it can be used to model various physical situations.

The matrix T is irreducible and it will be seen in Section 3 that the chain has a unique stationary probability distribution vector π^0 for each fixed p. Methods for computing π will be discussed in Section 4.

(2.4) Example Player I draws a card from a three-card deck consisting of two cards, one card marked W and one card marked L. If player I draws L, he pays player II \$1; while if he draws W, player II pays him \$1. At the outset, player I has \$1, and player II has \$3. The game is over when one of the players has won all the other player's money. We wish to find the probability of each player's emerging the winner. Also, what is the probability that a tenth draw will be required, i.e., that after each of the first nine draws both players have some money?

We will take player I to be the object and use the amount of money he has to describe various states. Perhaps the most natural way to do this would be to define s_1 to be the state "player I has $0," s_2 to be the state "player I has $1," s_3 to be the state "player I has $2," etc. This would then lead to the state transition matrix

$$\begin{bmatrix} 1 & 0 & 0 & 0 & 0 \\ \frac{1}{3} & 0 & \frac{2}{3} & 0 & 0 \\ 0 & \frac{1}{3} & 0 & \frac{2}{3} & 0 \\ 0 & 0 & \frac{1}{3} & 0 & \frac{2}{3} \\ 0 & 0 & 0 & 0 & 1 \end{bmatrix}.$$

However, it will be seen later that it is much more convenient to reorder the states, still in terms of player I's holdings. If the order is changed to

$$\$0, \$4, \$1, \$2, \$3$$

then the state transition matrix becomes

$$T = \begin{bmatrix} 1 & 0 & 0 & 0 & 0 \\ 0 & 1 & 0 & 0 & 0 \\ \frac{1}{3} & 0 & 0 & \frac{2}{3} & 0 \\ 0 & 0 & \frac{1}{3} & 0 & \frac{2}{3} \\ 0 & \frac{2}{3} & 0 & \frac{1}{3} & 0 \end{bmatrix}. \tag{2.5}$$

We now partition T into

$$T = \begin{bmatrix} I & 0 \\ B & C \end{bmatrix},$$

where

$$B = \begin{bmatrix} \frac{1}{3} & 0 \\ 0 & 0 \\ 0 & \frac{2}{3} \end{bmatrix}, \qquad C = \begin{bmatrix} 0 & \frac{2}{3} & 0 \\ \frac{1}{3} & 0 & \frac{2}{3} \\ 0 & \frac{1}{3} & 0 \end{bmatrix}.$$

It follows that

$$T^k = \begin{bmatrix} I & 0 \\ \sum_{i=0}^{k-1} C^i B & C^k \end{bmatrix}.$$

Moreover, $\rho(C) < 1$, so that

$$\lim_{k \to \infty} C^k = 0,$$

and by Lemma 6.2.1,

$$\lim_{k\to\infty} \sum_{i=0}^{k-1} C^i B = (I - C)^{-1} B = \frac{1}{15} \begin{bmatrix} 7 & 8 \\ 3 & 12 \\ 1 & 14 \end{bmatrix}.$$

Combining these results we have

$$\lim_{k\to\infty} T^k = \begin{bmatrix} 1 & 0 & 0 & 0 & 0 \\ 0 & 1 & 0 & 0 & 0 \\ \frac{7}{15} & \frac{8}{15} & 0 & 0 & 0 \\ \frac{3}{15} & \frac{12}{15} & 0 & 0 & 0 \\ \frac{1}{15} & \frac{14}{15} & 0 & 0 & 0 \end{bmatrix}.$$

Up to this point we have made no mention of the initial probability distribution vector. Since, by hypothesis, player I is initially in state s_3 (he has \$1), the initial probability distribution vector for the Markov chain is

$$\pi^0 = [0 \quad 0 \quad 1 \quad 0 \quad 0].$$

Then, by Lemma 1.2,

$$\pi = \lim_{k\to\infty} \pi^k = \lim_{k\to\infty} \pi^0 T^k = \pi^0 \lim_{k\to\infty} T^k$$
$$= [\tfrac{7}{15} \quad \tfrac{8}{15} \quad 0 \quad 0 \quad 0]$$

is the stationary distribution vector associated with the chain $\mathcal{M} = (T, \pi^0)$. Thus, the probability of player I eventually emerging the loser, i.e., being in state s_1 is $\frac{7}{15}$, while the probability of player I eventually emerging the winner, i.e., being in state s_2, is $\frac{8}{15}$. Moreover the analysis shows that the probability of the game's continuing endlessly is zero.

In the same way, it follows that there is no possibility of player I being in state s_3 or s_5 after nine draws ($k = 9$) and that the probability of his being in state s_4 is $(\frac{2}{3})^9$. Thus the probability that a tenth draw will be required is $(\frac{2}{3})^9$, which is approximately 0.026.

3 CLASSICAL THEORY OF CHAINS

As we have indicated earlier, it is imperative in Markov chain theory to classify states and chains of various kinds. In this respect we shall remain consistent with the notation and terminology developed in Chapter 2, concerning the Perron–Frobenius theory of nonnegative matrices and related topics.

Consider a Markov chain with state set $\{s_1, \ldots, s_n\}$ and transition matrix T. We first classify various types of states in conformity with the definitions given in Section 3 of Chapter 2.

(3.1) Definition We say that a state s_i has *access to* a state s_j, written $s_i \to s_j$, if it is possible in a finite number of steps for the object to move from state s_i to state s_j. If s_i has access to s_j and s_j has access to s_i, then s_i and s_j are said to *communicate* and we write $s_i \leftrightarrow s_j$.

The communication relation is an equivalence relation on the set of states and thus partitions the states into classes. With this in mind we give the following definition.

(3.2) Definition The *classes* of a Markov chain are the equivalence classes induced by the communication relation on the set of states. We say that a *class* α has *access to* a class β (written $\alpha \to \beta$) if $s_i \to s_j$ for some $s_i \in \alpha$ and $s_j \in \beta$. A class is called *final* if it has access to no other class. If a final class contains a single state then the state is called *absorbing.*

Clearly the states of a Markov chain with transition matrix T satisfy one of the definitions in (3.1) and (3.2) if and only if the matrix T satisfies the corresponding definition in (2.3.7) and (2.3.8). Thus, for example, $s_i \to s_j$ if and only if there is a path in the directed graph $G(T)$, associated with T (see Definition 2.2.4), from vertex i to vertex j, thus if and only if the (i,j) entry of some power of T is positive. In addition, a state s_i is absorbing if and only if $t_{ii} = 1$.

Next, suppose T is reduced to a triangular block form

$$PTP^t = \begin{bmatrix} T_{11} & 0 & \cdots & 0 \\ T_{21} & T_{22} & \cdots & 0 \\ \vdots & \vdots & & \vdots \\ T_{s1} & T_{s2} & \cdots & T_{ss} \end{bmatrix}, \tag{3.3}$$

where the diagonal blocks T_{ii} are square and either irreducible or 1×1 and null. Then these diagonal blocks T_{ii} correspond to the classes associated with the corresponding Markov chain. Moreover, the class associated with T_{ii} is final if and only if $T_{ij} = 0$ for $j = 1, \ldots, i-1$. From (3.3) it follows that every Markov chain has at least one final class and consists of exactly one (final) class if and only if T is irreducible.

We proceed now to further classify the states of a Markov chain.

(3.4) Definition A state s_i is called *transient* if $s_i \to s_j$ for some s_j but $s_j \nrightarrow s_i$, that is, s_i has access to some s_j which does not have access to s_i. Otherwise, the state s_i is called *ergodic*. Thus s_i is ergodic if and only if $s_i \to s_j$ implies $s_j \to s_i$.

It follows that if one state in a class of states associated with a Markov chain is transient (ergodic) then each state in that class is transient (ergodic). This leads to the next definition.

(3.5) Definition A class α induced by the access relation on the set of states of a Markov chain is called *transient* if it contains a transient state and is called *ergodic* otherwise. Thus a class is ergodic if and only if it is final.

As an illustration of these ideas we note that each state in Examples 2.1, 2.2, and 2.3 is ergodic since the transition matrix in each case is irreducible. Thus each of these chains consists of a single ergodic class. The situation is, however, different in Example 2.4. With the transition matrix T given by (2.5), it follows that each of the states s_1 and s_2 is absorbing; that is, if player I has either \$0 or \$4 then the game is over. In particular then, $\{s_1\}$ and $\{s_2\}$ are ergodic classes. However, states s_3, s_4, and s_5, corresponding to player I having \$2, \$3, and \$4, respectively, are transient, and the set $\{s_3, s_4, s_5\}$ forms a single transient class. As a further illustration we give this example.

(3.6) Example Consider the Markov chain with transition matrix

$$T = \begin{bmatrix} \frac{1}{2} & 0 & 0 & \frac{1}{2} & 0 & 0 \\ 0 & \frac{2}{3} & \frac{1}{3} & 0 & 0 & 0 \\ \frac{1}{3} & \frac{1}{12} & \frac{1}{3} & \frac{1}{4} & 0 & 0 \\ \frac{1}{2} & 0 & 0 & \frac{1}{2} & 0 & 0 \\ 0 & 0 & 0 & 0 & 1 & 0 \\ 0 & \frac{1}{2} & 0 & 0 & 0 & \frac{1}{2} \end{bmatrix}.$$

By permuting the second and fourth rows and the second and fourth columns of T we obtain the reduced form

$$PTP^{t} = \begin{bmatrix} \frac{1}{2} & \frac{1}{2} & 0 & 0 & 0 & 0 \\ \frac{1}{2} & \frac{1}{2} & 0 & 0 & 0 & 0 \\ \frac{1}{3} & \frac{1}{4} & \frac{1}{3} & \frac{1}{12} & 0 & 0 \\ 0 & 0 & \frac{1}{3} & \frac{2}{3} & 0 & 0 \\ 0 & 0 & 0 & 0 & 1 & 0 \\ 0 & 0 & 0 & \frac{1}{2} & 0 & \frac{1}{2} \end{bmatrix}.$$

Then states s_1, s_2, and s_5 are ergodic, and s_5 is also absorbing. Also, s_3, s_4, and s_6 are transient. In particular then, $\{s_1, s_2\}$ and $\{s_5\}$ are ergodic classes while $\{s_3, s_4\}$ and $\{s_6\}$ are transient classes.

We next use these concepts to classify certain types of chains. Once again, we do this in a manner consistent with the notation and terminology developed in Chapter 2.

(3.7) Definitions A Markov chain is called *ergodic* if it consists of a single ergodic class. It is called *regular* if at some fixed step k each state of the chain can move to each state of the chain with a positive probability and *periodic* if it is ergodic but not regular.

Clearly every regular Markov chain is ergodic. However, not every ergodic chain is regular. For example, a chain with transition matrix

$$T = \begin{bmatrix} 0 & 1 \\ 1 & 0 \end{bmatrix}$$

is ergodic but not regular and is thus periodic.

The reason for the term "periodic chain" will become apparent after Theorem 3.16. Our final classification of chains is the simplest.

(3.8) Definition A Markov chain in which each ergodic class consists of a single absorbing state is called an *absorbing chain.*

We now proceed to study the classification of chains in terms of fundamental properties of the state transition matrices T. Here, the material developed in Chapter 2, Section 2, for nonnegative matrices is vital.

(3.9) Theorem Let T be a state transition matrix for a Markov chain. Then the chain is

(1) ergodic if and only if T is irreducible.
(2) regular if and only if T is primitive.
(3) periodic if and only if T is irreducible and cyclic.

Proof By Theorem 2.2.7, the matrix T is irreducible if and only if the directed graph, $G(A)$ (see Definition 2.2.4), associated with A is strongly connected. But this is true if and only if all the states of the Markov chain have access to each other, thus if and only if the chain is ergodic. This establishes (1).

Part (2) follows from Theorem 2.1.7(c), and Definition 2.1.8. For a chain is regular if and only if some power of the transition matrix is positive.

Part (3) is immediate from Definition 3.7 and the definition of a cyclic matrix given in Section 2 of Chapter 2. ■

By applying Theorem 3.9, it is easily seen that the Markov chains given in Examples 2.1 and 2.2 are regular, while the Markov chain given in Example 2.3 is periodic.

We now investigate the existence and uniqueness of the stationary probability distribution vector in terms of the state transition matrix. This is accomplished in part by characterizing transition matrices associated with various types of chains. First some elementary facts are accumulated.

From Theorem 2.5.3 it follows that if T is stochastic then $\rho(T) = 1$ and by letting e denote the column vector of all ones

$$Te = e. \tag{3.10}$$

(3.11) Theorem Every Markov chain has a stationary probability distribution vector.

Proof If T is the state transition matrix associated with the chain, then since $\rho(T) = 1$, there exists a row vector $x > 0$ with $xT = x$, by Theorem 2.1.1. Normalizing x we obtain

$$\pi = \left(\sum_{i=1}^{n} x_i \right)^{-1} x,$$

and it follows that π is a stationary probability distribution vector for the chain. ■

Recall that from Definition 2.3.8, a class α associated with T is *basic* if and only if $\rho(T[\alpha]) = \rho(T)$, were $T[\alpha]$ is the submatrix of T associated with the indices in α. This leads to the following lemma.

(3.12) Lemma If T is a stochastic matrix then every final class associated with T is basic, and thus $v(t) = 1$.

Proof From the definition, the final classes of T are the classes corresponding to the ergodic classes of the associated Markov chain. Thus in the reduced triangular block form (3.3), if T_{ii} corresponds to a final class, then $T_{ij} = 0, j = 1, \ldots, i-1$. Then T_{ii} is stochastic and thus $\rho(T_{ii}) = 1 = \rho(T)$. ■

Next, we study the primary types of Markov chains in terms of characterizations of T.

(3.13) Theorem A stochastic matrix T is the transition matrix associated with an ergodic Markov chain if and only if

(a) one is a simple eigenvalue of T and
(b) there exists a row vector $x \gg 0$, unique up to positive scalar multiples, with

$$xT = x.$$

Proof By Corollary 2.3.15 it follows that T is irreducible if and only if one is a simple eigenvalue of T, and T has a positive left eigenvector x associated with one. But by Theorem 2.1.4, any left eigenvector of T associated with one is then a multiple of x. The proof is completed by applying Theorem 3.9(1). ■

As an immediate consequence we have the following corollary.

(3.14) Corollary A Markov chain is ergodic if and only if it has a unique positive stationary probability distribution vector.

Before giving the next result we remind the reader that

$$\delta(T) = \max\{|\lambda|; \lambda \in \sigma(T), \lambda \neq \rho(T)\}.$$

(3.15) Theorem A stochastic matrix T is the transition matrix associated with a regular Markov chain if and only if (a) and (b) of Theorem 3.13 hold and

(c) $\delta(T) < 1$.

Proof Since a regular Markov chain is ergodic, the proof follows by Theorem 3.13 and since T is primitive if and only if T is irreducible and $\delta(T) < 1$. ■

(3.16) Theorem A stochastic matrix T is the transition matrix for a periodic Markov chain if and only if (a) and (b) of Theorem 3.13 hold and

(d) there exists a permutation matrix P such that

$$PTP^{t} = \begin{bmatrix} 0 & T_1 & & \cdots & 0 \\ 0 & 0 & T_2 & \cdots & 0 \\ \vdots & \vdots & & \ddots & \vdots \\ & & & & T_{h-1} \\ T_h & 0 & & \cdots & 0 \end{bmatrix}, \tag{3.17}$$

where $h > 1$ and the zero diagonal blocks are square.

Proof The proof follows from Theorem 2.2.20, since the Markov chain is periodic if and only if T is irreducible but not primitive, so that T has $h > 1$ eigenvalues on the unit circle. ■

(3.18) Definition If a Markov chain with state transition matrix T is periodic then the *period* of the chain is the positive integer h given by (3.17), that is, the index of cyclicity of T (see Definition 2.2.26).

We remark that if the chain is periodic then an object in one of the states associated with T_i will be in the set of states associated with T_{i+1} after one step, for $i = 1, \ldots, h-1$. It will then move from a state in T_h to a state in T_1 in one step and the process will be repeated indefinitely.

We next develop a "standard form" for the transition matrix associated with an arbitrary Markov chain.

Let $S_1, \ldots, S_r$ denote the ergodic classes of states of an n-state Markov chain and let S_i consist of n_i states. Then the chain has $m = n_1 + \cdots + n_r$ ergodic states and $t = n - m$ transient states. Then by (3.3) and Lemma 3.12, the transition matrix associated with the chain can be permuted into the form

$$T = \begin{bmatrix} D_1 & 0 & \cdots & 0 & 0 \\ 0 & D_2 & \cdots & 0 & 0 \\ \vdots & & \ddots & \vdots & \vdots \\ & & & D_r & \\ B_1 & B_2 & \cdots & B_r & C \end{bmatrix}, \tag{3.19}$$

where D_i is an $n_i \times n_i$ irreducible state transition matrix associated with the class S_i, $i = 1, \ldots, r$, and where C is a square matrix of order t and all the states corresponding to C are transient. We shall call (3.19) the *standard form* for a state transition matrix.

Note that the matrix T given by (2.5) in Example 2.4 has been permuted into standard form. The importance of this will be apparent after the following results.

(3.20) Lemma Let T be a stochastic matrix in the standard form (3.19). Then

$$\rho(C) < 1.$$

Proof Letting $D = \text{diag}(D_1, \ldots, D_r)$, the kth power of T can be expressed in the form

$$T^k = \begin{bmatrix} D^k & 0 \\ B^{(k)} & C^k \end{bmatrix}$$

for some matrix $B^{(k)}$, depending upon $B^{(1)} = (B_1, \ldots, B_r)$ and the matrices C and D. Then since the process must be able to move from any transient state to an ergodic state in a finite number, say k_j, steps, each row of $B^{(k_j)}$ must have a positive entry. Thus the maximum row sum for C^{k_j} is less than one. Thus by Theorem 2.2.35 it follows that $\rho(C) < 1$. ∎

The preceding theorem will be used to obtain a canonical form for the transition matrix associated with an absorbing chain. This is given in the following theorem.

(3.21) Theorem A stochastic matrix is the transition matrix associated with an absorbing Markov chain if and only if it can be permuted into the form

$$T = \begin{bmatrix} I & 0 \\ B & C \end{bmatrix}, \tag{3.22}$$

where $\rho(C) < 1$.

Proof Since each ergodic class of an absorbing chain consists of a single absorbing state, the standard form (3.19) reduces to (3.22). The theorem then follows by applying Lemma 3.20. ■

We note that by Theorem 3.22, the Markov chain given in Example 2.4 is absorbing. Here the matrix T given by (2.5) is in the canonical form (3.22).

We conclude this section by investigating stationary probability distribution vectors associated with an arbitrary Markov chain.

(3.23) Theorem To each ergodic class of states S_i of a Markov chain there corresponds a unique stationary probability distribution vector $\pi(i)$ having the property that the positions of $\pi(i)$ corresponding to the states of S_i contain positive entries and all other entries are zero. Moreover, every stationary probability distribution vector π for the chain is a linear convex combination of the $\pi(i)$. That is,

$$\pi = \sum_{i=1}^{r} \lambda_i \pi(i), \qquad \lambda_i \geq 0, \qquad \sum_{i=1}^{r} \lambda_i = 1,$$

where r is the number of ergodic classes.

Proof Let T denote the state transition matrix associated with the Markov chain. Then without loss of generality we can assume that T has the standard form (3.19). Let $\theta(i)$ denote the unique positive stationary probability distribution vector for the Markov chain associated with state transition matrix D_i for $i = 1, \ldots, r$. Such vectors $\theta(i)$ exist by Corollary 3.14. Then the r,n-dimensional vectors

$$\begin{aligned}
\pi(1) &= (\theta(1),0, \ldots ,0,0) \\
\pi(2) &= (0,\theta(2), \ldots ,0,0) \\
&\vdots \\
\pi(r) &= (0,0, \ldots ,\theta(r),0)
\end{aligned}$$

are linearly independent stationary probability distribution vectors, and clearly every linear convex combination of the $\pi(i)$ is a stationary probability distribution vector.

Now let

$$\pi = (z_1, \ldots ,z_r,z_{r+1})$$

be any stationary probability distribution vector for the chain, partitioned conformally with T given by (3.19). Note first that

$$z_{r+1}C = z_{r+1}.$$

But by Lemma 3.20, $\rho(C) < 1$ so that one is not an eigenvalue of C, and thus

$$z_{r+1} = 0.$$

Moreover for each i,

$$z_i D_i = z_i$$

and consequently z_i is a positive multiple of $\theta(i)$, by Theorem 3.13(b). Hence

$$z_i = \alpha_i \theta(i), \qquad \alpha_i > 0, \qquad i = 1, \ldots, r.$$

Then since the sum of the entries in each of $\pi, \theta(1), \ldots, \theta(r)$ is one, it follows that

$$\sum_{i=1}^{r} \alpha_i = 1.$$

Thus π is a linear convex combination of the $\theta(i)$ and the theorem is proved. ■

The next section will be concerned will certain modern methods for analyzing Markov chains.

4 MODERN ANALYSIS OF CHAINS

One of the objectives of this book is to show how recent developments in the theory of generalized matrix inverses and singular M-matrices studied in Chapters 5 and 6, respectively, can be blended with the classical Perron–Frobenius theory of nonnegative matrices, given in Chapter 2, to produce modern, efficient methods for analyzing Markov chains. First, we review some of the concepts concerning singular M-matrices that will be needed here.

Recall that $Z^{n \times n}$ denotes the set of $n \times n$ real matrices with all nonpositive off-diagonal entries. Then $A \in Z^{n \times n}$ is an *M-matrix with "property c"* (see Definition 6.4.10) if A has a representation

$$A = sI - B, \qquad B \geq 0, \qquad s \geq \rho(B), \tag{4.1}$$

where the powers of B/s converge to some matrix. Equivalently, A is an M-matrix with "property c" if and only if (4.1) holds and rank A = rank A^2. This latter condition is then equivalent to the existence of the group generalized inverse, $A^{\#}$, of A defined by

$$A^{\#}x = y \qquad \text{if} \quad x \in R(A) \quad \text{and} \quad Ay = x$$

and

$$A^{\#}x = 0 \qquad \text{if} \quad x \in N(A).$$

The definition of $A^{\#}$ given here is consistent with Definition 5.4.11, since in this case R^n is the direct sum, $R^n = R(A) \oplus N(A)$. Recall also that in this case the group generalized inverse of A is the same as the Drazin generalized inverse (see Definition 5.4.10).

It will be seen in this chapter that for a finite homogeneous Markov chain, virtually everything that one would want to know about the chain can be determined by investigating a certain group inverse and a limiting matrix. It will also be demonstrated that their introduction into the theory provides practical advantages over the more classical analysis techniques and serves to unify the theory to a certain extent. We begin with the following theorem.

(4.2) Theorem If T is the transition matrix for a Markov chain then

$$A = I - T$$

is an M-matrix with "property c."

Proof Clearly A is an M-matrix, since $T \geq 0$ and $\rho(T) = 1$. To show that A has "property c," we need only show that $\operatorname{rank} A = \operatorname{rank} A^2$, and then apply Lemma 6.4.11. The proof of this is divided into two parts.

First, suppose that T is irreducible. Then by Theorem 3.9(1), the Markov chain associated with T is ergodic. Then by Theorem 3.13(a), one is a simple eigenvalue of T. Thus zero is a simple eigenvalue of $A = I - T$ and consequently the Jordan form for A can be expressed in the form

$$J_A = \begin{bmatrix} 0 & 0 \\ 0 & K \end{bmatrix},$$

where K is nonsingular. Thus $\operatorname{rank} A^2 = \operatorname{rank} A$ and hence A has "property c" by Lemma 6.4.11.

If T is reducible, then without loss of generality we may assume that T has the standard form (3.19). Then by part 1 of this proof, $I - D_i$ has "property c" for $i = 1, \ldots, r$. Moreover, $I - C$ is in fact a nonsingular M-matrix since $\rho(C) < 1$, by Lemma 3.20. It follows then, that $A = I - T$ has "property c." ■

As an immediate consequence we have the next corollary.

(4.3) Corollary If T is the transition matrix for a Markov chain and $A = I - T$, then $A^{\#}$, the group generalized inverse of A, exists.

Throughout this section we shall use the notation

$$L = I - AA^{\#}$$

whenever $A = I - T$ and T is the transition matrix for a Markov chain.

We are quite interested in analyzing the limiting behavior of a Markov chain. To this end we shall need the following results. Recall that from Definition 6.4.8, a matrix T is called *semiconvergent* whenever the powers of T converge to some matrix.

(4.4) Theorem Let T be a transition matrix, let $A = I - T$ and let $L = I - AA^{\#}$. Then T is semiconvergent if and only if $\delta(T) < 1$; that is, one is the only eigenvalue of T on the unit circle, in which case

$$\lim_{k \to \infty} T^k = L. \tag{4.5}$$

Proof Since $A = I - T$ is an M-matrix with "property c" by Theorem 4.2, it follows immediately that T is semiconvergent if and only if one is the only eigenvalue of T on the unit circle, by Exercise (6.4.9(3)). Equation (4.5) then follows from Lemma 7.6.11. ■

In Markov chain terminology we have the following corollary.

(4.6) Corollary The transition matrix T for an ergodic Markov chain is semiconvergent if and only if the chain is regular. The transition matrix for an absorbing chain is always semiconvergent.

Proof The proof of the first part follows immediately from Theorem 3.9(1) and Theorem 3.15. The second part is immediate from Theorem 3.21. ■

By considering the standard form (3.19), together with Theorem 4.4 and Corollary 4.6, we have the following theorem.

(4.7) Theorem Let T be a transition matrix for a Markov chain. Then T is semiconvergent if and only if there exists a permutation matrix P such that

$$PTP^{\mathrm{t}} = \begin{bmatrix} T_{11} & 0 & \cdots & 0 \\ 0 & T_{22} & \cdots & 0 \\ \vdots & \vdots & \ddots & \vdots \\ 0 & 0 & \cdots & T_{kk} \end{bmatrix}, \tag{4.8}$$

where each T_{ii} has the form (3.19) and each D_i in (3.19) is regular. In this case,

$$\lim_{k \to \infty} T^k = P^{\mathrm{t}} \begin{bmatrix} L_{11} & 0 & \cdots & 0 \\ 0 & L_{22} & \cdots & 0 \\ \vdots & \vdots & \ddots & \\ 0 & 0 & \cdots & L_{kk} \end{bmatrix} P,$$

where $L_{ii} = I - (I - T_{ii})(I - T_{ii})^{\#}$, $i = 1, \ldots, k$.

We shall see next that every transition matrix can be transformed into one that is semiconvergent by a simple eigenvalue shift. This technique will be used later for computing a stationary probability distribution vector by iterative methods.

(4.9) Theorem Let T be the transition matrix for a general Markov chain, let $A = I - T$ and let $L = I - AA^{\#}$. Then the matrix

$$T_\alpha = (1 - \alpha)I + \alpha T, \qquad 0 < \alpha < 1, \tag{4.10}$$

is semiconvergent and moreover,

$$\lim_{k\to\infty} T_\alpha^k = L. \tag{4.11}$$

Proof First note that for $0 < \alpha < 1$, T_α is stochastic; for clearly $T_\alpha \geq 0$ and for $e = (1, \ldots, 1)^t$,

$$T_\alpha e = (1 - \alpha)e + \alpha e = e.$$

Now by Exercise (6.4.3(b)), one is the only eigenvalue of T_α on the unit circle. Thus T_α is semiconvergent by Theorem 4.4 and moreover

$$\lim_{k\to\infty} T_\alpha^k = I - (I - T_\alpha)(I - T_\alpha)^{\#}.$$

But since $(I - T_\alpha) = \alpha(I - T)$,

$$(I - T_\alpha)(I - T_\alpha)^{\#} = \alpha(I - T)(1/\alpha)(I - T)^{\#} = AA^{\#},$$

and (4.11) follows. ■

A useful representation of the unique stationary distribution vector associated with an ergodic chain is given next.

(4.12) Theorem Let T be the transition matrix for an ergodic Markov chain, let $A = I - T$ and $L = I - AA^{\#}$. Then

$$L = \pi e = \begin{bmatrix} \pi \\ \vdots \\ \pi \end{bmatrix}, \tag{4.13}$$

where π is the stationary probability distribution (row) vector associated with the chain.

Proof We note first that from (4.11) of Theorem 4.9, L is a (row) stochastic matrix. Moreover,

$$LA = (I - AA^{\#})A = A - A = 0,$$

since $AA^{\#}A = A$. Thus

$$LT = L,$$

and (4.13) then follows from Corollary 3.14. ■

Notice that in Theorem 4.12, L is a nonnegative, rank 1, idempotent, stochastic matrix. For an arbitrary chain, L is a nonnegative, rank r idempotent, stochastic matrix where r is the number of ergodic classes associated with the chain. Such matrices were studied in detail in Section 3 of Chapter 3.

In the sequence of results to follow, T will denote the transition matrix for a Markov chain with state set $\{s_1, \ldots, s_n\}$, A will denote $I - T$, and L will denote the limit matrix $I - AA^{\#}$.

We will show next that virtually everything one might wish to know about the chain can be determined by analyzing the matrices $A^{\#}$ and L. Not only does L serve as the expression for the limiting matrix in all cases, it can also be used to classify the states of a general chain.

(4.14) Theorem For a general Markov chain, state s_i is a transient state if and only if the ith column of L is entirely zero; that is,

$$Le_i = 0,$$

where e_i is the ith unit vector. Equivalently, s_i is an ergodic state if and only if

$$Le_i \neq 0.$$

Proof Without loss of generality, we may assume that the states associated with the chain have been permuted so that all the ergodic states are listed before the transient states. In other words, without loss of generality, we may assume that the transition matrix T for the chain has been permuted into the form (3.19). In more compact form, we write T as

$$T = \begin{bmatrix} D & 0 \\ B & C \end{bmatrix}, \tag{4.15}$$

where $D = \text{diag}(D_1, \ldots, D_r)$ and the D_i correspond to the ergodic sets and where all the transient states correspond to the matrix C. Then $A = I - T$ has the form

$$A = \begin{bmatrix} A_{11} & 0 \\ A_{21} & A_{22} \end{bmatrix}, \qquad A_{11} = I - D, \quad A_{21} = -B, \quad A_{22} = I - C. \tag{4.16}$$

Here index $A_{11} = 1$, and by Lemma 3.20, $\rho(C) < 1$, so that A_{22} is nonsingular. Consequently, $A^{\#}$ is seen to have the form

$$A^{\#} = \begin{pmatrix} A_{11}^{\#} & 0 \\ * & A_{22}^{\#} \end{pmatrix}, \qquad A_{22}^{\#} = (I - C)^{-1}, \tag{4.17}$$

and so the matrix $L = I - AA^{\#}$ has the form

$$L = \begin{bmatrix} L_{11} & 0 \\ * & 0 \end{bmatrix}, \qquad L_{11} = I - A_{11}A_{11}^{\#}. \tag{4.18}$$

Finally, each column of L_{11} contains at least one nonzero entry by Theorem 4.12, since $A_{11} = I - D$ and $D = \operatorname{diag}(D_1, \ldots, D_r)$ and D_i is the transition matrix for an ergodic chain, for $i = 1, \ldots, r$. ∎

Even though L can provide a distinction between the ergodic states and the transient states, it does not completely determine the ergodic sets. It does, however, provide useful information concerning the ergodic sets.

(4.19) Theorem For a general Markov chain, if states s_i and s_j belong to the same ergodic set, then the ith and jth rows of L are equal; that is,

$$e_i^t L = e_j^t L.$$

Proof As in the proof of Theorem 4.14, we permute the states so that the transition matrix T for the chain has the form (4.15) and thus L has the form (4.18). Then since the ergodic classes correspond to the diagonal blocks D_i of D, if s_i and s_j are in the same ergodic set, then the ith and jth rows of L_{11} are equal, by Theorem 4.12. ∎

For a general chain the entries of L can be used to obtain the probabilities of eventual absorption into any one particular ergodic class, for each starting state.

(4.20) Theorem For a general Markov chain let s_k be any fixed state and let $[s_k]$ denote the ergodic class of the chain containing s_k. Let $\mathcal{T}_k$ denote the set of indices of the states belonging to $[s_k]$. Then for $L = (l_{ij})$ and $1 \leq i \leq n$,

$$\sum_{j \in \mathcal{T}_k} l_{ij} = \text{probability of eventual absorption of the process from } s_i \text{ into } [s_k]. \tag{4.21}$$

Proof We begin by permuting the states of the chain so that the transition matrix T has the form (4.15). Now replace D by I in (4.15) and let

$$\tilde{T} = \begin{bmatrix} I & 0 \\ B & C \end{bmatrix}$$

Then the modified chain with $\tilde{T}$ as transition matrix is clearly an absorbing chain. Also, the probability of the eventual absorption of the process from a state $\tilde{s}_i$ into a state $\tilde{s}_k$, corresponding to $(\tilde{T})_{kk} = 1$, is given by

$$\left(\lim_{m\to\infty} \tilde{T}^m\right)_{ik} = (\tilde{L})_{ik},$$

where ik denotes the entry in the ith row and kth column and where

$$\tilde{L} = I - (I - \tilde{T})(I - \tilde{T})^{\#}.$$

From this it follows that in the original chain the probability of eventual absorption into the set $[s_k]$ is simply the sum over $\mathscr{T}_k$ of the absorption probabilities, that is, for $\tilde{L} = (\tilde{l}_{ij})$,

(4.22) $\displaystyle\sum_{j\in\mathscr{T}_k} \tilde{l}_{ij}$ = probability of eventual absorption of the process from s_i into $[s_k]$.

Then by relating L to $\tilde{L}$, (4.21) follows from (4.22). ■

The entries of $A^{\#}$ also provide important information about the chain. We assume here that the chain has at least one transient state.

(4.23) Theorem For a general Markov chain with transient states s_i and s_j, with $A^{\#} = (a^{\#}_{ij})$,

(4.24) $a^{\#}_{ij}$ = expected number of times the process is in s_j when initially in s_i.

Proof As before, assume that the states of the chain have been permuted so that the transition matrix T has the form (4.15). Then s_i and s_j are associated with the matrix C. Now $\rho(C) < 1$ by Lemma 3.20, so that A_{22}^{-1} in (4.17) has the form

$$(I - C)^{-1} = \sum_{m=0}^{\infty} C^m$$

By using the fact that

$$\left(\sum_{m=0}^{k-1} C^m\right)_{ij} = \left(\sum_{m=0}^{k-1} T^m\right)_{ij}$$

is the expected number of times the process is in state s_j in k steps, when initially in state s_i, it follows that the expected number of times the process is in s_j when initially in s_i is given by

$$\lim_{k\to\infty}\left(\sum_{m=0}^{k-1} C^m\right)_{ij} = (I - C)^{-1}_{ij} = a^{\#}_{ij}. \quad ■$$

Frequently, the analyst wishes to determine the expected number of steps until absorption into an ergodic class when initially in a transient state. The matrix $A^{\#}$ can also provide this information. We assume again that the chain has at least one transient state.

(4.25) Theorem For a general Markov chain let $\mathcal{T}$ denote the set of indices corresponding to the transient states. If the chain is initially in state s_i, then for $A^{\#} = (a_{ij}^{\#})$,

(4.26)
$$\sum_{j \in \mathcal{T}} a_{ij}^{\#} = \text{expected number of times the chain will be in a transient state when initially in state } s_i.$$
$$= \text{expected number of steps for the chain to reach some ergodic state, when initially in state } s_i.$$

Proof The result follows from Theorem 4.23. For adopting the notation in that proof, it follows that

$$\sum_{j \in \mathcal{T}} a_{ij}^{\#} = \sum_{j \in \mathcal{T}} (I - C)_{ij}^{-1} = \text{expected number of times the chain is in a transient state when initially in } s_i. \quad \blacksquare$$

Now by applying Theorems 4.20, 4.23, and 4.25, we obtain the following immediate result for absorbing chains.

(4.27) Theorem If T is the transition matrix for an absorbing chain, then the following statements are true:

(1) If s_j is an absorbing state, then

$$l_{ij} = \text{probability of eventual absorption of the process from state } s_i \text{ to the class } [s_j] = \{s_j\}.$$

(2) If s_i and s_j are nonabsorbing (transient) states, then

$$a_{ij}^{\#} = \text{expected number of times the chain will be in } s_j \text{ when initially in } s_i.$$

(3) If $\mathcal{T}$ is the set of indices corresponding to the nonabsorbing (transient) states, then

$$\sum_{j \in \mathcal{J}} a_{ij}^{\#} = \text{expected number of steps until absorption when initially in a nonabsorbing state } s_i.$$

Proof The proof follows directly from Theorems 4.20, 4.23, and 4.25, by applying the standard form (3.22), of an absorbing chain. $\blacksquare$

We remark that in the classical methods for the analysis of absorbing chains, it is usually assumed that the states have been permuted or relabeled so that the transition matrix assumes a canonical form as in (4.15). However, when analyzing the chain by using the matrices $A^{\#}$ and L, the problem of first permuting the states may be completely avoided because all results involving $A^{\#}$ and L are independent of how the states are ordered or labeled. In other words, the results given thus far in this section help to perform this classification of states, rather than requiring that a classification previously exist.

We note also that the Markov chain in Example 2.4 is an absorbing chain. In fact, the transition matrix T for the chain, as given by (2.5), is already in the standard form (3.19).

A large majority of applications utilizing Markov chains involve ergodic chains and, in particular, regular chains. We now limit our discussion to these special kinds of Markov chains. Most of the important questions concerning ergodic chains can be answered by computing the matrix

$$Z = (I - T + L)^{-1}, \tag{4.28}$$

which is sometimes called the *fundamental matrix* associated with the chain (see Kemeny and Snell [1960]). It turns out that although $Z \neq A^{\#}$, the matrix $A^{\#}$ may be used in a manner similar to the way in which Z is used. In most cases, Z may be replaced directly by $A^{\#}$. We will not need to make further mention of the fundamental matrix Z except for the following exercise relating Z to $A^{\#}$.

(4.29) Exercise Let T be the transition matrix for an ergodic chain. Then

$$Z = A^{\#} + L = I + TA^{\#}.$$

(see Meyer [1975]).

We note that in order to apply many of the techniques for analyzing Markov chains developed in this section, it is often necessary to compute either the group generalized inverse $A^{\#}$ or the limiting matrix $L = I - AA^{\#}$, where $A = I - T$ and T is the transition matrix for the chain. An efficient direct method for computing $A^{\#}$ is given in Exercise (5.18). This section is concluded with a discussion of iterative methods for computing rows and/or columns of L.

Very often, the number of states in a Markov chain may be quite large, say, 10,000 or more. This situation, common in the study of Markovian queueing-theoretic models in computer systems, makes the use of an iterative process for calculating vectors associated with L quite attractive, especially the calculation of the stationary probability distribution vector for an ergodic chain (see Wallace [1974]).

As indicated in Theorems 4.12, 4.14, and 4.19, the computation of columns and rows of the limiting matrix, $L = I - AA^{\#}$, is quite important in the analysis of chains. For $\alpha > 0$ and $T_\alpha = (1 - \alpha)I + \alpha T$, consider the iterative process

(4.30) $$x^{k+1} = T_\alpha x^k, \qquad k = 0,1, \ldots,$$

and for row vectors x^j, consider the iterative process

(4.31) $$x^{k+1} = x^k T_\alpha, \qquad k = 0,1, \ldots.$$

From Theorem 4.9, we know that these processes will converge to the ith column and the ith row vector of L, respectively, for $x^0 = e_i$, the ith unit vector, for $i = 1, \ldots, n$.

We show next how, in certain cases, it is possible to choose α in such a way so as to optimize the asymptotic convergence rates of (4.30) and (4.31). For this purpose we will require that all the eigenvalues of T be real and we will drop the requirement that α satisfy $\alpha < 1$.

(4.32) Theorem Assume the transition matrix T for a general Markov chain has all real eigenvalues. Assume that $T \neq I$ and let s and t denote the smallest and largest eigenvalues of T, respectively, other than one. For $\alpha > 0$ let $T_\alpha = (1 - \alpha)I + \alpha T$. then for

(4.33) $$\alpha_b = \frac{2}{2 - (s + t)},$$

T_{α_b} is semiconvergent,

(4.34) $$\lim_{k \to \infty} T_{\alpha_b}^k = L,$$

and the asymptotic rate of convergence of each of (4.30) and (4.31) with $\alpha = \alpha_b$ is given by

(4.35) $$\operatorname*{Max}_{\alpha > 0} R_\infty(T_\alpha) = R_\infty(T_{\alpha_b}) = \ln \frac{t - s}{2 - (s + t)}$$

Proof Since the eigenvalues of T_α other than one have the form

$$1 - \alpha + \alpha\lambda, \quad \lambda \in \sigma(T), -1 \leq \lambda < 1,$$

it follows that

$$|1 - \alpha + \alpha\lambda| = |(1 - \lambda)\alpha - 1|$$

is minimal for $\alpha > 0$ if and only if

$$\alpha = \alpha_b \equiv \frac{2}{(1 - s) + 1 - t} = \frac{2}{2 - (s + t)}.$$

Moreover, in this case it follows that

$$\delta(T_{\alpha_b}) = \operatorname*{Max}_{\substack{\lambda \in \sigma(T) \\ \lambda \neq 1}} \left| \frac{2(1-\lambda)}{2-(s+t)} - 1 \right| = \frac{t}{2-(s+t)}$$

Thus T_{α_b} is convergent and (4.35) holds by (7.6.17). Clearly, (4.34) holds as in Theorem 4.4. ■

Note that α_b given by (4.3) may be greater than one. As an immediate consequence of Theorem 4.32 we have the following.

(4.36) Corollary Let T be the transition matrix for a Markov chain that is not a direct sum of regular and/or absorbing chains and suppose that all the eigenvalues of T are real. Let t be the largest eigenvalue of T, other than one. Then for

$$\alpha_b = \frac{2}{3-t}, \tag{4.37}$$

T_{α_b} is semiconvergent and nonnegative, (4.34) holds, and

$$\operatorname*{Max}_{\alpha > 0} R_\infty(T_\alpha) = R_\infty(T_{\alpha b}) = \ln \frac{t+1}{3-t}. \tag{4.38}$$

Proof By Theorem 4.7, T is not semiconvergent. Thus $-1 \in \sigma(T)$ since $\sigma(T)$ is real. Then $s = -1$ in Theorem 4.32 and consequently T_{α_b} is semiconvergent and (4.38) holds. Finally, the nonnegativity of T_{α_b} follows from the fact that $0 < \alpha_b < 1$. ■

Notice that in Theorem 4.32, $\alpha_b = |s| = |t|$ if and only if $t = -s$. Thus $R_\infty(T_{\alpha_b}) > R_\infty(T_\alpha)$, $\alpha > 0$, unless $s = -t$. Consequently, if s and t are bounds on the largest and smallest eigenvalues of T, respectively, given with reasonable accuracy, then α_b given by (4.33) may produce or accelerate the convergence of the iterative processes (4.30) and (4.31).

This section is concluded with a simple illustration of the significance of these procedures. We return to Example 2.3 and consider an n-state *random walk*. Suppose the particle in question moves to the left or right one step with equal probability, except at the boundary points where the particle always moves back one step. Then $p = q$ in Example 2.3 and the transition matrix associated with this Markov chain is the $n \times n$ matrix T given by

$$T = \begin{bmatrix} 0 & 1 & & & \\ \frac{1}{2} & 0 & \frac{1}{2} & \ddots & \\ & \frac{1}{2} & 0 & \ddots & \frac{1}{2} \\ & & \ddots & 1 & 0 \end{bmatrix}. \tag{4.39}$$

We now consider the problem of computing a stationary probability distribution vector π for this chain. We see that the chain is ergodic by Theorem 3.9, since T is irreducible. Thus π is unique and positive by Corollary 3.14. Since T is not primitive, the chain is periodic, also by Theorem 3.9. In particular, the chain has period 2, for starting from an even-numbered state, the process can be in even-numbered states only in an even number of steps, and in an odd-numbered state in an odd number of steps; hence, the even and odd states form two cyclic classes.

Now we wish to compute the stationary probability distribution vector π associated with this ergodic chain. The eigenvalues of T are easily seen to be the real numbers

$$\lambda_k = \cos[k\pi/(n-1)], \qquad k = 0,1,\ldots,n-1.$$

We know that T is not semiconvergent by Corollary 4.6, and we confirm this by noting that $\lambda_{n-1} = -1$ is an eigenvalue of T. Now the second largest eigenvalue of T is

$$\lambda_1 = \cos[\pi/(n-1)].$$

Then by Corollary 4.36, for

$$\alpha_b = \frac{2}{3 - \cos[\pi/(n-1)]}, \tag{4.40}$$

T_{α_b} is nonnegative and semiconvergent. Moreover,

$$R_\infty(T_{\alpha_b}) = -\ln \frac{\cos[\pi/(n-1)] + 1}{3 - \cos[\pi/(n-1)]} \tag{4.41}$$

is the optimal asymptotic convergence rate for T_α as a function of α, in applying the iteration (4.31) to compute the stationary probability distribution vector for the chain. Here we may choose the row vector $x^0 = e_i$, for any $1 \le i \le n$.

Finally, we note that even though λ_1 increases with n, the fact that T is sparse makes the iterative formulas (4.30) and (4.31) attractive for problems of this type. This concludes Section 4 and the text of this chapter.

5 EXERCISES

(5.1) A man is playing two slot machines. The first machine pays off with a probability $\frac{1}{2}$, the second with a probability $\frac{1}{4}$. If he loses, he plays the same machine again; if he wins, he switches to the other machine. Let s_i denote the state of playing the ith machine.

(a) Is the process a finite homogeneous Markov chain?
(b) If the answer to (a) is yes, what is the transition matrix?

(5.2) A sequence of experiments is performed, in each of which two coins are tossed. Let s_1 indicate that two heads come up, s_2 that a head and a tail come up, and s_3 that two tails come up.

(a) Show that this is a Markov chain and find the transition matrix.
(b) If two tails turn up on a given toss, what is the probability of two tails turning up three tosses later?
(c) Classify the states of the Markov chain; that is, determine which states are ergodic and which are transient.

(5.3) The following problem describes a common model used to explain diffusion in gases. We begin with two urns, each of which contains five balls. The balls in one urn are all white, and those in the other urn are all black. Each second, one ball is selected from each urn and moved to the other urn.

(a) Show that this is a Markov chain and find the transition matrix and the initial probability distribution vector.
(b) What is the probability that both urns will have at least one white ball after 3 sec, after 5 sec?
(c) Classify the states of the chain.

(5.4) Prove that if T is a nonsingular transition matrix for a Markov chain and $T^{-1} \geq 0$, then T^{-1} is the transition matrix for a Markov chain.

(5.5) Show that if both T and T^{-1} are transition matrices for Markov chains then T is a permutation matrix.

(5.6) Consider the Markov chains with transition matrices

$$(1)\quad \begin{bmatrix} 0 & 1 & 0 & 0 & 0 \\ 0 & 0 & \frac{1}{2} & \frac{1}{4} & \frac{1}{4} \\ 0 & 0 & 0 & 1 & 0 \\ \frac{1}{2} & \frac{1}{2} & 0 & 0 & 0 \\ 0 & 0 & 0 & 1 & 0 \end{bmatrix}, \qquad (2)\quad \begin{bmatrix} \frac{1}{2} & \frac{1}{2} & 0 & 0 & 0 \\ \frac{2}{3} & 0 & 0 & \frac{1}{6} & \frac{1}{6} \\ 0 & \frac{1}{2} & \frac{1}{2} & 0 & 0 \\ 1 & 0 & 0 & 0 & 0 \\ 0 & 0 & 1 & 0 & 0 \end{bmatrix}.$$

(a) Determine the ergodic classes and the transient classes for each chain.
(b) Determine the standard form (3.19) for each transition matrix.

(5.7) Let T denote the transition matrix for a Markov chain and let $\tilde{T} = \frac{1}{2}(I + T)$. Show that the ergodic classes and the transient classes are the same for T and $\tilde{T}$.

(5.8) Give the transition matrix T for a two-state periodic Markov chain. Is T unique?

(5.9) Consider a Markov chain with transition matrix.

$$T = \begin{bmatrix} \frac{1}{2} & 0 & \frac{1}{2} \\ 1 & 0 & 0 \\ 0 & 1 & 0 \end{bmatrix}.$$

(a) Show that the chain is regular.

(b) Determine the stationary probability distribution vector associated with the chain by solving $\pi T = \pi$ and using $\pi_1 + \pi_2 + \pi_3 = 1$.

(c) Determine

$$L = \lim_{k \to \infty} T^k.$$

(5.10) Classify the Markov chains associated with the following transition matrices according to whether they are (a) ergodic, (b) regular, (c) periodic, (d) absorbing, or (e) none of these.

$$(1) \begin{bmatrix} \frac{1}{2} & \frac{1}{2} \\ \frac{1}{2} & \frac{1}{2} \end{bmatrix}, \qquad (2) \begin{bmatrix} 1 & 0 \\ \frac{1}{2} & \frac{1}{2} \end{bmatrix},$$

$$(3) \begin{bmatrix} 0 & \frac{1}{2} & \frac{1}{2} \\ 1 & 0 & 0 \\ 1 & 0 & 0 \end{bmatrix}, \qquad (4) \begin{bmatrix} 0 & 1 & 0 \\ 0 & 0 & 1 \\ \frac{1}{2} & \frac{1}{2} & 0 \end{bmatrix},$$

$$(5) \begin{bmatrix} 0 & \frac{1}{3} & 0 & \frac{1}{3} & \frac{1}{3} \\ 0 & \frac{1}{2} & 0 & \frac{1}{4} & \frac{1}{4} \\ 0 & 0 & 1 & 0 & 0 \\ \frac{1}{2} & 0 & \frac{1}{2} & 0 & 0 \\ 0 & 0 & 0 & 0 & 1 \end{bmatrix}.$$

(5.11) Put each of the transition matrices in Exercise (5.10) in the standard form (3.19).

(5.12) Show that for any probability vector π, there is a regular Markov chain with π as its stationary probability distribution vector.

(5.13) Consider arbitrary Markov chains with state transition matrices T, and let $T^m = (t_{ij}^{(m)})$. Prove or give a counterexample to each of the following:

(a) If state s_i is transient and $t_{ij}^{(m)} > 0$ for some m, then state s_j is transient.

(b) If state s_j is transient and $t_{ij}^{(m)} > 0$ for some m, then state s_i is transient.

(c) If T_1 and T_2 are transition matrices associated with regular Markov chains then so is the product $T_1 T_2$.

(d) If T_1 and T_2 are defined as in part (c), then for any $0 < \lambda < 1$, the matrix $T = \lambda T_1 + (1 - \lambda)T_2$ is a transition matrix associated with a regular Markov chain.

(5.14) Show that if the transition matrix associated with a regular Markov chain with n states is doubly stochastic, then each component of the stationary probability distribution vector associated with the chain has the value $1/n$.

(5.15) Let T be the transition matrix associated with a Markov chain with n states. Show that this chain is ergodic if and only if the chain associated with $\tilde{T} = \frac{1}{2}(I + T)$ is regular. Show that in this case $\tilde{T}^n \gg 0$. (See Exercise (5.7).)

(5.16) Show that the period of a periodic Markov chain, whose transition matrix is nonsingular, divides the number of states in the chain (see Theorem 2.2.27).

(5.17) Let T denote the transition matrix associated with an ergodic Markov chain and let $A = I - T$. Show that $A^\# = A^+$, the Moore–Penrose generalized inverse of A, if and only if T is doubly stochastic. (See Meyer [1975].)

(5.18) For an n-state ergodic Markov chain with transition matrix T, let $A = I - T$ be partitioned as

$$A = \begin{bmatrix} A_{11} & c \\ d^t & a_{nn} \end{bmatrix}$$

where A_{11} is $(n-1) \times (n-1)$. Adopt the following notation:

$$h^t = d^t A_{11}^{-1},$$

$$\delta = -h^t A^{-1} e, \qquad e = (1, \ldots, 1)^t,$$

$$\beta = 1 - h^t e,$$

$$F = A_{11}^{-1} - (\delta/\beta) I.$$

Show that the group generalized inverse, $A^\#$, of A is then given by

$$A^\# = \left[\begin{array}{c|c} A_{11}^{-1} + \dfrac{A_{11}^{-1} e h^t A_{11}^{-1}}{\delta} - \dfrac{F e h^t F}{\delta} & -\dfrac{Fe}{\beta} \\ \hline \dfrac{h^t F}{\beta} & \dfrac{\delta}{\beta^2} \end{array}\right].$$

(See Meyer [1975].)

(5.19) Show that in the notation of Exercise (5.18), the stationary probability distribution vector π is given by

$$\pi = \frac{1}{1 - d^t A_{11}^{-1} e}(-d^t A_{11}^{-1}, 1) = \frac{1}{\beta}(-h^t, 1),$$

and that the limiting matrix $L = I - AA^{\#}$ is given by

$$L = \frac{1}{\beta}\begin{bmatrix} -eh^t & e \\ -h^t & 1 \end{bmatrix}.$$

(See Meyer [1975].)

(5.20) Use Exercises (5.18) and (5.19) to compute $A^{\#}$, π, and L for the Markov chain specified in Exercise (5.9).

(5.21) Write a computer program to use the iterative formula (4.31) to approximate the stationary probability distribution vector for the random walk with transition matrix T given by (4.39), for $n = 4$, 10, 20, and 100 states. Use the optimum α_b in (4.39), where α_b is specified by (4.40) and compute $R_\infty(T_{\alpha_b})$ in each case. What conclusions can you draw concerning n?

6 NOTES

(6.1) The fundamental topic discussed in this application chapter is named after A. A. Markov, who laid the foundations for the theory in a series of papers beginning in 1908. His main contributions to the subject concerned the finite state space case, and matrix theory played a primary role in his work. For a detailed account of the early literature on this topic see Frechet [1938].

(6.2) The theory of Markov chains can be approached in two quite different ways. We chose here to emphasize the algebraic aspects of the subject in order to make strong use of the theory of nonnegative matrices and the theory of M-matrices. The second approach is to treat Markov chains as a special type of stochastic process and to rely on probability theory rather than matrix theory. This approach is taken, for example, by Chung [1967]. Our development of the subject here is, of course, no more than an introduction to the theory. There are many problems of a probabilistic nature which we have not touched upon. For a more detailed treatment of the subject we refer to the texts: Kemeny and Snell [1960], Cinlar [1975], or Isaacson and Madsen [1976], and others.

(6.3) The examples in Section 2 represent adaptations of examples from a variety of sources. Many of them can be found in several places. The literature is rich in applications of the theory of Markov chains.

(6.4) The classical theory of chains developed in Section 3 includes many of the standard theorems concerning Markov chains. They can be found in various books including Kemeny and Snell [1960], Cinlar [1975], and Isaacson and Madson [1976]. Our treatment, however, is along the lines of the developed in Pearl [1973, Chapter 3].

(6.5) Much of the material concerning the modern analysis of Markov chains, given in Section 4, is due to C. D. Meyer, Jr. Corollary 4.3, Theorem 4.4, and Corollary 4.6 are in Meyer [1975], while most of the other results are in Meyer [1978]. Finally, Theorem 4.32 and Corollary 4.36 are essentially in Neumann and Plemmons [1978].

(6.6) We remark again that the study of Markov chains provides some of the most beautiful and elegant applications of the classical Perron–Frobenius theory of nonnegative matrices. In addition, many of the modern methods for analyzing these chains make strong use of recent developments in the theory of singular M-matrices and generalized matrix inverses, making the material developed in this chapter quite appropriate for the book.

CHAPTER 9

INPUT–OUTPUT ANALYSIS IN ECONOMICS

1 INTRODUCTION

In this chapter we study another important discipline in the mathematical sciences in which the theory of nonnegative matrices finds elegant applications. In particular, we will see that the primary mathematical tools in the study of linear economic models involve nonnegativity and, especially, the theory of M-matrices developed in Chapter 6. In the present chapter we show how many of the results on M-matrices given earlier can be used to greatly simplify the construction and the analysis of Leontief's input–output models in economics.

It has been said by Miernyk [1965] that: "When Wassily Leontief published his 'Quantitative input–output relations in the economic system of the United States' in *The Review of Economics and Statistics* [1936], he launched a quiet revolution in economic analysis that has steadily gained momentum." It was only a matter of timing that the article, which represents a turning point in the development of economic thought, did not at first attract wide acclaim. The nations of the noncommunist world were in the midst of the Great Depression. Moreover, John Maynard Keynes had just published his *General Theory of Employment, Interest, and Money* [1936], a treatise that immediately attracted worldwide attention since it was focused on the problems of chronic unemployment in the capitalist economics of that day. It turns out that, unlike Keynes, Leontief was not concerned with the causes of disequilibrium in a particular type of economic system during a particular phase of its development; he was primarily interested in the structure of economic systems. In particular, he was interested in the way the component parts of an economy fit together and influence one another. He developed an analytical model that can be applied to any kind of economic system during any phase of its development. As he noted himself, input–output is above all a mathematical tool. It can be used in the analysis of a wide variety of economic problems and as a guide for the implementation of various kinds of economic policies.

Leontief's input–output analysis deals with this particular question: *What level of output should each of n industries in a particular economic situation produce, in order that it will just be sufficient to satisfy the total demand of the economy for that product*? With this in mind we now give an overview of Leontief's models. These concepts will be made more precise in later sections of this chapter.

In Leontief's approach, production activities of an economy are disaggregated into n sectors of industries, though not necessarily to individual firms in a microscopic sense, and the transaction of goods among the sectors is analyzed. His basic assumptions are as follows:

(1) Each of the n sectors produces a single kind of commodity. Broadly interpreted, this means that the n sectors and n kinds of commodities are in one-to-one correspondence. The sector producing the ith good is denoted by i.

(2) In each sector, production means the transformation of several kinds of goods in some quantities into a single kind of good in some amount. Moreover this pattern of input–output transformation is assumed to be stable.

Intuitively, in a Leontief system this pattern assumes the following form. To produce one unit of the jth good, t_{ij} units of the ith good are needed as inputs for $i = 1, \ldots, n$ in sector j, and λ units of output of the jth good require λt_{ij} units of the ith good. The magnitudes t_{ij} are called *input coefficients* and are usually assumed to be constant. In the economist's terminology, the ratios of inputs are constant, and constant returns to scale prevail.

Let x_i denote the *output of the ith good* per fixed unit of time. Part of this *gross output* is consumed as the input needed for production activities of the n sectors. Thus

$$\sum_{j=1}^{n} t_{ij}x_j$$

units of the ith good is consumed in production activities, leaving

$$d_i = x_i - \sum_{j=1}^{n} t_{ij}x_j$$

units of the ith good as the net output. This net output d_i is normally called the *final demand* of the ith good. Alternatively, d_i can be thought of as the contribution of the *open sector* of the economy, in which labor costs, consumer purchases leading to profits, etc., are taken into account.

Thus letting x and d denote the n-vectors with components x_i and d_i, respectively, we obtain the system of linear equations

$$(I - T)x = d. \tag{1.1}$$

The coefficient matrix

$$A = I - T \tag{1.2}$$

of this system of linear equations is obviously in $Z^{n \times n}$. It will be seen later that the economic situation is "feasible" if and only if A is a nonsingular M-matrix; in which case the system can be solved for the gross output vector $x = A^{-1}d$, which is necessarily nonnegative. Thus the system (1.1) has the characteristic feature that for the obvious economic reason, the relevant constants t_{ij} and d_i, as well as the solutions x_i, should satisfy the nonnegativity constraint. From the economic point of view, *the solvability of* (1.1) *in the nonnegative unknowns* $x_i \geq 0$ *means the feasibility of the Leontief model*, as previously mentioned.

The model just described is called the *open Leontief model*, since the open sector lies outside the system. If this open sector is absorbed into the system as just another industry, the model is called a *closed Leontief model*. In this situation, final demand does not appear; in its place will be the input requirements and the output of the newly conceived industry. All goods will now be intermediate in nature, for everything that is produced is produced only for the sake of satisfying the input requirements of the industries or sectors of the model. Mathematically, the disappearance of the final demands means that we now have a homogeneous system of linear equations, where the coefficient matrix is again in $Z^{n \times n}$. The problem here is to determine when this matrix is a (singular) M-matrix and when the system has nonnegative solutions. This will be discussed later in this chapter.

Thus far we have considered only the *static Leontief models*; that is, models in which the input coefficients and the demands from the open sector are held constant. However, a dynamic version of, say, the open Leontief model can be constructed as follows. Let x_i^k denote the output of the ith good at time k, let t_{ij} denote the amount of output of industry i per unit of input of industry j at the next time stage, and let α, $0 < \alpha < 1$, be the proportion of output, which is the same for each industry, that is available for internal use in the economy. Then if the final demands d_i are held constant, we have the difference equation

$$x^{k+1} = \alpha T x^k + d,$$

which can then be studied in terms of the framework of Chapter 7 on iterative methods. However, we shall be concerned only with the static models in the present chapter.

The input–output models just described have wide ranging important applications. One of the original purposes was to study national economies. Consider the following example. The American economy is known for its high wage rates and intensive use of capital equipment in production. Labor is commonly regarded as a scarce factor and capital as an abundant factor. Therefore it was generally believed that America's foreign trade is based on exchanging capital-intensive goods for labor-intensive goods. Thus it was quite a surprise when Leontief [1953] published his finding, by use of input–output analysis, that the pattern of American trade is just opposite to the common view. In his study on American exports and imports, Leontief calculated the total requirements of labor and capital for the production of both exports and imports. It turned out that the United States actually exported labor-intensive goods and imported capital-intensive goods, a remarkable discovery made possible by input–output analysis. More recently Almon *et al.* [1974] have used such methods to produce an interindustry forecast of the American economy for the year 1985. In addition, Sarma [1977] has described an input–output model developed by IBM whose purpose is to forecast the industrial implications in the overall American economy.

Although input–output models were originally introduced by Leontief in order to model the structure of a national economy, at the present time they have been used, and are being used, in several other areas as well. For example, they have been used in connection with certain cooperate planning problems (see Stone [1970] or Sandberg [1974a]), they serve as a basis for solving some cost evaluation problems (see Hibbs [1972]), and they play a central role in studies of environmental pollution (see Gutmanis [1972]).

In the present chapter, Section 3 is devoted to an extensive treatment of the open input–output model while the closed model is discussed in Section 4. In each of these sections we shall primarily be concerned with nonnegative solutions to appropriate systems of linear equations that are derived from the models.

As usual, the last two sections are devoted to the exercises and the notes, respectively.

Next, in Section 2, we lay the framework for the material to be presented later by illustrating many of the important concepts by a simple example, that of analyzing the flow of goods in an economy with three aggregated industries—agriculture, manufacturing, and services.

2 A SIMPLE APPLICATION

There are four goals of this section: (1) to acquaint the reader with the way in which economists construct and analyze an input–output table, (2) to illustrate, by this application, the way in which input–output models can be used in economic forecasting, (3) to acquaint the reader with certain notation and terminology conventions, and most importantly, (4) to provide material related to nonnegativity for reference in later sections of this chapter.

We begin by discussing an example of an input–output table showing the flow of goods and services among different branches of an economic system during a particular time period. For an alternate example, see Sarma [1977].

In order to engage in production, each sector or industry must obtain some *inputs*, which might include raw materials, semifinished goods, and capital equipment bought from other industries. In addition, business taxes must be paid and labor hired. Very often, some *intermediate products*, which are used as inputs to produce other goods and services as against final or *outside products* which do not reenter the production processes, are purchased from other industries. The output produced by each industry is sold either to outside users or to other industries or sectors which use the goods as inputs. A table summarizing the origin of all the various inputs and the destination of all the various outputs of all the industries in an economy is called an *input–output table*.

As an example, we consider a simple hypothetical economy consisting of three sectors: (1) agriculture, (2) manufactures, and (3) services. Each of the sectors produces precisely one kind of output: agricultural goods, manufactured goods, or services. These three sectors are to be interdependent, in that they purchase inputs from and sell outputs to each other. No government, foreign imports, or capital equipment will be involved here. All finished goods and services which do not reenter into production processes are used by the outside sector consisting of consumers, etc.

Following the previously stated assumptions, we give a hypothetical table summarizing the flow of goods and services, measured in dollars.

(2.1) Table Input–Output Table (in Dollars)

Output to / Input from	(1) Agriculture	(2) Manufactures	(3) Services	Outside demand	Gross output
(1) Agriculture	15	20	30	35	100
(2) Manufactures	30	10	45	115	200
(3) Services	20	60	—	70	150

Here, the data in any row show the distribution of input to various sectors and users while the data in a column indicate the sources of inputs needed for production. For example, reading across the first row (agriculture), we find that, of the gross output of $100 of agricultural goods produced, $15 is to be used in further agricultural production, \$20 of goods is sold to manufactures, \$30 of goods is sold to services, and finally \$35 of agriculture goods go to satisfy the outside demand. Similarly, reading down the second column, we see that in order to produce \$200 of gross output, manufactures has to input \$20 of agriculture goods, \$10 of its own goods, and \$60 of services.

In order to analyze Table 2.1, we will need the following notation.

Subscripts: (1) agriculture, (2) manufactures, (3) services.
x_i: Gross output of sector i.
x_{ij}: Sales of sector i to sector j.
d_i: Final demand on sector i.

Then the basic row relation of Table 2.1 is

(2.2) $$x_i = x_{i1} + x_{i2} + x_{i3} + d_i, \qquad i = 1,2,3.$$

This of course says that the gross output of a sector consists of the *intermediate product* sold to various production sectors and the final outside product which takes into account the consumers and the open sector.

As usual, we collect the gross output produced by all sectors in the vector

$$x = \begin{bmatrix} x_1 \\ x_2 \\ x_3 \end{bmatrix}$$

and the final demand in the vector

$$d = \begin{bmatrix} d_1 \\ d_2 \\ d_3 \end{bmatrix}.$$

Now it is much more convenient in the analysis if the input–output table in question is converted into one indicating the input requirements for the production of one *unit of output* for each sector. This table is often called a technical input–output table and its entries are called the *input coefficients* for the economy, which were mentioned in the introduction to this chapter. In order to construct the technical input–output table associated with Table 2.1, the various inputs of each sector are divided by the gross output of that sector. For example, each entry of the first column of Table 2.1 is divided by the gross output of agriculture, which is \$100.

Letting t_{ij} denote the input coefficient indicating the amount of product i needed to produce one unit output of product j, we have

(2.3) $$t_{ij} = x_{ij}/x_j, \qquad 1 \leq i,j \leq 3.$$

The technical table thus derived from Table 2.1 is then the following.

(2.4) Table Technical Input–Output Table

Input from \ Output to	(1) Agriculture	(2) Manufactures	(3) Services
(1) Agriculture	0.15	0.10	0.20
(2) Manufactures	0.30	0.05	0.30
(3) Services	0.20	0.30	0.00

Next, rewriting (2.3) as

(2.5) $$x_{ij} = t_{ij}x_j, \qquad 1 \leq i,j \leq 3,$$

and substituting (2.5) into (2.2) we obtain

(2.6) $$x_i = t_{i1}x_1 + t_{i2}x_2 + t_{i3}x_3 + d_i, \qquad i = 1,2,3.$$

Then letting $T = (t_{ij})$, we have

(2.7) $$T = \begin{bmatrix} 0.15 & 0.10 & 0.20 \\ 0.30 & 0.05 & 0.30 \\ 0.20 & 0.30 & 0.00 \end{bmatrix},$$

so that the matrix $A = I - T$ is given by

(2.8) $$A = \begin{bmatrix} 0.85 & -0.10 & -0.20 \\ -0.30 & 0.95 & -0.30 \\ -0.20 & -0.30 & 1.00 \end{bmatrix}.$$

As we have seen, Table 2.1, or alternatively Table 2.4, summarizes the gross purchases or sales of products among the three sectors of the hypothetical economy. It also describes the technology of production. From Table 2.1, we see, for example, that the production of \$150 of services requires the input of \$30 of agricultural goods and \$45 of manufactured goods. Here \$70 of the services go to satisfy the outside demand. This leads us to ask what the gross output of services should be if the technical input coefficients are held fixed while the outside demand is allowed to change. This *economic forecasting* is based upon the assumption that when the level of output is changed, the amounts of all inputs required are also changed proportionately. This is called the assumption of *fixed proportion of factor inputs*. Thus we are assuming that Table 2.4 remains fixed even though the outside demand and, accordingly, the gross output columns of Table 2.1 may change. Thus in order to forecast what level of output x_i each of the three sectors should have in order to satisfy the input requirements and the outside demands d_i, we need only solve the linear system

(2.9) $$Ax = d,$$

where A is given by (2.8). If the economic system is to be *feasible*, then (2.9) must have a nonnegative solution for each outside demand vector d. It will be seen in Section 3, that this is just the requirement that A be a nonsingular M-matrix. By condition (N_{38}) of Theorem 6.2.3 then, we require that A^{-1} exist and satisfy $A^{-1} \geq 0$. In our case, the inverse of the matrix A given by (2.8) is

$$A^{-1} = \begin{bmatrix} 1.3459 & 0.2504 & 0.3443 \\ 0.5634 & 1.2676 & 0.4930 \\ 0.4382 & 0.4304 & 1.2167 \end{bmatrix}, \tag{2.10}$$

where the entries are rounded to four decimal places. Then since $A^{-1} \geq 0$, the output vector $x = A^{-1}d$ is nonnegative for each outside demand vector $d \geq 0$. Thus this particular economic system is feasible. For example, if the outside demand on agriculture is \$100, that on manufactures is \$200 and that on services is \$300, then $d = (100, 200, 300)^t$ so that the output vector x is computed using (2.10) to be

$$x = A^{-1}d = \begin{bmatrix} 287.96 \\ 457.76 \\ 494.91 \end{bmatrix}.$$

Thus \$287.96 of agriculture, \$457.76 of manufactures, and \$494.91 of services are required to satisfy the input demands of each of the three sectors and the outside sector.

Now suppose the demand for commodity (1), agriculture, is increased to \$300, so that $\hat{d} = (300, 200, 300)$. Then a new output vector $\hat{x}$ is given by

$$\hat{x} = A^{-1}\hat{d} = \begin{bmatrix} 557.14 \\ 570.44 \\ 582.55 \end{bmatrix}.$$

Here we see that the output of each sector increases, but that the output of commodity (1), agriculture, increases by the greater amount. It will be shown in Section 3 that this will always be the case if the row sums of the production matrix T satisfy

$$\sum_{j=1}^{n} t_{ij} < 1,$$

and T is irreducible.

Under these conditions it will be shown that if only the demand of commodity j increases, then the output of commodity j increases by the greatest amount, although all outputs may increase. In our example, each row sum of the production matrix T given by (2.7) is, of course, strictly less than one, and T is irreducible.

3 THE OPEN MODEL

As we have mentioned earlier, interindustry analysis of the Leontief type is concerned primarily with systems in which the products of economic factors (machines, materials, labor, etc.) are themselves used as factors to produce further goods. Various Leontief-type models have been presented, the simplest of which is the open model which will be discussed in the present section.

As before, we assume that the economy is divided into n sectors, each producing one commodity to be consumed by itself, by other industries and by the outside sector. Then identifying the ith sector with the ith commodity, we have the following notation:

x_i: Gross output of sector i.
x_{ij}: Sales of sector i to sector j.
d_i: Final demand on sector i.
t_{ij}: Input coefficient, the number of units of commodity i required to produce one unit of commodity j.

Then the overall input–output balance of the entire economy can be expressed in terms of the n equations:

$$x_i = \sum_{j=1}^{n} x_{ij} + d_i, \qquad 1 \le i \le n. \tag{3.1}$$

Now assuming that if the level of output is changed then the amounts of all inputs required are also changed proportionally; that is, assuming a *fixed proportion of factor inputs*, the input coefficients t_{ij} are constant and satisfy

$$t_{ij} = x_{ij}/x_j, \qquad 1 \le i,j \le n. \tag{3.2}$$

The system of linear equations (3.1) then becomes

$$x_i = \sum_{j=1}^{n} t_{ij}x_j + d_i, \qquad 1 \le i \le n. \tag{3.3}$$

Letting $T = (t_{ij})$ and

$$A = I - T \tag{3.4}$$

as before, the overall input–output balance of the entire economy is expressed in terms of the system of n linear equations in n unknowns:

$$Ax = d, \tag{3.5}$$

where $x = (x_i)$ is the *output vector* and $d = (d_i)$ is the *final demand vector*.

The model just described is called the *open Leontief model* and the matrix $T = (t_{ij})$ is called the *input matrix* for the model. Then A, given by (3.4), is in $Z^{n\times n}$; that is, $a_{ij} \le 0$ for all $i \ne j$. Matrices in $Z^{n\times n}$ are often called *matrices of Leontief type*, or sometimes *essentially nonpositive matrices* by economists. Clearly the features of the open Leontief model are completely determined by the properties of A.

The economic model just described also has an associated price–valuation system, which gives the pricing or value side of the input–output relationship. Our notation will be

p_j: Price of the jth commodity.
v_j: Value added per unit output of the jth commodity.

Then

$$\sum_{i=1}^{n} t_{ij} p_i, \qquad 1 \le j \le n,$$

is the unit cost of the jth commodity, so that

$$p_j - \sum_{i=1}^{n} t_{ij} p_i$$

is the net revenue per unit output of the jth commodity; that is, the value added per unit output, v_j. Then letting p denote the vector with entries p_j and v denote the vector with entries v_j, the relationship just stated is described by the system of linear equations

$$p^t - p^t T = v^t \tag{3.6}$$

or

$$p^t A = v^t, \tag{3.7}$$

where $A = I - T$, as before. The vector p is called the *price vector* and the vector v is called the *value added vector* of the associated open Leontief model. Obviously $v \ge 0$ and the economist is interested in solving the system (3.6), or (3.7), for price vectors $p \ge 0$.

A link between the two systems (3.5) and (3.6) is given by the relation

$$\sum_{i=1}^{n} v_i x_i = \sum_{j=1}^{n} p_j d_j,$$

which may be interpreted by the following economic statement: The "national income" and the "national product" are equal.

While the solvability of the original output system (3.5) for the nonnegative outputs x_i, $1 \le i \le n$, means the *feasibility* of the model, that of solving the price system (3.7) in the nonnegative prices p_j, $1 \le j \le n$, means *profitability*. This leads to the following definitions.

(3.8) Definitions An open Leontief model with input matrix T is said to be *feasible* if the system (3.5) has a nonnegative solution for the output vector x, for each open demand vector d. It is said to be *profitable* if the system (3.7) has a nonnegative solution for the price vector p, for each value added vector v.

The duality of these concepts is made apparent in the following elementary theorem, which is based upon the theory of M-matrices developed in Chapter 6. Recall that A is a nonsingular M-matrix if and only if $A = sI - B$, $s > 0$, $B \geq 0$, with $s > \rho(B)$, the spectral radius of B.

(3.9) Theorem Consider an open Leontief model with input matrix T and let $A = I - T$. Then the following statement are equivalent:

(1) The model is feasible.
(2) The model is profitable.
(3) A is a nonsingular M-matrix.

Proof We show first that (1) and (3) are equivalent. If (1) holds, then by choosing the demand vector d to be positive, it follows that A satisfies:

(3.10) $$\text{There exists } x > 0 \text{ with } Ax = d \gg 0.$$

But this is just condition (I_{28}) of Theorem 6.2.3, which characterizes nonsingular M-matrices. Thus (3) holds since $A \in Z^{n \times n}$. Conversely, if (3) holds, then by condition (N_{38}) of Theorem 6.2.3, $A^{-1} \geq 0$. Thus (3.5) has the nonnegative solution $x = A^{-1}d$ for each $d \geq 0$, so that (1) holds. The equivalence of (2) and (3) is established in a similar manner by using the fact that A is a nonsingular M-matrix if and only if the same is true for A^t. ■

Recall that 50 characterizations for $A \in Z^{n \times n}$ to be a nonsingular M-matrix were given in Theorem 6.2.3. Thus we can state the following corollary.

(3.11) Corollary Consider an open Leontief model with input matrix T and let $A = I - T$. Then the following statements are equivalent:

(1) The model is feasible.
(2) The model is profitable.
(3) A satisfies one (and thus all) of the conditions (A_1)–(Q_{50}) of Theorem 6.2.3.

Now condition (A_1) of Theorem 6.2.3, which states that $A \in Z^{n \times n}$ is an M-matrix if and only if all the principal minors of A are positive, was established in the economics text by Hawkins and Simon [1949]. Their result has consequently been known in the economics literature as the *Hawkins–Simon condition* for the feasibility (or profitability) of the open Leontief model. Actually, condition (A_1) was established much earlier for nonsingular M-matrices by Ostrowski [1937].

We remark also that if the input matrix T of an open Leontief model is irreducible, then by Theorem 6.2.7, the model is feasible (or profitable) if and only if

$$A^{-1} \gg 0 \tag{3.12}$$

or equivalently

$$Ax > 0 \qquad \text{for some} \quad x \gg 0. \tag{3.13}$$

In this case, (3.12) is equivalent to the statement that the model has all positive inputs for any nonzero demand vector and/or the model has all positive prices for any nonzero value added vector. In this regard, we note that the input matrix T, given by (2.7) for the example discussed in Section 2, is irreducible. Thus the economy discussed there is feasible, if and only if $A^{-1} \gg 0$, where $A = I - T$; but this in turn is verified by (2.10).

We also mention in passing that if T is the input matrix for a feasible (profitable) open Leontief model, then since $\rho(T) < 1$, the output and price vectors may be computed from the series

$$\sum_{k=0}^{\infty} T^k d \qquad \text{and} \qquad \sum_{k=0}^{\infty} v^t T^k,$$

respectively, by Lemma 6.2.1. However, these methods are usually not practical.

This section is concluded with a discussion of the effects that changes in the open demands of a feasible model may have on the final outputs, and the effects that changes in the value added requirements on a feasible model may have on the prices.

First we shall need the following technical lemma.

(3.14) Lemma Let A be a nonsingular M-matrix of order n whose row sums are all nonnegative; that is, $Ae \geq 0$ where $e = (1, \ldots, 1)^t$. Then the entries of A^{-1} satisfy

$$(A^{-1})_{ii} \geq (A^{-1})_{ki}, \qquad 1 \leq i,k \leq n. \tag{3.15}$$

Proof Since

$$A^{-1} = \frac{1}{\det A} \operatorname{Adj} A,$$

where Adj A is the transposed matrix of the cofactors of A, it suffices to show that if C_{ij} is the cofactor of the ith row and jth column of A, then

$$C_{ii} \geq C_{ik}, \qquad 1 \leq i,k \leq n.$$

For that purpose we write $A = sI - B$, $s > 0$, $B \geq 0$. Then $s > \rho(B)$ and, by assumption, $s \geq \max_i \sum_{j=1}^n b_{ij}$. We consider two cases. Assume that $s > \max_i \sum_{j=1}^n b_{ij}$. Now we can replace any zero entries of B by a small positive number δ to form a new positive matrix $\hat{B} = (\hat{b}_{ij})$, but that still

$$s > \max_i \sum_{j=1}^n \hat{b}_{ij}.$$

Thus if we can prove $\hat{C}_{ii} \geq \hat{C}_{ik}$ for all i and k in this case, then $C_{ii} \geq C_{ik}$ for all i and k in the original case, by continuity in δ, letting δ approach zero. As a result, it suffices to assume that $B \gg 0$, which we shall do.

Consider $s \neq k$ with k fixed (but arbitrary). Then replace all the elements of the ith row of B by zeros, except the (i,i)th and (i,k)th which we replace by $s/2$. Denote the new matrix by W; it is clearly irreducible, and moreover has all row sums not exceeding s, and all but the ith less than s. Thus $\rho(W) < s$ by the Perron–Frobenius theorem and Theorem 1.1.11. Then

$$-\frac{s}{2} C_{ik} + \frac{s}{2} C_{ii} = \det(sI - W) > 0,$$

since $sI - W$ is a nonsingular M-matrix and recalling that the cofactors remain the same as for $sI - B$. Therefore $C_{ii} \geq C_{ik}$.

Second, if $s = \max_i \sum_{j=1}^n b_{ij}$, then we take any $\varepsilon > 0$ and consider $s + \varepsilon$ in place of s for A. Case one then applies and (3.15) follows from continuity by letting $\varepsilon \to 0$. Thus the lemma is proved. ■

Now consider an open Leontief input–output model with input matrix T. Then even if the model is feasible, so that $A = I - T$ is a nonsingular M-matrix, it does not always follow that the row sums of T are all at most one. But if this is the case we can prove the following theorem.

(3.16) Theorem Let T be the input matrix for a feasible open Leontief model and suppose that

$$Te \leq e, \qquad e = (1, \ldots, 1)^t.$$

Then if the demand for commodity i alone increases, none of the outputs decrease and the output of commodity i increases and increases by the greatest amount, although other outputs may increase by the same amount.

Proof As before we let $A = I - T$, and let x and d denote the output and demand vectors, respectively. Then A is a nonsingular M-matrix by Theorem 3.9. Moreover

$$Ae = (I - T)e = e - Te \geq 0$$

by assumption, so that Lemma 3.14 applies.

Now suppose that the ith term of the demand vector d is increased by the amount δ. Then the resulting demand vector becomes

$$\hat{d} = d + \delta e_i,$$

where e_i denotes the ith unit vector. From (3.5), $x = A^{-1}d$ and the new output vector $\hat{x}$ becomes

$$\begin{aligned}\hat{x} = A^{-1}\hat{d} &= A^{-1}(d + \delta e_i)\\ &= A^{-1}d + \delta A^{-1}e_i = x + \delta(A^{-1})_i,\end{aligned}$$

where $(A^{-1})_i$ denotes the ith column of A^{-1}. Then since $A^{-1} \geq 0$ by condition (N_{38}) of Theorem 6.2.3, it follows from

$$\hat{x}_k = x_k + \delta(A^{-1})_{ki}, \qquad 1 \leq k \leq n,$$

that $\hat{x}_i > x_i$ and, also, none of the outputs decrease. Moreover by Lemma 3.14,

$$\delta(A^{-1})_{ki} \leq \delta(A^{-1})_{ii}, \qquad 1 \leq k \leq n,$$

so that x_i increases by the greatest amount, although other outputs may increase by the same amount. ■

(3.17) Corollary If the input matrix T for an open model is irreducible and $Te \ll e$, then if the demand of commodity i alone increases all the outputs increase and the output of commodity i increases by the greater amount.

Proof The result follows since $A^{-1} \gg 0$ and since strict inequality holds in (3.15) for each i and k, $k \neq i$. ■

Now since the linear systems (3.5) and (3.8) have dual analyses, one would expect that a similar relationship between the values added and the prices will hold, except in this case column sums rather than row sums are considered. But if an open Leontief model is profitable, then no sector of the economy operates at a loss and moreover at least one sector operates at a profit. In terms of the input matrix T, this means that

$$\sum_{i=1}^{n} t_{ij} \leq 1, \qquad 1 \leq j \leq n, \tag{3.18}$$

with strict inequality in at least one j. This leads to the following.

(3.19) Theorem If the value added in commodity i alone of a feasible open Leontief model is increased, then none of the prices decrease and the price of commodity i increases by the greatest amount, although other prices may increase by the same amount.

Proof By Theorem 3.9 the model is profitable so that by (3.18) the column sums of the input matrix T for the model are at most one. Now letting $A = I - T$, and letting p and v denote the price and value added vectors, respectively, the linear system (3.7) can be written in the form

$$A^t p = v.$$

The proof is then completed in a manner similar to the proof of Theorem 3.16, by replacing A by A^t. ■

For the irreducible case, we give a result which is dual to Corollary 3.17. The proof is similar to that of Corollary 3.17 and is omitted.

(3.20) Corollary If the input matrix T for an open model is irreducible and $T^t e \ll e$, then if the value added in commodity i alone is increased, all of the prices increase and the price of commodity i increases by the greater amount.

Two remarks concerning Theorems 3.16 and 3.19 are now in order: (i) It is not necessarily true that an increase in demand for a single commodity forces a greatest increase in the supply of that commodity compared to others; and (ii) it is always true that an increase in the value added in a single commodity forces a greatest increase in the price in that commodity compared to others.

To illustrate (i), consider an open Leontief model with input matrix T given by

$$T = \begin{bmatrix} \frac{3}{7} & \frac{5}{7} \\ \frac{2}{7} & \frac{1}{7} \end{bmatrix}.$$

Then since the maximum column sum of T is less than one, the model is profitable and thus feasible, and the matrix

$$A = I - T = \tfrac{1}{7}\begin{bmatrix} 4 & -5 \\ -2 & 6 \end{bmatrix}$$

is a nonsingular M-matrix. Then

$$A^{-1} = \tfrac{1}{2}\begin{bmatrix} 6 & 5 \\ 2 & 4 \end{bmatrix},$$

which agrees with our assertions in this section. Note also that the sum of the entries in the first row of T is greater than one. Thus the model does not satisfy Theorem 3.16. In fact, if the demand for commodity 2 increases, by a single unit, then the supply vector increases by

$$\begin{bmatrix} \frac{5}{2} \\ 2 \end{bmatrix},$$

which is a greater increase in the supply of commodity 1 than of commodity 2.

On the other hand, Theorem 3.19 is obviously satisfied here since the column sums of T are less than one. Also, since T is irreducible, Corollary 3.20 holds, so that an increase in a single value added component results in an increase in each price component, with a greater increase in the given component. This is clearly illustrated by the fact that A^{-1} is strictly row diagonally dominant here.

Now consider the example developed in Section 2. The matrix T, given by (2.7), has all row sums as well as all column sums less than one. Then Corollary 3.17, as well as Corollary 3.20, is satisfied by this model, since T is irreducible.

4 THE CLOSED MODEL

If the outside sector of the open input–output model is absorbed into the system as just another industry, then the system will become a closed model. In such a model, final demand and primary input do not appear; in their place will be the input requirements and the output of the newly conceived industry. Here all commodities will now be intermediate in nature, since everything that is produced is produced only for the input requirements of the sectors or industries within the model itself.

Thus in the closed Leontief input–output system, the consumer or open sector will be regarded as a production sector. In the particular formulation of the closed model discussed here, the open sector's "inputs" are various consumption goods and services, and its "output" is labor. In this system, final demand and items such as employment and the wage rate are all treated as unknown and their equilibrium values are solved for simultaneously with the rest of the variables.

Because there is no variable determined from outside, we ask a different question in the closed system. *Given the technology of production, what are the equilibrium output and price levels such that there is no unsatisfied demand*? To analyze this problem we begin with an equation that looks similar to Eq. (3.5), $Ax = d$, where x_i is the output of the ith sector and d_i is the outside demand on the ith sector, and where $A = I - T$, where T is the input matrix for the model. However, final demands are now considered to be inputs of the consumer sector. Each component d_i of d is now related to a level of employment $\mathscr{E}$, so that consumers purchase an amount d_i which varies with the level of employment $\mathscr{E}$. To achieve this we define fixed technical coefficients c_i by

$$c_i = d_i/\mathscr{E},$$

so that

$$d_i = c_i\mathscr{E}.$$

Then assuming that the original open model contains $n - 1$ sectors, the input–output relations in the closed model are described by the $n - 1$ simultaneous linear equations

$$x_i = \sum_{j=1}^{n-1} t_{ij}x_j + c_i\mathscr{E}, \qquad i = 1,\ldots,n-1. \tag{4.1}$$

Now the total labor supply; that is, the level of employment $\mathscr{E}$, is the sum of the labor used by each sector,

$$\mathscr{E} = \sum_{i=1}^{n} L_i$$

where L_i is the amount of labor used by sector i, and where $i = n$ is the consumer sector. Now defining fixed labor input coefficients l_i by

$$l_i = L_i/x_i, \quad i = 1,\ldots,n-1 \qquad \text{and} \qquad l_n = L_n/\mathscr{E},$$

the total labor supply can be expressed in terms of a linear equation

$$\mathscr{E} = \sum_{j=1}^{n} l_jx_j + l_n\mathscr{E}. \tag{4.2}$$

Then attaching (4.2) to the linear system (4.1) as the last equation, we have

$$\begin{aligned}
x_1 &= t_{11}x_1 + \cdots + t_{1,n-1}x_{n-1} + c_1\mathscr{E},\\
x_2 &= t_{21}x_1 + \cdots + t_{2,n-1}x_{n-1} + c_2\mathscr{E},\\
&\;\;\vdots\\
x_{n-1} &= t_{n-1,1}x_1 + \cdots + t_{n-1,n-1}x_{n-1} + c_{n-1}\mathscr{E},\\
\mathscr{E} &= l_1x_1 + \cdots + l_{n-1}x_{n-1} + l_n\mathscr{E}.
\end{aligned}$$

To write this linear system in matrix notation, we let

$$T = \begin{bmatrix} t_{11} & t_{12} & \cdots & t_{1,n-1} & c_1 \\ t_{21} & t_{22} & \cdots & t_{2,n-1} & c_2 \\ \vdots & & & \vdots & \vdots \\ t_{n-1,1} & t_{n-1,2} & \cdots & t_{n-1,n-1} & c_{n-1} \\ l_1 & l_2 & \cdots & l_{n-1} & l_n \end{bmatrix}$$

Now for notational convenience we define

$$t_{in} = c_i, \quad i = 1,\ldots,n-1, \qquad \text{and} \qquad t_{nj} = l_j, \quad j = 1,\ldots,n.$$

We also define

$$x^t = (x_1,\ldots,x_{n-1},\mathscr{E})^t,$$

and let $x_n = \mathscr{E}$. Then the input–output relationships for the closed model just described are given by the system of linear equations

$$x = Tx, \tag{4.3}$$

and T is called the *input matrix* for the model. Or letting $A = I - T$, the linear system (4.3) can be written as

$$Ax = 0, \tag{4.4}$$

where 0 denotes the zero vector of dimension n. Since the linear system (4.4) is homogeneous, it either has only the trivial solution $x = 0$ or it has infinitely many solutions. As before, we shall be interested in solutions where the output coefficients x_i are nonnegative and, in this case, positive.

This closed Leontief input–output system takes into account the demand factors as well as supply factors. In a certain sense, the *level of employment*, $\mathscr{E}$, *represents the final demand on the system.* This final demand is not given but is determined with the other supply variables, the total outputs x_i of each of the $n - 1$ original industries. Thus by solving for the final demand, $\mathscr{E}$, and output requirements x_i simultaneously, the closed model takes into account both the impact of demand on supply and that of supply on demand. Thus the equilibrium output levels calculated from a closed model incorporate not only the outputs required to meet a given final demand but also the outputs required to meet the change in the final demand which is induced by changes in production.

Now let x be a vector of outputs produced by this system. Then since

$$\sum_{j=1}^{n} t_{ij} x_j$$

is the quantity of the ith input necessary to produce this output bundle, $y = Tx$ is the vector of inputs. But since outputs x_i produce the only source of inputs, the system cannot operate unless $y \leq x$. Moreover, production processes will be assumed to be irreversible so that x is necessarily nonnegative and, in fact, positive is our case. This leads to the following definition.

(4.5) Definition A closed Leontief model with input matrix T is said to be *feasible* if there exists some x such that

$$Tx \leq x, \qquad x \gg 0. \tag{4.6}$$

Such an x, if it exists, is called a *feasible output solution to the model.*

Now if the model is feasible, we search for some solution to the model, which we are willing to consider an output equilibrium. With this in mind we give the following.

(4.7) Definition A vector x is called an *output equilibrium vector* for a closed Leontief model with input matrix T if

$$Tx = x, \qquad x \gg 0. \tag{4.8}$$

It follows from Definition 4.7 then that a closed input–output model has an output equilibrium vector if and only if the homogeneous system (4.4) has a positive solution; that is,

$$Ax = 0, \qquad x \gg 0, \tag{4.9}$$

is consistent. In this case the model is then feasible. However, we will see that a feasible model does not necessarily possess an output equilibrium vector.

It turns out that the theory of singular M-matrices, developed in Chapter 6, is quite useful in analyzing the feasibility of a closed Leontief model. Recall that A is an M-matrix with "*property* c" provided that A has a representation $A = sI - B$, $s > 0$, $B \geq 0$, where the powers of B/s converge to some matrix (see Definition 6.4.10).

(4.10) Theorem A closed Leontief model with input matrix T is feasible only if $A = I - T$ is an M-matrix with "property c."

Proof By Definition 4.5, the model is feasible only if (4.6) holds, so that $Ax \geq 0$ for some $x \gg 0$. But since $A \in Z^{n \times n}$, it follows that A is an M-matrix with "property c," by Exercise (6.4.14). ∎

Now by applying Theorem 6.4.12 and Exercise 6.4.13, we have the following.

(4.11) Corollary A closed Leontief model with input matrix T is feasible only if $A = I - T$ satisfies one, and consequently all, of conditions (A_1)–(F_{13}) of Theorem 6.4.12, and there exists a symmetric positive definite matrix W such that

$$AW + WA^t$$

is positive semidefinite.

However, the converse of Theorem 4.10 does not hold; that is, $A = I - T$ may be an M-matrix with "property c" while there is no $x \gg 0$ with $Ax \geq 0$. Such is the case for a closed Leontief model with input matrix

$$T = \begin{bmatrix} 1 & 1 \\ 0 & 0 \end{bmatrix},$$

for then setting

$$A = I - T = \begin{bmatrix} 0 & -1 \\ 0 & 1 \end{bmatrix}$$

it follows that there is no $x \gg 0$ with $Ax \geq 0$.

In order to further study feasible Leontief models and to investigate the existence of an output equilibrium vector, it will be convenient to permute T into a special block triangular form. If T is any square, reducible matrix, then there is a permutation matrix P for which $P\,TP^t$ is in *reduced triangular block form*:

$$(4.12)\qquad PTP^t = \begin{bmatrix} T_{11} & & & 0 \\ T_{21} & T_{22} & & \\ \vdots & \vdots & \ddots & \\ T_{k1} & T_{k2} & \cdots & T_{kk} \end{bmatrix},$$

where each block T_{ii} is either square and irreducible, or a 1×1 null matrix. For our purposes here, it is convenient to define a 1×1 null matrix to be irreducible, and if T is irreducible, we write simply $T = (T_{11})$. We first establish the following.

(4.13) Lemma Suppose the input matrix T of a closed Leontief model is irreducible and let $A = I - T$. Then the following statements are equivalent:

(1) The model is feasible.

(2) The model has an output equilibrium vector x which is unique up to positive scalar multiples.

(3) A is an M-matrix with "property c."

Proof That (1) implies (3) was established in Theorem 4.10. Now assuming (3), it follows that there exists $x \gg 0$ with $Ax = 0$ by Theorem 6.4.16 (2). But since T is irreducible and $Tx = x$, it follows that x is unique up to positive scalar multiples by the Perron–Frobenius theorem, Theorem 2.1.4. Thus (2) holds. Finally, the implication (2) implies (1) is clear from Definitions 4.5 and 4.7. ∎

Next, we investigate feasible Leontief models in terms of the reduced normal form (4.12) of the input matrix.

(4.14) Theorem Let T be the input matrix for a closed Leontief model and suppose that PTP^t has the reduced triangular block form (4.12). Let $A = I - PTP^t$ be partitioned conformally with PTP^t. Then the model is feasible if and only if A is an M-matrix and for each i, $1 \le i \le k$,

$$(4.15)\qquad A_{ii} \text{ singular} \quad \text{implies} \quad T_{ij} = 0 \qquad \text{for all} \quad i \ne j.$$

Proof Assume first that the model is feasible. Then A is an M-matrix by Theorem 4.10, and there exists a partitioned vector $x = (y_1, \ldots, y_k)^t \gg 0$ such that $PTP^tx \le x$, with $PAP^tx \ge 0$, so that for $A_{ij} = I - T_{ij}$,

$$(4.16)\qquad \sum_{j<i} A_{ij} y_j \ge 0 \qquad \text{for all} \quad 1 \le i \le k.$$

Because $A_{ij} \leq 0$ for all $j < i$, it follows that $A_{ii} y_i \geq 0$ for all $1 \leq i \leq k$. But then if A_{ii} is singular, it follows that $A_{ii} y_i = 0$ from Theorem 6.4.16 (5), since A_{ii} is irreducible. It follows further that $A_{ij} y_j = 0$ for all $j < i$, since $A_{ij} \leq 0$. Thus $A_{ij} = 0$ for all $j \neq i$ and consequently $T_{ij} = -A_{ij} = 0$ for all $j \neq i$, so that (4.15) holds.

Conversely assume that $A = I - PTP^t$ is an M-matrix and that (4.15) holds. If A_{ii} is singular then by construction, A_{ii} is a singular irreducible M-matrix, and as such, there exists $y_i \gg 0$ for which $A_{ii} y_i = 0$ by Theorem 6.4.16(2). Similarly, if A_{ii} is nonsingular, then A_{ii} is a nonsingular M-matrix since A is an M-matrix, so that there exists $y_i \gg 0$ for which $A_{ii} y_i \gg 0$ by Theorem 6.2.3, part (I_{27}). Recalling that (4.15) holds, it follows that the y_i's can be scaled so that $x = (y_1, \ldots, y_k)$ satisfies $x \gg 0$, and (4.16) holds. Thus $Ax \geq 0$ so that $PTP^t x \leq x$ and consequently, $T(P^t x) \leq P^t x$, $P^t x \gg 0$ and so the model is feasible. ■

As mentioned earlier, a feasible closed Leontief model does not necessarily have an output equilibrium vector. For example, if T is the input matrix and $A = I - T$ is a nonsingular M-matrix, then the model is feasible since there exists $x \gg 0$ with $Ax \geq 0$, by Theorem 6.2.3, part (I_{27}); but of course there is no x with $Ax = 0$.

Our final result characterizes those closed Leontief models possessing an output equilibrium vector, in terms of the reduced triangular block form (4.12) for the input matrix T.

(4.17) Theorem Let T and A be as in Theorem 4.14. Then the model has an output equilibrium vector if and only if A is an M-matrix and for each i, $1 \leq i \leq k$,

$$(4.18) \qquad A_{ii} \text{ is singular} \quad \text{if and only if} \quad T_{ij} = 0 \qquad \text{for all} \quad j \neq i.$$

Proof If the model has an output equilibrium vector then the model is feasible; thus by Theorem 4.14, A is an M-matrix and A satisfies (4.15). Now suppose that i is an index such that $T_{ij} = 0$ for all $i \neq j$. Then by assumption there is a partitioned vector $x = (y_1, \ldots, y_k) \gg 0$ such that $Ax = 0$. Then $A_{ii} y_i = 0$ and thus A_{ii} is singular, establishing (4.18).

For the converse, assume that A is an M-matrix and that (4.18) holds. Suppose first that i is an index such that A_{ii} is singular. Then we can choose $y_i \gg 0$ so that $A_{ii} y_i = 0$, since A_{ii} is a singular irreducible M-matrix. In addition, $A_{ij} = 0$ for all $i \neq j$ by (4.18). We now assume, without loss of generality, that the blocks A_{ii} have been ordered so that all the singular blocks come first; that is, $A_{11}, \ldots, A_{gg}$ are all the singular diagonal blocks of A. Let $y_i \gg 0$ be such that $A_{ii} y_i = 0$, $i = 1, \ldots, g$. Now if $g = k$, the result holds.

If $k > g$ define

$$y_h = A_{hh}^{-1} \sum_{j=1}^{h-1} A_{hj} y_j, \qquad h = g+1, \ldots, k.$$

Now $A_{hh}^{-1} \gg 0$ since A_{hh} is a nonsingular irreducible M-matrix. Moreover $A_{hj} \neq 0$ for some $1 \leq j \leq h-1$ by (4.18). Thus by induction, all the vectors y_j are positive, and moreover,

$$\sum_{j=1}^{h} A_{hj} y_j = 0, \qquad h = g+1, \ldots, k.$$

Consequently, letting $x = (y_1, \ldots, y_k)$, we see that $x \gg 0$ and $Ax = 0$. Thus $TP^t x = P^t x$ so that $P^t x$ is an output equilibrium vector for the model, completing the proof. ■

We remark that an alternate proof of Theorem 4.17 can be given by using the concepts of basic and final classes of a nonnegative matrix (see Theorem 2.3.10).

This section is concluded by noting that if the input matrix T for a feasible closed Leontief model is irreducible, then the output equilibrium vector x can be computed by the methods discussed in Chapter 8 for computing the stationary distribution vector associated with an ergodic Markov chain. In particular, for any $\alpha > 0$, the powers of $T_\alpha = (1 - \alpha)I + \alpha T$ converge, by Theorem 8.4.9, to a matrix L. Moreover in this case

$$L = xe, \qquad e = (1, \ldots, 1),$$

where

$$Tx = x, \qquad x \gg 0.$$

Then the iterative procedure

$$x^{k+1} = Tx^k, \qquad k = 0, 1, \ldots, \tag{4.19}$$

converges to x for any $x^0 = e_i$, the ith unit vector. Here Theorem 8.4.32 can often be used to choose α in such a way as to optimize the asymptotic convergence rate of (4.19).

Moreover, we note that the direct computational methods discussed in Exercises 8.5.18 and 8.5.19 can also be modified in this case in order to compute an output equilibrium vector for a closed Leontief model, where the input matrix T is once again assumed to be irreducible.

Finally, suppose that the input matrix T for a closed Leontief model is irreducible and has the property that

$$\sum_{i=1}^{n} t_{ij} = 1, \qquad 1 \leq j \leq n. \tag{4.20}$$

Then since T^t is then a stochastic matrix, the entire analysis developed in Chapter 8 for the stationary distribution vector associated with an ergodic Markov chain carries over here to the study of the output equilibrium vector, with T replaced by T^t. In economic terms, condition (4.20) means simply that each sector of the economy is in equilibrium in terms of internal factors.

5 EXERCISES

(5.1) Consider an open Leontief model with input matrix T given by

$$T = \begin{bmatrix} 0.2 & 0.3 & 0.2 \\ 0.4 & 0.1 & 0.2 \\ 0.1 & 0.3 & 0.2 \end{bmatrix}.$$

Show that if $A = I - T$ then A^{-1} is approximated by

$$A^{-1} \approx \frac{1}{0.38} \begin{bmatrix} 0.66 & 0.30 & 0.24 \\ 0.34 & 0.64 & 0.24 \\ 0.21 & 0.27 & 0.60 \end{bmatrix},$$

where the entries are rounded to two decimal places.

(5.2) Show that the model with input matrix T described in Exercise 5.1 is feasible and compute the outputs x_1,x_2, and x_3 associated with the open demands

$$d_1 = 10, \qquad d_2 = 5, \qquad d_3 = 6.$$

(5.3) Show that Corollaries 3.17 and 3.20 apply to the model given in Exercise 5.1.

(5.4) Determine if the open Leontief model with input matrix T given by

$$T = \begin{bmatrix} 0.5 & 0.6 \\ 0.3 & 0.7 \end{bmatrix}$$

is feasible.

(5.5) Show that if an open Leontief model with input matrix T is feasible, then the sum of the entries in at least one column of T is less than one.

(5.6) Consider an open Leontief model in which each sector supplies exactly one other sector and in which each sector has exactly one supplier. Show by example that the input matrix for such a model is not necessarily irreducible.

(5.7) Consider a closed Leontief model whose input–output relationships are described by the simultaneous equations

$$x_1 = 0.2x_1 + 0.2x_2,$$

$$x_2 = 0.3x_2 + 0.3\mathscr{E},$$

$$\mathscr{E} = 0.8x_1 + 0.5x_2 + 0.7\mathscr{E},$$

where x_1 and x_2 are the outputs of the two internal sectors and where $\mathscr{E}$ is the level of employment.

Show that the model described here is feasible.

(5.8) Show that the closed model described in Exercise 5.7 has an output equilibrium vector, $(x_1, x_2, \mathscr{E})^t$, which is unique up to positive scalar multiples. Compute the output equilibrium vector where $x_1 + x_2 + \mathscr{E} = 1$.

(5.9) Give an example of a closed Leontief model having at least two linearly dependent output equilibrium vectors.

(5.10) Show that a closed Leontief model having a symmetric input matrix T is feasible if and only if $A = I - T$ is an M-matrix. Show by example, however, that such a model need not have an output equilibrium vector and give necessary and sufficient conditions for such a vector to exist in terms of the reduced normal form (4.12) for T.

(5.11) Let A be a nonsingular M-matrix and consider the linear systems $Ax_i = d_i$, for vectors $d_i > 0$, $i = 1,2$, and let

$$\Delta d = d_2 - d_i, \qquad \Delta x = x_2 - x_1.$$

Then $x_i > 0$, $i = 1,2$, and the following assertions hold

(a) $\Delta d \neq 0 \leftrightarrow \Delta x \neq 0$;
(b) $\Delta d > 0 \rightarrow \Delta x > 0$;
(c) $\Delta d < 0 \rightarrow \Delta x < 0$;

and moreover $\Delta x \gg 0$ and $\Delta x \ll 0$, respectively, in (b) and (c) if A is irreducible. In view of the open Leontief model,

(a) means that the production of at least one commodity changes if and only if the demand for at least one commodity changes,

(b) means that if the demand for at least one commodity increases then the production of at least one commodity increases (all increase if A is irreducible), and

(c) means the same as (b) for a decrease in demand (Sierksma [1979]).

(5.12) In the notation of Exercise 5.11, assume in addition that A is irreducible and let $\Delta x_j, \Delta d_j$ denote the jth components of Δx and Δd, respectively.

Show that

(a) $\Delta x \gg 0 \leftrightarrow \Delta x_j > 0$ for each j with $\Delta d_j < 0$,
(b) $\Delta x \gg 0 \leftrightarrow \Delta x_j < 0$ for each j with $\Delta d_j > 0$.

In terms of the open Leontief model

(a) means that if the production increases of all commodities for which the demand decreases, then all the productions increase and
(b) means that if the production decreases of all commodities for which the demand increases, then all the productions decrease.

Note also that in view of Exercise 5.11, if the production increases of an industry in which the demand decreases, these must be an industry for which the demand increases (Sierkesma [1979]).

(5.13) Let $\alpha \in R$. Then a matrix $A \in R^{n \times n}$ will be called a matrix of class $M(\alpha)$ if

$$a_{jj} - a_{ij} = \alpha \qquad \text{for all} \quad i \neq j.$$

Thus a 3×3 matrix $A \in M(\alpha)$ has the form

$$A = \begin{bmatrix} a_{11} & a_{22} - \alpha & a_{33} - \alpha \\ a_{11} - \alpha & a_{22} & a_{33} - \alpha \\ a_{11} - \alpha & a_{22} - \alpha & a_{33} \end{bmatrix}.$$

Show that if $A \in M(\alpha)$ for some $\alpha \in R$, then

(a) $\det A = \alpha^{n-1} \left(\sum_{i=1}^{n} a_{ii} - (n-1)\alpha\right)$;
(b) $\alpha^{n-1} Ae = (\det A)e$ for $e = (1, \ldots, 1)^t$.

Show in addition that if $\alpha \neq 0$, then

$$A \in M(\alpha) \leftrightarrow \mathrm{Adj}(A) \in M((\det A)/\alpha)$$

(Sierksma [1979]).

(5.14) Let A be an irreducible, nonsingular M-matrix satisfying $Ae \gg 0$. Show that in the notation of Exercises 5.11–5.13, the following statements are equivalent:

(a) $A \in M(\alpha)$ for some $\alpha \in R$;

(b) $\Delta x_i < \Delta x_j \leftrightarrow \Delta d_i < \Delta d_j$ for each Δd with $\Delta d_i \neq 0$ and $\Delta d_j \neq 0$ and each i and j;

(c) $\Delta x_m = \max_i \Delta x_i \leftrightarrow \Delta d_m = \max_i \Delta d_i$, for each $\Delta d \neq 0$.

(d) $x_m = \min_i \Delta x_i \leftrightarrow \Delta d_m = \min_i \Delta d_i$, for each $\Delta d \neq 0$.

In terms of the open Leontief model, (c) and (d) assert that if the input matrix T is irreducible and satisfies $Te \ll e$, then the maximal and minimal changes in the demand and production will always occur in the same industry if and only if $A \in M(\alpha)$, for some $\alpha \in R$; where $A = I - T$. Note also that this exercise relates Theorems 3.16 and 3.19 (Sierksma [1979]).

(5.15) Verify Exercise (5.14) for the input matrix

$$T = \begin{bmatrix} 0 & \frac{3}{10} & \frac{3}{5} \\ \frac{1}{5} & \frac{1}{10} & \frac{3}{5} \\ \frac{1}{5} & \frac{3}{10} & \frac{2}{5} \end{bmatrix}$$

by showing that $A = I - T$ is in $M(\alpha)$ for some $\alpha \in R$ and by computing Δx for $\Delta d = (12, -12, 24)^t$ and noting that (b),(c), and (d) then hold for this pair.

6 NOTES

(6.1) Leontief presented the first version of his input–output model in 1936, but developed the topic more fully in *The Structure of the American Economy* [1941]. A dynamic version was later presented by Leontief [1953], and a brief summarizing presentation of it was given by him [1966, Chapter VII]. The input–output literature is voluminous and an early bibliography has been compiled by Taskier [1961]. For elementary descriptions of input–output models, see Miernyk [1965] and Yan [1969]. A sophisticated analytical treatment of the subject has been given by Schumann [1968].

(6.2) The notation and terminology adopted in this chapter are along the lines of that used by Sarma [1977] and Yan [1969]. Many of the results in Sections 3 and 4 are based upon properties of nonsingular and singular M-matrices, respectively, developed in Chapter 6. Lemma 3.14 is essentially given in Seneta [1973, Theorem 2.3, p. 29], for the irreducible case. Also Theorem 3.16 is his Exercise 2.2, p. 34. Other results in Section 2 are fairly obvious in the terminology of nonsingular M-matrices.

The first input–output model presented by Leontief [1936] was a closed model. The particular "form" of the closed model developed in Section 4 is essentially along the lines of that given by Yan [1969, Chapter III]. However, Theorems 4.10, 4.14, and 4.17, concerning positive solutions to singular M-matrix equations, are believed to be new. For excellent papers on related topics, see Schneider [1956] and Carlson [1963].

(6.3) Other approaches to the analysis of input–output models have been given in the literature. For example, Dorfman *et al.* [1958] have taken a linear programming approach to the analysis of open Leontief models. Here the optimizing solution is the one and only efficient solution possible for the output vector. In addition, Kemeny and Snell [1960, Section 7.7] have studied closed Leontief models by associating with them a finite homogeneous Markov chain and then studying the model by investigating the properties of this chain. (See the concluding remarks to Section 4.) On the other hand, a nonlinear version of the linear input–output open model of Leontief has been developed and studied by Sandberg [1974a,b]. In the nonlinear version, each product $t_{ij}x_j$ of the linear system (3.3) is replaced by a possibly nonlinear function $t_{ij}(x_j)$ of x_j which is continuously differentiable. This approach enables one to give a comparative analysis of certain classes of input–output models.

(6.4) Input–output analysis is perhaps the most significant, but certainly not the only, area in economics in which nonnegative matrices and M-matrices play an important role. Another important application is to the study of the *von Neumann model.* The model was developed by von Neumann [1945/46] to introduce the concept of equilibrium growth. It provided the first proof that a solution to a general equilibrium model existed, and it was the first programming model. The main difference between von Neumann's model and the closed Leontief model is that in Leontief's model, the methods of production are given a priori, and in von Neumann's model, a larger number of possible production methods are available; one of the problems, then, is to choose the best technique. These production possibilities can be described by two nonnegative matrices A and B, and one of the mathematical problems is to find the maximum scalar $\alpha > 0$ such that

$$\alpha Az \leq Bz, \qquad z \gg 0.$$

As expected, the Perron–Frobenius theory of nonnegative matrices given in Chapter 2 plays an important role in the study of such problems. (See Hansen [1970, Chapter 16].)

An interesting application of the theory of M-matrices to the study of world goods prices has been developed. In a special reference to the protection of domestic factor owners by tarriff, Stolper and Samuelson [1941] have studied the effect on factor rewards of a change in world goods prices. Here the noted *Stolper–Samuelson theorem asserts that a rise in the price of a good entails a rise in the price of its associated factor and a fall in the prices of all other factors.* A matrix $A \in \mathscr{R}^{n \times n}$ is said to satisfy the *Stolper–Samuelson condition* if A^{-1} is an M-matrix. Thus in this case, $A \geq 0$ and all the principal minors of A are positive (see Theorem 6.2.3). It turns out that the factor price situation just discussed is ensured whenever the Jacobian of a certain nonlinear function satisfies the Stolper–Samuelson condition. (For further discussions of this and related problems and an extensive bibliography, see Uekawa *et al.* [1972].)

We mention also that additional applications of nonnegativity to economics can be found in Nikaido [1968] and in survey papers by Maybee and Quirk [1969] and by Johnson [1974b].

(6.5) Since the completion of the main body of this chapter the authors have become aware of several new and interesting applications of nonnegative matrices to the open Leontief model, by Sierksma [1979]. Some of these applications have been included in the exercises.

CHAPTER 10

THE LINEAR COMPLEMENTARITY PROBLEM

1 INTRODUCTION

As an example of the role of the matrices studied in this book in mathematical programming we study the *linear complementarity problem* (LCP): for a given $r \in R^n$ and $M \in R^{n \times n}$ find (or conclude there is no) $z \in R^n$ such that

(1.1) $$w = r + Mz,$$

(1.2) $$w \geq 0, \qquad z \geq 0, \qquad z^t w = 0.$$

This problem will be denoted by the symbol (r,M).

Problems in (linear and) convex quadratic programming, the problem of finding a Nash equilibrium point of a bimatrix game (e.g., Cottle and Dantzig [1968] and Lemke [1965]), and also a number of free boundary problems of fluid mechanics (e.g., Cryer [1971]) can be posed in the form of Eqs. (1.1) and (1.2). In the convex quadratic programming problems the matrix M of (1.1) is (not necessarily symmetric) positive semidefinite. In the bimatrix game problems it can be taken to be nonnegative and in the free boundary problems it belongs to the class $Z^{n \times n}$, or briefly Z, of matrices with nonpositive off-diagonal elements.

The main two algorithms which have been developed for solving (r,M) are the principal pivoting method of Dantzig and Cottle [1967] and the complementary pivot algorithm of Lemke [1968]. The LCP (r,M) has a unique solution for every $r \in R^n$ if and only if all the principal minors of M are positive ($M \in P$). We prove this result in Section 2. For $M \in P$ both algorithms just mentioned converge to the unique solution. We prove the convergence of the principal pivoting method in Section 2. The complementary pivot algorithm is described, as are additional results and generalizations of LCP, in Section 7.

A matrix M is a Q-matrix ($M \in Q$), if (r,M) has a solution for every $r \in R^n$. ($P \subseteq Q$.) In Section 3 it is shown that a nonnegative matrix is in Q if and only if its diagonal elements are positive.

The set of feasible vectors z associated with (r,M) is

$$X(r,M) = \{z \in R^n_+ ; r + Mz \geq 0\}.$$

In Section 4 it is shown that the off-diagonal elements of M are nonpositive ($M \in Z$) if and only if $X(r,M)$ has a least element which is a solution of (r,M) for each r such that $X(r,M) \neq \phi$ and that M is a nonsingular M-matrix if and only if $X(r,M)$ has a least element which is the unique solution of (r,M). The theory of least elements is related to the question of whether the LCP (r,M) can be solved by one linear program.

The parametric linear complementarity problem (PLCP) and linear complementarity problem with upper bounds are defined in Section 5. In terms of these problems, characterizations are given for matrices in P to be nonsingular M-matrices.

The exercises are grouped in Section 6.

2 P-MATRICES

A square matrix is a *P*-matrix if all its principal minors are positive. We denote by P the class of P-matrices. It includes the positive definite matrices, the totally positive matrices (Exercise 2.6.25), the matrices that satisfy condition (H_{24}) in Theorem 6.2.3 (*"diagonally" stable matrices*), and the nonsingular M-matrices. Note that P is the class of matrices which satisfy (A_1)–(A_6) in Theorem 6.2.3.

A very interesting characterization of P-matrices can be added to those listed in Theorem 6.2.3, a characterization stated in terms of LCP. We start by describing the *principal pivoting method* and showing that it solves (r,M) if $M \in P$.

We rewrite (1.1) as

$$Iw - Mz = r. \tag{2.1}$$

(2.2) Definitions Variable pairs (w_i,z_i) are called *complementary*. The variables w_i and z_i are the complements of each other. A pair (w,z) of vectors in R^n is a *complementary solution* of (2.1) provided $z_i w_i = 0$, $i = 1, \ldots, n$. A *basic set of variables* consists of any ordered set of n variables w_i and z_i, such that their coefficient matrix in (2.1), called the *basis*, is nonsingular. A *complementary set of variables* is one which contains exactly one variable of each complementary pair (w_i,z_i). A *basic solution* is the one found by solving for the values of a given set of basic variables when the nonbasic variables are set equal to zero.

(2.3) Examples $\{w_1,\ldots,w_n\}$ is a complementary basic set of variables. The corresponding basic solution is $z = 0$, $w = r$. The variables $\{z_1,\ldots,z_n\}$ form a complementary basic set if and only if M is nonsingular. If this is the case, the basic solution is $w = 0$, $z = M^{-1}r$.

(2.4) Definitions Let B be a basis corresponding to a *complementary* basic set of variables $\{y_1,\ldots,y_n\}$ (where y_i and x_i are complementary). Multiplying through (2.1) on the left by B^{-1} yields a *principal pivot transform* of (2.1) of the form $y = Ax + b$. A is called a *principal pivot transform* of M.

The validity of the algorithm, to be described, depends on the fact that the class P is invariant under principal pivotal transforms. We leave it for the reader to prove (Exercise 6.1) that

(2.5) Theorem If A is a principal pivot transform of a P-matrix, then $A \in P$.

From this we conclude the following.

(2.6) Theorem Let M in (2.1) be a P-matrix. Let y be a complementary basic set of variables of (2.1), expressed in terms of the nonbasic variables $x,y = Ax + b$. Then the component y_s is a strictly increasing function of x_s, $s = 1,\ldots,n$.

Proof The diagonal element a_{ss} is positive since A, like M, is a P-matrix. ■

(2.7) Theorem Let $M \in P$. If increasing the value of the nonbasic variable x_s causes the basic variable y_r to decrease, then after replacing y_r by x_r in the basic set (which can be done since $M \in P$), the rth basic variable will increase with the sth nonbasic variable.

Proof Let $y = Ax + b$ be the relation between the basic and nonbasic variables before the change of x_r and y_r. Then $a_{rs} < 0$ and $a_{rr} > 0$. Let $\hat{y} = \hat{A}\hat{x} + \hat{b}$ be the relation between basic and nonbasic variables after the change of x_r and y_r. Then $\hat{a}_{rr} = (1/a_{rr}) > 0$ and $\hat{a}_{rs} = -(a_{rs}/a_{rr}) > 0$. ■

(2.8) Definition A basic solution is *degenerate* if it has zero as a basic variable.

(2.9) Example If in (2.1), $r_i = 0$, for some i, then the basic solution $z = 0$, $w = r$ is degenerate.

To avoid problems of degeneracy, we shall use the standard device of replacing the vector r in (2.1) by the matrix $(r|I)$ and regarding the components of w and z as vectors.

(2.10) Definition A nonzero vector is *lexico-positive* (*lexico-negative*) if its first nonzero component is positive (negative).

The vector values of any set of basic variables of (2.1) with basis B are given by the rows of $(B^{-1}r|B^{-1})$. Since the rows of B^{-1} are linearly independent this implies the next theorem.

(2.11) Theorem All basic solutions of (2.1) are nondegenerate in the lexicographic sense.

All ordering relations in the sequel should be interpreted in the lexicographic sense.

(2.12) Theorem Consider the matrix system

$$w = Mz + (r|I) \qquad (w \text{ and } z \text{ are } n \times (n+1) \text{ matrices}). \tag{2.13}$$

If all but the first components of all but one nonbasic, say x_s, variables are zeros, then at most one basic variable can be a zero vector.

Proof Let $y = Ax + b$ be a principal pivot transform of (2.13) (y, x, and b are $n \times (n+1)$ matrices). By the argument preceding Theorem 2.11, the rows of the matrix obtained from b by deleting the first column are linearly independent. By the assumption on x, the rows of the matrix obtained from Ax by deleting the first column are proportional. Thus y can have at most one zero row. ■

(2.14) Theorem If $M \in P$, then (r,M) has a solution for every $r \in R^n$.

Proof The proof is constructive. It is based on the *principal pivot method* which will now be described. The algorithm consists of *major cycles.* Such a cycle is initiated with a complementary basic solution $(y,x) = (b,0)$ of (2.1). (In the first major cycle $(y,x) = (w,z) = (r,0)$.) *If* $y \geq 0$, (y,x) *is the solution of* (r,M). If not, suppose that $y_s < 0$. By Theorem 2.6 an increase of x_s increases y_s as well. However the values of positive basic variables may decrease. The major cycle has two steps:

Step I Increase x_s until it is *blocked* by a (lexico-) positive basic variable decreasing to zero or by the negative y_s increasing to zero.

Step II Replace the blocking variable in the basic set by its complement. If the blocking variable is y_s, initiate a new major cycle with the new complementary basic solution. Otherwise return to Step I.

To prove the validity of the algorithm notice that, by Theorem 2.12, x_s cannot be blocked at the same time by y_r and y_t, $r \neq t$, and that if x_s is blocked by y_r, $r \neq s$, then by Theorem 2.7, interchanging x_r and y_r permits the further increase of x_s. Thus during a major cycle, the number of negative basic variables is decreased and so there are at most n such cycles. Finally, the number of iterations within a major cycle is finite since the number of basic sets is finite and no basis can be repeated with a larger value of x_s. This proves the convergence of the algorithm to a solution of (r,M). ■

If $M \in P$ the solution of (r,M) is unique. In fact the following is true.

(2.15) Theorem M is a P-matrix if and only if (r,M) has a unique solution for every $r \in R^n$.

Proof If $M \in P$ then by the previous theorem, (r,M) has a solution for every r. To show that the solution is unique we use the equivalence of conditions (A_1) and (A_6) in Theorem 6.2.3, namely, an $n \times n$ matrix M is *not* a P-matrix if there exists a nonzero vector x such that $x_i(Mx)_i \leq 0$, $i = 1, \ldots, n$. Suppose that $(w^{(1)}, z^{(1)})$ and $(w^{(2)}, z^{(2)})$ are two distinct solutions of (r,M). Then

$$w^{(1)} - w^{(2)} = M(z^{(1)} - z^{(2)}), \quad z^{(1)} - z^{(2)} \neq 0$$

and

$$w^{(1)t}z^{(1)} = w^{(2)t}z^{(2)} = 0.$$

Let $x = z^{(1)} - z^{(2)}$. Then $x_i(Mx)_i \leq 0$, $i = 1, \ldots, n$, contradicting the assumption that $M \in P$.

Conversely, suppose M is not a P-matrix. Then there exists $x \neq 0$ such that $x_i y_i \leq 0$, $i = 1, \ldots, n$, where $y = Mx$. Denoting

$$v_i^+ = \operatorname{Max}\{v_i, 0\}, \quad v_i^- = -\operatorname{Min}\{v_i, 0\}, \quad y^+ = (y_i^+), \quad y^- = (y_i^-), \quad x^+ = (x_i^-),$$

and

$$x^- = (x_i^-),$$

we see that

$$y = y^+ - y^-, \quad x = x^+ - x^-, \quad y_i^+ y_i^- = 0, \ i = 1, \ldots, n, \quad x_i^+ x_i^- = 0, \ i = 1, \ldots, n,$$

and

$$y^+ \geq 0, \qquad y^- \geq 0, \qquad x^+ \geq 0, \qquad x^- \geq 0. \tag{2.16}$$

Since $x_i y_i \leq 0$, $i = 1, \ldots, n$, it follows that $x_i^+ y_i^+ = x_i^- y_i^- = 0$, $i = 1, \ldots, n$. So

$$y^{+t}x^+ = y^{-t}x^- = 0. \tag{2.17}$$

Since $y = Mx$,

$$y^+ - Mx^+ = y^- - Mx^-. \tag{2.18}$$

Call this common value $\hat{r}$, then by (2.16)–(2.18), (y^+,x^+) and (y^-,x^-) solve $(\hat{r},M)$, but since $x \neq 0$ these are two distinct solutions. ■

The previous characterization has the following geometric interpretation.

(2.19) Definitions A *complementary set of column vectors* consists of the columns of the matrix $(I|-M)$ of (2.1) which correspond to a complementary set of variables. The polyhedral cone generated by these columns is a *complementary cone* (*of* M).

A complementary cone has a nonempty interior if and only if its generators are the columns of a basis. Thus all the 2^n complementary cones of M have nonempty interiors if and only if all the principal minors of M are nonzero. (This is equivalent to "(r,M) has a finite number of solutions for every $r \in R^n$"; see Exercise 6.3.) A solution of (r,M) represents r as a nonnegative linear combination of the generators of some complementary cone and every such combination defines a solution of (r,M). Thus the union of all the complementary cones is the set of all r for which (r,M) has a solution. If this union is all of R^n, the cones have nonempty interiors and no two cones have a common interior point, then the cones form a parition of R^n.

(2.20) Example The four complementary cones of

$$\begin{bmatrix} 1 & 1 \\ 0 & 1 \end{bmatrix}$$

form a partition of R^2 as seen in the accompanying figure.

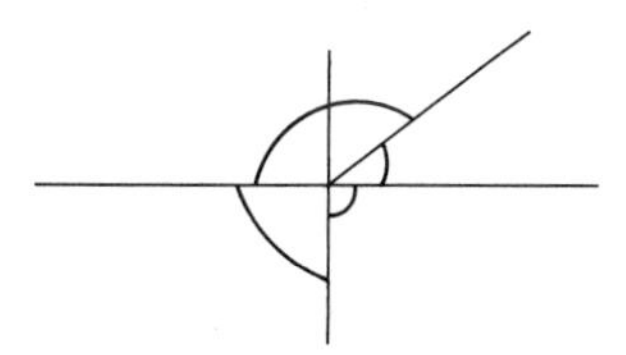

The translation of Theorem 2.15 to this geometric terminology is the following.

(2.21) Theorem The set of complementary cones of M forms a partition of R^n if and only if M is a P-matrix.

3 Q-MATRICES

In this section we study matrices M for which (r,M) has at least one solution for every $r \in R^n$.

(3.1) Definitions A square matrix M is a Q-matrix ($M \in Q$), if the LCP (r,M) has a solution for every $r \in R^n$. M is *regular* ($M \in R$) if the system

$$x > 0, \qquad t \geq 0,$$
$$x_k > 0 \rightarrow (Mx)_k + t = 0,$$
$$x_k = 0 \rightarrow (Mx)_k + t \geq 0$$

is inconsistent. M is *strictly semimonotone* ($M \in S$) if for every $x > 0$ there exists an index k such that $x_k > 0$ and $(Mx)_k > 0$.

Since M is a P-matrix if and only if for every $x \neq 0$, $x_k(Mx)_k > 0$ for some k, ($(A_1) \leftrightarrow (A_5)$ in Theorem 6.2.3), the inclusions

$$P \subseteq S \subseteq R$$

follow immediately from the definitions.

(3.2) Examples The matrices

$$\begin{bmatrix} 1 & 1 \\ 1 & 1 \end{bmatrix} \quad \text{and} \quad \begin{bmatrix} -1 & 1 \\ -2 & 1 \end{bmatrix}$$

demonstrate that the previous inclusions are proper.

We shall show that $R \subseteq Q$ and that here too the inclusion is proper. We start with two results on vector functions.

(3.3) Theorem Let S be a nonempty, compact and convex subset of R^{n+1} and let $F: S \rightarrow R^{n+1}$ be a continuous map on S. Then there exists $\bar{x} \in S$ such that

$$(x - \bar{x})^t F(\bar{x}) \geq 0 \qquad \text{for all} \quad x \in S. \tag{3.4}$$

Proof Consider the map Γ from S into the class of its subsets defined by

$$\Gamma(x) = \{u \in S;\ u^t F(x) = \operatorname*{Min}_{V \in C} v^t F(x)\}.$$

For each $x \in S$, $\Gamma(x)$ is a nonempty convex subset of S. The map Γ is upper semicontinuous on S and so by Kakutani's fixed point theorem (Kakutani [1941]), Γ has a fixed point $\bar{x}$, that is, $\bar{x} \in \Gamma(\bar{x})$; but this means exactly that $\bar{x}$ satisfies (3.4). ■

(3.5) Corollary Let $S = \{x \in R_+^{n+1};\ \sum_{i=1}^{n+1} x_i = 1\}$ and let $F: S \rightarrow R^{n+1}$ be continuous on S. Then there exists $\bar{x} \in S$ that satisfies (3.4) and

$$\bar{x}_k > 0 \rightarrow G_k(\bar{x}) = \operatorname*{Min}_i F_i(\bar{x}) = m, \tag{3.6}$$

$$\bar{x}_k = 0 \rightarrow F_k(\bar{x}) \geq m. \tag{3.7}$$

Proof The set S in the corollary satisfies the assumptions of the theorem. Inequality (3.7) follows trivially from (3.6). (3.6) follows from (3.4) since

$$\bar{x}^t F(\bar{x}) = \underset{x \in S}{\text{Min}}\, x^t F(\bar{x}) = \underset{i}{\text{Min}}\, F_i(\bar{x}) = m.$$

If for some $\bar{x}_k > 0$, $F_k(\bar{x}) > m$, then

$$\bar{x}^t F(\bar{x}) = \sum_{i=1}^{n+1} \bar{x}_i F_i(\bar{x}) > m,$$

a contradiction. ■

(3.8) Theorem If M is a regular matrix, then (r,M) has a solution for every $r \in R^n$.

Proof Consider the map $F : R_+^{n+1} \to R^{n+1}$ defined by

$$F\begin{bmatrix} z \\ s \end{bmatrix} = \begin{bmatrix} Mz + sr + se \\ s \end{bmatrix},$$

where e is a vector of ones in R^n ($z \in R_+^n$, $s \in R_+$). F is continuous on the set S defined in Corollary 3.5. Thus there exists

$$\bar{x} = \begin{bmatrix} \bar{z} \\ \bar{s} \end{bmatrix} \in S$$

satisfying (3.6) and (3.7). Now $\bar{s} > 0$ because $\bar{s} = 0$ implies

$$0 = \bar{s} = \bar{x}_{n+1} \geq m,$$

by (3.7), and

$$(M\bar{z})_i = m \leq 0 \qquad \text{if} \quad \bar{z}_i > 0,$$

$$(M\bar{z})_i \geq m \qquad \text{if} \quad \bar{z}_i = 0$$

contradicting the regularity of M. Thus $m = \bar{s} > 0$ and

$$\bar{z}_i > 0 \to (M\bar{z})_i + sr_i + s = s,$$

$$\bar{z}_i = 0 \to (M\bar{z})_i + sr_i + s \geq s,$$

so that $z = \bar{z}/\bar{s}$ and $w = Mz + r$ solve (r,M). ■

Notice that this is a (nonconstructive) proof of Theorem 2.14.

(3.9) Example Let

$$M = \begin{bmatrix} -1 & 2 & 2 \\ 2 & -1 & 2 \\ 2 & 2 & -1 \end{bmatrix}.$$

By verifying that the union of all eight complementary cones of M is R^3 it can be shown that $M \in Q$. However $M \notin R$ as is shown by choosing

$$x = \begin{bmatrix} 1 \\ 0 \\ 0 \end{bmatrix} \qquad \text{and} \qquad t = 1.$$

A characterization of Q-matrices can be given when the matrices in question are nonnegative.

(3.10) Theorem Let $M \geq 0$. M is a Q-matrix if and only if $m_{ii} > 0$ for each $i = 1, \ldots, n$.

Proof If all the diagonal elements are positive, then $M \in S \subset R \subset Q$. (This also follows from the complementary pivot algorithm; see Section 7.)

Suppose, conversely, that M has a zero diagonal element, say m_{11}. Let $\hat{r} = (-1,1,1,\ldots,1)^t$. Then since $M \geq 0$, $w_i > 0$, $i = 2, \ldots, n$, in every solution of $(\hat{r},M)$ so that $z_i = 0$, $i = 2, \ldots, n$. But then

$$w_1 = \sum_{i=1}^{n} m_{1i} z_i + \hat{r}_1 = -1 < 0$$

meaning that $(\hat{r},M)$ has no solution. ■

(3.11) Corollary If $M \geq 0$ is a Q-matrix then all principal submatrices of M are also Q-matrices.

(3.12) Corollary If $M \geq 0$, then the union of the complementary cones of M is the whole space R^n, if and only if $m_{ii} > 0$, $i = 1, \ldots, n$.

4 Z-MATRICES, LEAST ELEMENTS, AND LINEAR PROGRAMS

It is fairly obvious from the definition of LCP that if it has a solution it has one which is an extreme point of its "feasible set" (see Exercise 6.2). It is thus natural to ask when does this feasible set has a "least element" and when does this element solve the LCP. We define these concepts and use them in characterizing *Z-matrices* (matrices in class Z denoted by $Z^{n \times n}$ in Chapter 6).

(4.1) Definition A vector z is a *least element* of $X \subseteq R^n$ if $z \in X$ and $x \in X \rightarrow z \leq x$. (If X has a least element then $X \neq \varnothing$ and the least element is unique.)

(4.2) Examples Let

(a) $r^{(1)} = \begin{bmatrix} -1 \\ 0 \end{bmatrix}, \quad M^{(1)} = \begin{bmatrix} -2 & -3 \\ 1 & 1 \end{bmatrix},$

(b) $r^{(2)} = \begin{bmatrix} -1 \\ 2 \end{bmatrix}, \quad M^{(2)} = \begin{bmatrix} -2 & 1 \\ 3 & -1 \end{bmatrix},$

(c) $r^{(3)} = \begin{bmatrix} -1 \\ 4 \end{bmatrix}, \quad M^{(3)} = \begin{bmatrix} 3 & -1 \\ -2 & 1 \end{bmatrix},$

(d) $r^{(4)} = \begin{bmatrix} -1 \\ 6 \end{bmatrix}, \quad M^{(4)} = \begin{bmatrix} 1 & 1 \\ -2 & -3 \end{bmatrix},$

(e) $r^{(5)} = \begin{bmatrix} -1 \\ 8 \end{bmatrix}, \quad M^{(5)} = \begin{bmatrix} 0 & 1 \\ 1 & 0 \end{bmatrix}.$

The feasible set $X(r^{(1)},M^{(1)})$ is empty. The feasible sets of the other problems are drawn in the accompanying figures.

(b)

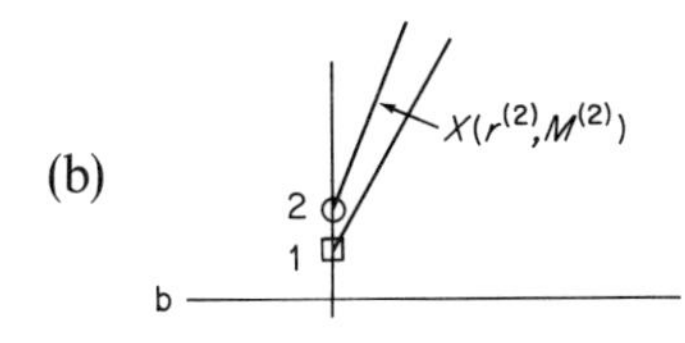

(c)

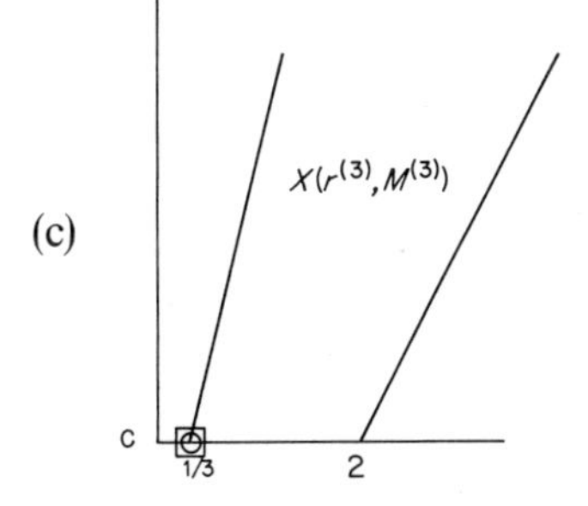

(d)

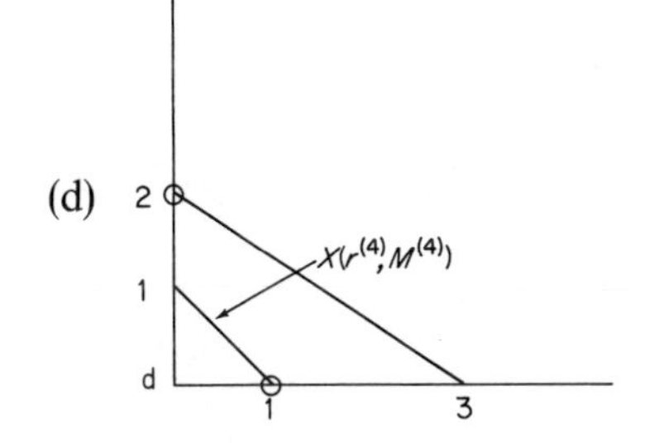

(e)

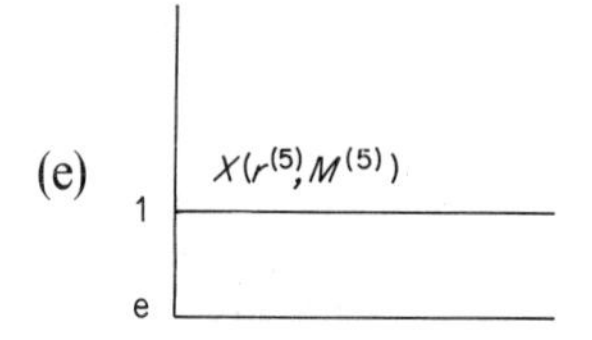

The solutions of the problems are circled and the least elements are in squares. Thus $X(r^{(4)},M^{(4)})$ does not have a least element. The least element of $X(r^{(2)},M^{(2)})$ does not solve $(r^{(2)},M^{(2)})$, the least element of $X(r^{(3)},M^{(3)})$ is a solution of $(r^{(3)},M^{(3)})$, and $(r^{(5)},M^{(5)})$ has no solution even though $X(r^{(5)},M^{(5)})$ is not empty.

We first consider polyhedral sets which have a least element. A polyhedral set in R^n is of the form

$$X_{A,b} = \{x \in R^n,\ Ax \geq b\}, \qquad \text{where} \quad A \in R^{m \times n} \quad \text{and} \quad b \in R^m.$$

(4.3) Definition If B is a *basis*, i.e., a nonsingular submatrix of A of order n (see Definition 2.2) and b_B is the corresponding subvector of b, then x is *determined by* B if $x = B^{-1}b_B$.

A vector $\bar{x}$ is the least element of $X_{A,b}$ if and only if it is an optimal solution, for every $c \geq 0$, of the linear program

$$\text{(4.4)} \qquad \text{minimize} \quad c^t x \qquad \text{subject to} \quad Ax \geq b.$$

Recall that the dual of (4.4) is

$$\text{(4.5)} \qquad \text{maximize} \quad b^t y \qquad \text{subject to} \quad A^t y = c \quad \text{and} \quad y \geq 0.$$

The existence of a least element of $X_{A,b}$ is related to the program (4.5) and to A being monotone in the sense of Chapter 5.

(4.6) Theorem (i) Equivalent conditions for

(a) the vector $\bar{x}$ is the least element of $X_{A,b}$,

are

(b) $\bar{x} \in X_{A,b}$ and A has a nonnegative left inverse C ($C \geq 0$, $CA = I$, implying rank A = rank $C = n$) such that $Cb = \bar{x}$;

(c) there exists an $n \times m$ matrix C such that $Cb = \bar{x}$ and $y(c) \equiv C^t c$ is an optimal solution of (4.5);

(d) There exists an $n \times m$ matrix C such that $Cb = \bar{x}$, $y(c)$ is a feasible solution of (4.5) for all $c \geq 0$, and $y(c)$ is an optimal solution of (4.5) for some positive c.

(ii) Sufficient conditions for (a) are

(e) $\bar{x} \in X_{A,b}$ and is determined by a monotone basis, (that is, $B^{-1} \geq 0$);
(f) $\bar{x}$ is determined by a monotone basis and the corresponding basic solution of (4.5) is optimal for some positive c.

(iii) If $A\bar{x} - b$ has at most n zero elements, then (e) and (f) are also necessary conditions.

Proof (a)$\leftrightarrow$(b): Let e_i be the ith row of the identity matrix of order n. A vector $\bar{x}$ is the least element of $X_{A,b}$ if, and only if, for each i, it minimizes $e_i x$, over $x \in X_{A,b}$. By the duality theorem, this is so if and only if $\bar{x} \in X_{A,b}$ and for each i there exists $y_i^t \in R_+^m$ such that $y_i A = e_i$ and $y_i b = \bar{x}_i$.

$$C = \begin{bmatrix} y_1 \\ \vdots \\ y_n \end{bmatrix}$$

clearly satisfies (b) so (a) $\rightarrow$ (b). The proof is clearly reversible.

(b) $\rightarrow$ (c): $\bar{x}$ and $y(c)$ are feasible solutions of (4.4) and (4.5), respectively, and $c^t\bar{x} = b^t y(c)$ for all $c \geq 0$. Thus by the duality theorem, $y(c)$ is an optimal solution of (4.5) for all $c \geq 0$.

(c) $\rightarrow$ (d): Trivial.

(d) $\rightarrow$ (b): $C \geq 0$ since $y(c) \geq 0$ for all $c \geq 0$. $CA = I$ since $y(c)A = c$ for all $c \geq 0$. For some positive vector $\bar{c}$, $y(\bar{c})$ is an optimal solution of (4.5). Thus there is, by the duality theorem, an x that is an optimal solution of (4.4) with $c = \bar{c}$. By complementary slackness, $(y(\bar{c}))^t(Ax - b) = 0$. Since $C \geq 0$ and $\bar{c}$ is positive, $(y(c))^t(Ax - b) = 0$ for all $c \geq 0$. This implies $c^t(x - Cb) = 0$ for all $c \geq 0$, so necessarily $x = Cb = \bar{x}$ whence $\bar{x} \in X_{A,b}$ completing the proof of (b).

We leave the proof of parts (ii) and (iii) for the exercises. ■

The classes of matrices defined next should not be confused with input–output matrices studied in Chapter 9. See also Note 7.5.

(4.7) Definition $A \in R^{m \times n}$ is *pre-Leontief* if each column of A has at most one positive element. A pre-Leontief matrix is *Leontief* if Ax is positive for some $x \geq 0$ and *totally Leontief* if, in addition, yA is positive for some $y \geq 0$.

If $A \in R^{m \times n}$ is a Leontief matrix, then rank $A = m$ and A has an $m \times m$ Leontief submatrix. Every square Leontief matrix is totally Leontief and monotone (Exercise 6.12).

The nonsingular M-matrices of order m are the $m \times m$ Leontief matrices with positive diagonal elements.

(4.8) Theorem (i) A necessary and sufficient condition for

(a) $X_{A,b}$ has a least element for every b such that $X_{A,b} \neq \varnothing$;

is

(b) there is a basis B such that $c^{t}B^{-1} \geq 0$ for some positive vector c and each such basis is monotone.

(ii) A sufficient condition for (a) is

(c) A^{t} is Leontief.

(iii) If the identity matrix of order n is a submatrix of A, then condition (c) is also necessary.

Proof (i) For every A, $X_{A,0} \neq \varnothing$. Thus if (a) holds then $X_{A,0}$ has a least element and by Theorem 4.6 this element is zero and A has a nonnegative left inverse C. Let $c \gg 0$. Then $y = C^{t}c$ is a feasible solution of (4.5). Since rank $A = n$ (which follows from $CA = I$), there is a basis B such that $c^{t}B^{-1} \geq 0$. To show that B^{-1} is nonnegative assume, for notational convenience, that B consists of the first rows of A. Let

$$b = \begin{bmatrix} b_B \\ b_N \end{bmatrix}, \qquad b_B = 0, \qquad b_N \ll 0$$

be partitioned in conformity with A and let

$$\bar{y} = \begin{bmatrix} \bar{y}_B \\ \bar{y}_N \end{bmatrix} = \begin{bmatrix} B^{-t}c \\ 0 \end{bmatrix}$$

By complementary slackness, $x = 0$ and $\bar{y}$ are optimal solutions of (4.4) and (4.5), respectively. Thus, since $X_{A,b}$ has a least element, this element must be zero. Since $A0 - b$ has exactly n zero elements, the monotonicity of B follows from (e) in Theorem 4.6.

Conversely, if $X_{A,b} \neq \varnothing$ and $c \gg 0$, then there is an optimal basis B for (4.5). B and $\bar{x} = B^{-1}b$ satisfy (f) of Theorem 4.6 so (b) $\rightarrow$ (a).

Part (ii) is included in the remark following Definition 4.7. To prove (iii) suppose a is a row of A, which has a positive element δ, which is not a row of I. For notational convenience assume that $a = (\delta,d)$. To prove (c) we have to show that $d \leq 0$. Indeed A contains a basis

$$B = \begin{bmatrix} \delta & d \\ 0 & I \end{bmatrix},$$

here I is of order $n - 1$. The inverse of this basis is

$$B^{-1} = \begin{bmatrix} \delta^{-1} & -\delta^{-1}d \\ 0 & I \end{bmatrix}.$$

If c is positive and its first element is small enough, $cB^{-1} \geq 0$. Then by (b), $B^{-1} \geq 0$. Hence $-\delta^{-1}d \geq 0$ or $d \leq 0$, completing the proof. ■

Let $X_{A,b,a} = \{x \in X_{A,b};\, x \geq a\}$. (For square matrices M, $X(r,M) = X_{M,-r,0}$.)

(4.9) Corollary (a) $X_{A,b,0}$ has a least element for each b such that $X_{A,b,0} \neq \varnothing$

if and only if

(b) A^t is pre-Leontief.

Proof Since $X_{A,b,0} = a + X_{A,b-Aa,0}$, (a) is equivalent to $X_{A,b,0}$ having a least element for all a,b for which $X_{A,b,0} \neq \varnothing$. By Theorem 4.8 this is equivalent to $(A^t|I)$ being Leontief or, what is the same thing, A^t being pre-Leontief. ■

(4.10) Corollary (a) $X_{A,b,0}$ has a least element for every b

if and only if

(b) $(A^t|I)$ is totally Leontief.

Proof $X_{A,b,a} \neq \varnothing$ for every a and b if and only if there is a positive vector x such that Ax is positive. This, combined with the previous corollary, proves the desired equivalence. ■

Let Z denote the class of matrices with nonpositive off-diagonal elements (we used $Z^{n \times n}$ when the order n was to be specified), and let K denote $Z \cap P$, the nonsingular M-matrices. The relation between the theory of least elements and LCP is described in the following two theorems.

(4.11) Theorem The following statements are equivalent.

(a) For every r such that $X(r,M) \neq \varnothing$, $X(r,M)$ has a least element which is a solution of (r,M).

(b) $M \in Z$.

Proof (a) → (b): Suppose m_{ij} is positive. We have to show that $i = j$. Let $r_i = -m_{ij}$ and $r_k = 1 - m_{kj}$ if $k \neq i$. Then $I^j \in X(r,M)$ where A^j is the jth column of A. Thus $X(r,M)$ has a least element but if $0 \leq z \leq I^j$ satisfies $r + Mz \geq 0$ then $z = I^j$ so I^j is the least element of $X(r,M)$. Thus I^j solves (r,M) so $(I^j, r + M^j) = r_j + m_{jj} = 0$. By the way r was defined, this implies that $i = j$.

(b) → (a) Since A is pre-Leontief, $X(r,M)$ has, by Corollary 4.9, a least element if it is not empty. It is easy to verify that, since $A \in Z$, this least element indeed solves (r,M). ■

(4.12) Theorem The following statements are equivalent.

(a) $X(r,M)$ has a least element which is the unique solution of (r,M), for all r.

(b) $M \in K$.

Proof The proof follows from Theorems 4.11 and 2.15 or from Corollary 4.10 and Theorem 2.15. ■

This may be a good time to mention a characterization of nonsingular M-matrices, assuming they are in Z, which will supplement Theorem 6.2.3.

(4.13) Theorem Let e be the vector of ones and let $M \in Z$. Then

(a) $M \in K$.

if and only if

(b) The LCP (r,M) has a unique solution for $r = 0$ and $r = e$.

Proof (a) → (b) since $K \subset P$.

(b) → (a): If $(0,M)$ has only the trivial solution then there is no $x > 0$ such that

$$x_i > 0 \rightarrow (Mx)_i = 0 \qquad \text{and} \qquad x_i = 0 \rightarrow (Mx)_i \geq 0.$$

If (e,M) has only the trivial solution then there are no $x > 0$ and $t > 0$ such that

$$x_i > 0 \rightarrow (Mx)_i = -t \qquad \text{and} \qquad x_i = 0 \rightarrow (Mx)_i = t.$$

Thus M is regular (Definition 3.1) and, by Theorem 3.8, (r,M) has a solution for every r. In particular, $(-e,M)$ has a solution so there exists $z \geq 0$ such that $Mz \gg 0$. So M which is in Z, satisfies condition (I_{28}) of Theorem 6.2.3 and thus is a nonsingular M-matrix. ■

Suppose $M \in Z^{n \times n}$ and $X(r,M) \neq \varnothing$. Then since the existence of a least element of $X(r,M)$ is guaranteed by Theorem 4.11 it follows that for every positive vector π this least element is the optimal solution of

$$\text{(4.14)} \qquad \text{minimize} \quad \pi^t z \qquad \text{subject to} \quad r + Mz \geq 0, \quad z \geq 0.$$

We now extend the class of matrices M for which a linear program solves (r,M).

(4.15) Definition Let the triple (s,r,M) denote the linear program

$$\text{minimize} \quad s^t z \qquad \text{subject to} \quad r + Mz \geq 0, \quad z \geq 0.$$

The LCP (r,M) is *LP-solvable* if one can find a vector s such that each solution of the linear program (s,r,M) solves (r,M).

(4.16) Notation Let C denote the class of square matrices that satisfy

$$MX = Y \qquad \text{where} \quad X, Y \in Z \tag{4.17}$$

and

$$p^tX + q^tY \gg 0 \qquad \text{for some} \quad (p,q) \geq 0. \tag{4.18}$$

We shall show that if $X(r,M) \neq \varnothing$ and $M \in C$ then (r,M) is LP-solvable. We shall use the following characterization of matrices in C.

(4.19) Theorem An $n \times n$ matrix M satisfies conditions (4.17) and (4.18) if and only if X and Y are in Z, X is nonsingular,

$$\begin{bmatrix} M_{11} & M_{12} \\ M_{21} & M_{22} \end{bmatrix} \begin{bmatrix} X_{11} & X_{12} \\ X_{21} & X_{22} \end{bmatrix} = \begin{bmatrix} Y_{11} & Y_{12} \\ Y_{21} & Y_{22} \end{bmatrix}, \tag{4.20}$$

where

$$\begin{bmatrix} M_{11} & M_{12} \\ M_{21} & M_{22} \end{bmatrix} = P^tMP, \quad \begin{bmatrix} X_{11} & X_{12} \\ X_{21} & X_{22} \end{bmatrix} = P^tXP, \quad \text{and} \quad \begin{bmatrix} Y_{11} & Y_{12} \\ Y_{21} & Y_{22} \end{bmatrix} = P^tYP$$

for some permutation matrix P, X_{11}, Y_{11}, and M_{11} are square of the same order and

$$\begin{bmatrix} X_{11} & X_{12} \\ Y_{21} & Y_{22} \end{bmatrix} \in K. \tag{4.21}$$

Proof First we show that if X and Y are in $Z^{n \times n}$ then (4.18) and (4.21) are equivalent. By Exercise 1.3.7, (4.18) is equivalent to $Xu \leq 0$, $Yu \leq 0$, $u \geq 0 \rightarrow u = 0$. Suppose (4.21) holds and $u \geq 0$ satisfies $Xu \leq 0$ and $Yu \leq 0$. Then

$$\begin{bmatrix} X_{11} & X_{12} \\ Y_{21} & Y_{22} \end{bmatrix} u \leq 0,$$

but since

$$\begin{bmatrix} X_{11} & X_{12} \\ Y_{21} & Y_{22} \end{bmatrix}$$

is monotone, $u \leq 0$. Thus $u = 0$.

Conversely, suppose (4.18) holds, that is

$$[X^t \mid Y^t] \begin{bmatrix} p \\ q \end{bmatrix} \gg 0 \qquad \text{for some} \quad \begin{bmatrix} p \\ q \end{bmatrix} \geq 0.$$

Since X and Y are Z-matrices and satisfy (4.18), $A = (X^t \mid Y^t)$ is a Leontief matrix. Thus it has a monotone submatrix of order n which has exactly one positive element in each column. Hence, by permuting the columns of B, if necessary, we may assume that the ith column of B is the ith row of X or the ith row of Y. This suggests the permutation and partitioning for which (4.21) holds.

The equivalence just given proves the if part of the theorem. For the necessity part it remains to show that X is nonsingular. Condition 4.21 implies that X_{11} is nonsingular. By Schur's determinantal formula (Exercise 6.1),

$$\det X = \det X_{11} \det(X_{22} - X_{21}X_{11}^{-1}X_{12}).$$

By (4.20)

$$(4.22) \qquad M_{22}(X_{22} - X_{21}X_{11}^{-1}X_{12}) = Y_{22} - Y_{21}X_{11}^{-1}X_{12}.$$

The right-hand side is the Schur complement of Y_{22} in the nonsingular M-matrix in (4.21), so it is nonsingular. By (4.22), $X_{22} - X_{21}X_{11}^{-1}X_{12}$ is nonsingular, completing the proof. ■

To prove that (r,M) is LP-solvable if $M \in C$ and $X(r,M) \neq \varnothing$ we use least element arguments. The set $X(r,M)$ need not have a least element so the argumentation is done via a transformation to a polyhedral set which has such an element.

(4.23) Lemma Let $X,Y \in Z^{n\times n}$ and let $q,r \in R^n$. Suppose that the polyhedral set

$$V = \{v \in R^n; r + Yv \geq 0, q + Xv \geq 0\}$$

is nonempty and bounded below. Then there exists a least element $\overline{v} \in V$ satisfying $(r + Yv, q + Xv) = 0$. This least element can be obtained by solving the linear program.

$$(4.24) \qquad \text{minimize } \pi^t v \qquad \text{subject to} \quad r + Yv \geq 0, \quad q + Xv \geq 0$$

for any positive vector π.

Proof The constraint set V is bounded below and closed so the linear program (4.24) has a solution, say $\overline{v}$, with $\pi = e$ the vector of ones. For every $v \in V$ let v' be the vector with $v_i' = \text{Min}\{v_i,\overline{v}_i\}$. It is easy to see that $v' = \overline{v}$. Thus $\overline{v}$ is the least element of V and this implies that it solves (4.24) for every positive vector π. It remains to verify that $\overline{v}$ satisfies the complementarity condition. Suppose $(r + Y\overline{v})_i > 0$ and $(q + X\overline{v})_i > 0$. Let ε be a positive number which satisfies

$$\varepsilon < (r + Y\overline{v})_i/y_{ii} \qquad \text{if} \quad y_{ii} > 0,$$

$$\varepsilon < (q + X\overline{v})_i/x_{ii} \qquad \text{if} \quad x_{ii} > 0,$$

and let $v = \bar{v} - \varepsilon I^i$, where I^i is the ith unit vector. Then $v \in V$, contradicting the fact that $\bar{v}$ is the least element. ■

(4.25) Theorem Let $M \in C$ and suppose $X(r,M) \neq \emptyset$. Let V be the image of $X(r,M)$ under the map $v = X^{-1}z$, where X satisfies (4.17) and (4.18). Then v has a least element $\bar{v}$, $\bar{z} = X\bar{v}$ solves (r,M), and $\bar{z}$ can be obtained by solving the linear program (s,r,M) with $s = X^{-t}\pi$ where π is any positive vector.

Proof The feasible set $X(r,M) = \{z \geq 0; r + YX^{-1}z \geq 0\}$ is transformed via $v = X^{-1}z$ to

$$V = \left\{v; \begin{bmatrix} X \\ Y \end{bmatrix} v \geq \begin{bmatrix} 0 \\ -r \end{bmatrix}\right\}$$

Partitioning

$$r = \begin{bmatrix} r_1 \\ r_2 \end{bmatrix}$$

according to Theorem 4.19, we see that $v \in V$ implies that

$$\begin{bmatrix} X_{11} & X_{12} \\ Y_{21} & Y_{22} \end{bmatrix} v \geq \begin{bmatrix} 0 \\ -r_2 \end{bmatrix}.$$

Since

$$\begin{bmatrix} X_{11} & X_{12} \\ Y_{21} & Y_{22} \end{bmatrix}$$

is a nonsingular M-matrix,

$$v \in V \rightarrow v \geq \begin{bmatrix} X_{11} & X_{12} \\ Y_{21} & Y_{22} \end{bmatrix}^{-1} \begin{bmatrix} 0 \\ -r_2 \end{bmatrix};$$

i.e., the set V is bounded below. Since $X(r,M) \neq \emptyset$, so is V and by Lemma 4.2.3 V has a least element $\bar{v}$ satisfying $(r + Yv)^t Xv = 0$ and $\bar{v}$ can be obtained by solving the linear program

minimize $\pi^t v$ subject to $r + Yv \geq 0$, $Xv \geq 0$ for any positive vector π.

Letting $\bar{z} = X\bar{v}$, one sees that $\bar{z}$ is a solution of (r,M) and it can be obtained by solving the linear program $X^{-t}\pi,r,M)$. ■

(4.26) Example In Example 4.2(d), $M \in C$ as is shown by the factorization

$$M = \begin{bmatrix} 1 & 1 \\ -2 & -3 \end{bmatrix} = \begin{bmatrix} \frac{1}{2} & -2 \\ -\frac{1}{2} & 3 \end{bmatrix} \begin{bmatrix} 1 & -3 \\ -\frac{1}{2} & 1 \end{bmatrix}^{-1} = YX^{-1}.$$

Here $X(r,M)$ has no least element. The LCP (r,M) has two solutions (0,2) and (1,0). It is the first solution that is obtained via the program in Theorem 4.2.5, since the map $X(r,M) \to V = X^{-1}(X(r,M))$ maps it to $(-12,-4)$, the least element of V as shown in the accompanying figure.

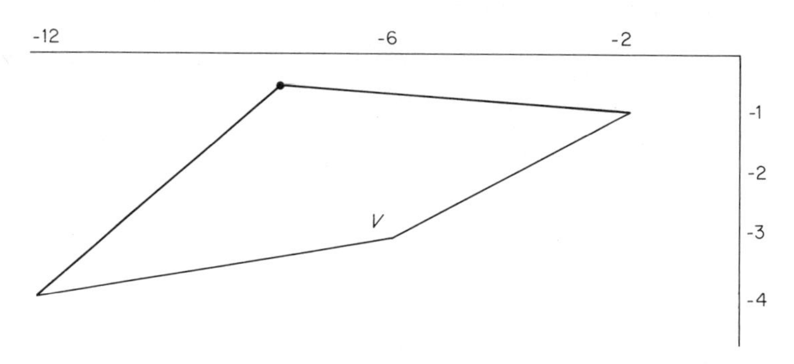

To conclude we describe several subclasses of C.

(4.27) Theorem Let

(a) $M = YX^{-1}$, $X \in K$, $Y \in Z$, and $s \geq 0$ such that $s^tX \gg 0$, or
(b) $M = YX^{-1}$, $X \in Z$, $Y \in K$, and $s = M^tt$, where $t \geq 0$ and $s^tY \gg 0$, or
(c) $M \in Z$ and s be positive, or
(d) $M^{-1} \in Z$ and $s = M^tt$ where t is positive.

Then $M \in C$ and if $X(r,M) \neq \varnothing$, any solution of the linear program (s,r,M) solves the LCP (r,M).

Proof The matrices in (a) and (b) satisfy the conditions of Theorem 4.19. The factorization $M = YI$ for M in (c) shows that M satisfies conditions (4.17) and (4.18). Observe that in this case (s,r,M) is the program (4.14). Factorizing M in (d) to $M = IX^{-1}$ completes the proof. ■

(4.28) Theorem Let

(a) $M = Y + ab^t$, $Y \in K$, $a \geq 0$, $b \geq a$, or
(b) $M = 2A - B$, $A \in Z$, $B \in K$, where $A \geq B$, or
(c) $m_{ii} > \sum_{j \neq i} |m_{ij}|$, for all i or
(d) $m_{jj} > \sum_{i \neq j} |m_{ij}|$, for all j.

Then $M \in C \cap P$ and (r,M) is LP-solvable.

Proof If M is the sum of a nonsingular M-matrix and the product of two nonnegative vectors then so is every principal submatrix of M. Thus to prove that $M \in P$ it is enough to show that det M is positive. Clearly,

$$\det(Y + ab^t) = (\det Y)(\det(I + (Y^{-1}a)b^t)).$$

Using the formula

$$\det(xy^t - \lambda I) = (-1)^n \lambda^{n-1}(\lambda - x^t y)$$

with $\lambda = -1$, $x = Y^{-1}a$, and $y = b$, yields

$$\det(Y + ab^t) = (\det Y)(1 + (Y^{-1}a)^t b) \geq \det Y > 0$$

because $Y^{-1} \geq 0$ and $a,b \geq 0$.

To prove that $M \in C$, factorize it as

$$Y + ab^t = Y(I + Y^{-1}ab^t).$$

By the first part of the proof, $I + Y^{-1}ab^t \in P$ because $Y^{-1}a \geq 0$. Letting $X = (I + Y^{-1}ab^t)^{-1}$ it follows that

$$X = I + \frac{1}{1 + b^t Y^{-1} a} Y^{-1}ab^t \in Z$$

so that $M = Y + ab^t$ is in the form (b) of Theorem 4.27. In fact we saw that $X^{-1} \in P$, thus $X \in P$, so $X \in K$. Thus both X and Y in the factorization $M = YX^{-1}$ are nonsingular M-matrices.

In part (b), factorize $M = YX^{-1}$ with $X = A^{-1}B$ and $Y = 2B - BA^{-1}$. We leave the details for the exercises.

If M or M^t is strictly diagonally dominant and has positive diagonal elements, then all its principal submatrices are positive stable so all the principal minors are positive. To show that $M \in C$ let $B = (b_{ij})$ be the comparison matrix of M (Definition 6.2.8).

$$b_{ij} = \begin{cases} m_{ii} & \text{if } \ j = i \quad \text{(here } m_{ii} > 0), \\ m_{ij} & \text{if } \ j \neq i \ \text{ and } \ m_{ij} \leq 0, \\ -m_{ij} & \text{if } \ j \neq i \ \text{ and } \ m_{ij} > 0. \end{cases}$$

Let $2A = B + M$. Then $M = 2A - B$ and $A \geq B$. By their construction A and B belong to Z. If M satisfies (c) then $B^t e \gg 0$, where e is the vector of ones. If M satisfies (d) then $Be \gg 0$. In both cases M satisfies condition (K_{33}) of Theorem 6.2.3 and thus is a nonsingular M-matrix. This presents M in the form (b), proving that $M \in C$ and reproving that $M \in P$. Since in all cases M is a P-matrix, there is no need to *assume* that $X(r,M)$ is nonempty and the unique solution of the LCP (r,M) can be found by solving a linear program. ■

5 CHARACTERIZATIONS OF NONSINGULAR M-MATRICES

In this section we state additional characterizations of nonsingular M-matrices under the assumption that they are P-matrices. They are described in terms of the following problems.

(5.1) Definition Let $p,q,r,a \in R^n$, $a \gg 0$, and $M \in R^{n \times n}$. *The parametric linear complementarity problem* (PLCP) is the family of LCP's of the form

$$\{(q + \alpha p, M);\ \alpha \geq 0\}. \tag{5.2}$$

If $M \in P$, $z(\alpha; q,p,M)$ denotes the unique solution of (5.2). *The linear complementarity problem with upper bounds,* $(r,M)_a$, is that of finding z such that

$$r + Mz \geq 0, \qquad a \geq z \geq 0, \qquad z^t(r + Mz) = 0.$$

The set $\{r \in R^n;\ (r,M)_a \text{ has a solution}\}$ is denoted by $R(a,M)$.

Both PLCP and the LCP with upper bounds arise in certain problems in *structural mechanics* addressed by Maier [1970]. Considering a PLCP with a P-matrix M and a positive vector q, Maier observed that the *regular progression hypothesis* in structural analysis is valid if and only if $z(\alpha; q,p,M)$ is isotone (i.e., monotone nondecreasing) componentwise. He posed (Maier [1972]) the problem of determining conditions on M under which $z(\alpha; q,p,M)$ is isotone. Cottle [1972] solved this problem by proving that for a P-matrix M, $z(\alpha; q,p,M)$ is isotone in $\alpha \geq 0$ for every $q \geq 0$ and every p if and only if M is a nonsingular M-matrix. Maier has also shown that the *safety factor* of the structures that he studied may be determined by solving the maximization problem.

(5.3) Find $\alpha^* = \text{Max}\{\alpha \geq 0;\ q + \alpha p \in R(a,M)\}$ where $a \gg 0$, $q \geq 0$, $p \ngeq 0$ (if $p \geq 0$, the maximum does not exist).

Cottle [1975] showed that if $R(a,M)$ is convex then (5.3) can be solved by the *restricted basis simplex method,* the ordinary simplex method with pivot choice that keeps the linear programming bases complementary, and that $R(a,M)$ is convex for every $a \gg 0$ if and only if M is a nonsingular M-matrix.

(5.4) Definition A set $S \subseteq R^n$ is *star-shaped* with respect to $s \in S$ if for every $s' \in S$ and $0 \leq \lambda \leq 1$

$$(1 - \lambda)s + \lambda s' \in S.$$

S is *star-shaped on* $T \subseteq S$ if it is star-shaped with respect to every s in T.

Note that a set is convex if and only if it is star-shaped on itself. For $r \geq 0$, $(r,M)_a$ has the trivial solution so $R^n_+ \subseteq R(a,M)$. Kaneko [1978] showed that the convexity of $R(a,M)$ is equivalent to its star-shapedness on R^n_+. In other works of Kaneko he characterized "$M \in K$" in terms of convexity (of each component) of $z(\alpha; q,p,M)$ for every p and q and isotonicity of z for every $q \geq 0$ and $p \leq 0$. Another variation of these characterizations arises in a *portfolio selection problem* studied by Pang (see Pang, *et al.* [1976])

where in the PLCP (5.2), p is nonpositive and there is no sign condition on q. In this case, too, the isotonicity of z is equivalent to $M \in K$. Note that $z(\alpha; q,p,M)$ cannot be isotone in $\alpha \geq 0$ for every p if $q \ngeq 0$, or for every q if $p \nleq 0$, for let $q = -p$. Then zero cannot be a solution of the LCP (q,M) if $q \ngeq 0$ or $p \nleq 0$ but it is a solution of $(q + \alpha p,M)$ for $\alpha = 1$.

We sum up the preceding in the following theorem.

(5.5) Theorem Let $M \in R^{n \times n}$ be a P-matrix. The following statements are equivalent:

(a) $M \in K$.
(b) $z(\alpha; q,p,M)$ is convex in $\alpha \geq 0$ for every q and every p.
(c) $R(a,M)$ is convex for every $a \in \operatorname{int} R^n_+$.
(d) $R(a,M)$ is convex for *some* $a \in \operatorname{int} R^n_+$.
(e) $R(a,M)$ is star-shaped on R^n_+ for *some* $a \in \operatorname{int} R^n_+$.
(e′) $R(a,M)$ is star-shaped on $\operatorname{int} R^n_+$ for *some* $a \in \operatorname{int} R^n_+$.
(f) $z(\alpha; q,p,M)$ is isotone in $\alpha \geq 0$ for every q and every $p \in -R^n_+$.
(f′) $z(\alpha; q,p,M)$ is isotone in $\alpha \geq 0$ for every q and every $p \in -\operatorname{int} R^n_+$.
(g) $z(\alpha; q,p,M)$ is isotone in $\alpha \geq 0$ for every $q \in R^n_+$ and every $p \in -R^n_+$.
(g′) $z(\alpha; q,p,M)$ is isotone in $\alpha \geq 0$ for every $q \in \operatorname{int} R^n_+$ and every $p \in -\operatorname{int} R^n_+$.
(h) $z(\alpha; q,p,M)$ is isotone in $\alpha \geq 0$ for every $q \in R^n_+$ and every p.
(h′) $z(\alpha; q,p,M)$ is isotone in $\alpha \geq 0$ for every $q \in \operatorname{int} R^n_+$ and every p.
(i) $R(a,M)$ is star-shaped on R^n_+ for every $a \in \operatorname{int} R^n_+$.
(i′) $R(a,M)$ is star-shaped on $\operatorname{int} R^n_+$ for every $a \in \operatorname{int} R^n_+$.

Proof We refer the reader to Kaneko [1978] where the plan of the proof is

$$
\begin{array}{ccccccccc}
(a) & \rightarrow & (b) & \rightarrow & (h) & & & & \\
 & & \downarrow & & \downarrow & & & & \\
 & & (c) & \rightarrow & (i) & & & & \\
 & & \downarrow & & \downarrow & & & & \\
 & & (d) & \rightarrow & (e) & \rightarrow & (f) \rightarrow (g) & \rightarrow & (a).
\end{array}
$$

The equivalence of (x) and (x'), $x \in \{e,f,g,h,i\}$, follows from the continuity of the solution of the LCP (r,M), as a function of r, with $M \in P$. ■

6 EXERCISES

(6.1) Suppose M is cogredient to

$$
\begin{bmatrix} M_{11} & M_{12} \\ M_{21} & M_{22} \end{bmatrix}.
$$

Let

$$A = \begin{bmatrix} M_{11}^{-1} & -M_{11}^{-1}M_{12} \\ M_{21}M_{11}^{-1} & M_{22} - M_{21}M_{11}^{-1}M_{12} \end{bmatrix}$$

be a principal pivot of M. The matrix $M/M_{11} \equiv M_{22} - M_{21}M_{11}^{-1}M_{12}$ is called the *Schur complement* of M_{11} (in M).

(a) Prove *Schur's formula*: $\det M = \det M_{11} \det(M/M_{11})$
(b) Prove that if $M \in P$ then $A \in P$.
(c) Prove that if $M \in K$ then $M/M_{11} \in K$.

(6.2) Show that if (r,M) has a solution then it has a solution which is an extreme point of $X(r,M)$.

(6.3) A vector $r \in R^n$ is *nondegenerate with respect to a matrix* $M \in R^{n \times n}$ if for every w and z such that $w = r + Mz$, at most n of the $2n$ variables $\{w_j,z_j\}$ are zero. A square matrix M is *nondegenerate* if all its principal submatrices are nonzero.

(a) Prove that the number of solutions of (r,M) is finite if and only if r is nondegenerate with respect to M (Lemke [1965]).

(b) Prove that the number of solutions of (r,M) is finite *for every* r if and only if M is nondegenerate (Murty [1972]).

(6.4) Let $M \in P$.

(a) Show that the solution of the LCP (r,M) is continuous in r.

(b) Show that $z(\alpha; q,p,M)$, the solution of the PLCP $(q + \alpha p,M)$, is a continuous, piecewise linear function in $\alpha \geq 0$.

(c) Let $\bar{r}^t = (\bar{r}_1,r_2, \ldots ,r_n)^t$ and $\hat{r}^t = (\hat{r}_1,r_2, \ldots ,r_n)^t$. Let $z(r)$ be the solution of (r,M). Suppose $\hat{r}_1 > \bar{r}_1$. Prove that if $z_1(\bar{r}) > 0$ then $z_1(\hat{r}) < z_1(\bar{r})$ and that if $z_1(\bar{r}) = 0$ then $z_1(\hat{r}) = 0$.

(6.5) Solve (r,M) where

(a)

$$(r|M) = \begin{bmatrix} 1 & 0 & 0 & 0 \\ -1 & 1 & 0 & 0 \\ 3 & -1 & -1 & -1 \end{bmatrix},$$

(b)

$$(r|M) = \begin{bmatrix} 1 & 0 & 0 & 0 & 0 \\ -1 & 1 & 0 & 0 & 1 \\ 3 & -1 & -1 & -1 & -1 \\ 1 & 0 & 0 & 0 & -1 \end{bmatrix},$$

(c)

$$(r|M) = \begin{bmatrix} 1 & 0 & 0 & 0 & 1 \\ -1 & 1 & 0 & 0 & 1 \\ 3 & 1 & -1 & -1 & 1 \\ 1 & 0 & 0 & 0 & -1 \end{bmatrix},$$

(6.6) Solve (r,M) for all possible values of r and

(a)

$$M = \begin{bmatrix} 2 & -1 \\ -1 & 1 \end{bmatrix}$$

(b) $M = J$, a matrix with all elements equal to one. (Here J is square.)

(6.7) Prove that $M \in P$ if and only if (r,M) has a unique solution for every r which is a column of I, J, M, and $-M$. (Thus it is enough to check only $3n + 1$ values of r; see Tamir [1974].)

(6.8) Suppose the number of solutions of (r,M) is a constant for all $r \neq 0$. Prove that $M \in P$ (so the constant is one; see Murty [1972]).

(6.9) Show that if $JX \in Z$, then $JX \leq 0$ and $X \notin P$. Conclude that $J \notin C$ (See 4.16).

(6.10) Using the notation of Section 4, let

$$S_1 = \{s \in R^n;\ s = p + M^t q \text{ for some } (p,q) \geq 0 \text{ such that } p^t X + q^t Y \gg 0\}$$

and

$$S_2 = \{s \in R^n;\ s = X^{-t}\pi \text{ for some } \pi \gg 0\}.$$

(S_2 is the set of vectors s for which (s,r,M) solves (r,M) if $X(r,M) \neq \phi$ and $M \in C$.) Prove that $S_1 = S_2$.

(6.11) Complete the proof of Theorem 4.6 (Cottle and Veinott [1972]).

(6.12) Show that Leontief matrices have the properties described after Definition 4.7.

(6.13) (a) Prove part (b) of Theorem 4.28 (Cottle and Pang [1978]).
(b) For each of the conditions satisfied by M in Theorem 4.28, determine conditions on s so that (r,M) can be solved via (s,r,M) (Mangasarian [1976b]).

(6.14) Prove that $Z \cap Q = K$ (Mohan [1976]).

7 NOTES

(7.1) An *almost complementary* solution of (1.1) is one in which $z_i w_i = 0$, but for one value of i. It is *feasible* if $z_i \geq 0$, $w_i \geq 0$, $i = 1, \ldots, n$. Under the assumption of nondegeneracy, the extreme points of $X(r,M)$ are in one-to-one correspondence with the basic feasible solutions of (1.1). In an almost complementary basic feasible solution (a.c.b.f.s.) of (1.1) there is exactly one index, say β, such that w_β and z_β are basic variables and exactly one index, say v, such that w_v and z_v are nonbasic. We shall refer to $\{w_\beta, z_\beta\}$ and $\{w_v, z_v\}$ as the basic and nonbasic pairs, respectively.

In the complementary pivot algorithm of Lemke one moves from an a.c.b.f.s. by increasing one of the members of the nonbasic pair, say z_v, holding the other nonbasic variables at value zero.

If the value of z_v can be made arbitrarily large without forcing any basic variable to become negative the process *terminates in a ray* of $X(r,M)$.

If a member of the basic pair blocks the increase of z_v, then the new basic solution is complementary and thus solves (r,M).

If z_v is blocked by another basic variable then a new a.c.b.f.s. is obtained. In this case, the next nonbasic variable to be increased is the complement of the (now nonbasic) blocking variable.

Problem (a) in Exercise (6.5) has no solution. If Lemke's algorithm is initiated at $z^t = (1,0,2)$ by increasing z_2 then it returns to this point after four iterations. Problem (b) is solved by $z^t = (1,0,1,0)$, but here again, starting from $z^t = (3,0,0,0)$ by increasing z_3, one returns to the starting point after four iterations.

If the procedure is initiated at an a.c.b.f.s. which is the endpoint of an *almost complementary ray* (a.c.r.) of $X(r,M)$ then it terminates in a ray or in a solution of (r,M). An a.c.b.f.s. which is an endpoint of an a.c.r. can be obtained by introducing an artificial variable to (1.1). Even when the procedure is so initiated its termination in a ray does not imply that (r,M) has no solution. For example, problem (c) in Exercise (6.5) is solved by $z^t = (0,1,0,1)$ but starting at $z^t = (1,0,0,0)$, which is the endpoint of the a.c.r. $w^t = (1, w_2, 4 + w_2, 1)$, $z^t = (1 + w_2, 0,0,0)$, by increasing z_2, the procedure terminates in a ray. There are, however, special classes of matrices for which termination in a ray implies that (r,M) has no solution.

A square matrix M is *copositive* if $u \geq 0 \rightarrow u^t M u \geq 0$ and *copositive plus* ($M \in C_+$) if, in addition, $u \geq 0$, $u^t M u = 0 \rightarrow (M + M^t)u = 0$. M is *strictly copositive* ($M \in SC$) if $u > 0 \rightarrow u^t M u > 0$. If $M \in S$ (Definition 3.8) then for every r, Lemke's algorithm solves (r,M). Notice that

$$PD \subset (SC \cap P) \subset (SC \cup P) \subset S.$$

The reader is refered to Cottle and Dantzig [1968] for the proofs of the previous statements and for the description of the device for initiating the procedure. Finally Eaves [1971] defined M to be an *L-matrix* ($M \in L$) if for every $x > 0$ there exists an index k such that $x_k > 0$ and $M_k x \geq 0$ and if for some $x > 0$, $Mx \geq 0$ and $x^t M x = 0$, there exist nonnegative diagonal matrices D and E such that $DE \neq 0$ and $(EM + M^t D)x = 0$; and he proved that if $M \in L$ and Lemke's algorithm for solving (r,M) terminates in a ray, then (r,M) has no solution. Observe that

$$PSD \subset (SC \cap C_+) \qquad \text{and} \qquad (C_+ \cup S) \subset L.$$

(7.2) The complementary pivot algorithm described above is an extension of an iterative technique of Lemke and Howson [1964] for finding equilibrium points of bimatrix games. As was just mentioned, this algorithm can be used if $M \in$ PSD (see Exercise 1.5.12) which is the case in (1.1) arising in convex quadratic programming. A modified Dantzig–Cottle algorithm can also be used for $M \in$ PSD. Graves [1967] proposed another principal pivot algorithm. A parametric version of this algorithm is used in Cottle [1972] to solve PLCP.

Saigal [1971] proved that Lemke's algorithm also solves (r,M) where $M \in Z$. Mohan [1976] observed that in this case the sequence of almost complementary solutions is exactly the same as the one resulting from applying the ordinary simplex method to

$$\text{minimize } z_0 \qquad \text{subject to} \quad r + Mz + z_0 e \geq 0, \quad (z_0, z) \geq 0.$$

Another algorithm to solve (r,M), $M \in Z$, was devised by Chandrasekaran [1970].

Special algorithms were developed by Cottle and Sacher [1977], and Cottle *et al.* [1978] to solve large-scale LCP s efficiently when M is a nonsingular M-matrix having additional structures as tridiagonality and block tridiagonality. They are based on the Dantzig–Cottle principal pivoting algorithm, Chandrasekaran's algorithm, and successive overrelaxation procedures for solving a linear system.

(7.3) Theorem 2.5 is due to Tucker [1963]. Theorem 2.15 in its geometric version 2.21 goes back to Samelson *et al.* [1958]. The proof of Theorem 2.15, given in this chapter, is due to Gale and is taken from Murty [1972].

(7.4) The concept of regularity, Theorems 3.3 and 3.8, and Corollary 3.5 are taken from Karamardian [1972]. The characterization of nonnegative Q matrices and its corollaries are due to Murty [1972].

(7.5) The theory of least elements including Theorem 4.12 is based on Cottle and Veinott [1972]. Some authors define a Leontief matrix as one which has *exactly* one positive element in each column (and such that $Ax \gg 0$ for some $x \geq 0$). Notice that in Definition 4.7 the requirement is that the matrix has *at most* one positive element in each column. Theorem 4.11 is due to Tamir [1974] and Theorem 4.13 to Kaneko [1978].

(7.6) The second part of Section 4 is based on the work of Cottle and Pang [1978a]. Pang [1976] proved that the solution $\bar{z}$ of (r,M), obtained in Theorem 4.25, is the least element of $X(r,M)$ under the partial ordering induced by the polyhedral cone $\{r \in R^n;\ X^{-1}r \geq 0\}$, where $M = YX^{-1}$ with X and Y satisfying conditions (4.17) and (4.18).

Theorems (4.27) and (4.28) were proved by Mangasarian [1976a] and [1976b], respectively. (See also Mangasarian [1978].) The first proof is based on the duality theorem. In the second proof (r,M) is extended to the "slack LCP"

$$\left(\begin{bmatrix} r \\ 0 \end{bmatrix}, \begin{bmatrix} M & B \\ 0 & A \end{bmatrix}\right).$$

In Mangasarian's results the vector s, of Theorem 4.25, belongs to the class $\{s \in R^n;\ s = p + M^t q$ for some $(p,q) \geq 0$ such that $p^t X + s^t Y \gg 0\}$. By Exercise 6.10, this is the same class as in Theorem 4.25. As a final remark we point out that Mangasarian proved Theorem 4.28 with the additional assumption that M is nonnegative. This, of course, is a reasonable assumption to make in a book on nonnegative matrices, but as was shown in Section 4, it is enough to assume that the diagonal elements of M are positive.

(7.7) Theorem 5.5 is taken from Kaneko [1978]. Conditions (c) and (h) were proved by Cottle [1972] and [1975], respectively. The remaining conditions are due to Kaneko. Applications of the problems mentioned in Section 5 are described in Pang, *et al.* [1976] and in the references cited there.

(7.8) Example 4.2(e) is an example of "$X(r,M) \neq \varnothing$ but (r,M) has no solution." The question of characterizing matrices M for which "$X(r,M) \neq \varnothing$ implies (r,M) has a solution" is open. Sufficient conditions are, of course, $M \in Q$ and $M \in C$. There is no inclusion relation between these two classes. For example,

$$\begin{bmatrix} 1 & -1 \\ -1 & 1 \end{bmatrix} \in Z \cap C \text{ and } \notin Q, \qquad \left(X\left(\begin{bmatrix} -1 \\ -1 \end{bmatrix}, \begin{bmatrix} 1 & -1 \\ -1 & 1 \end{bmatrix}\right) = \phi\right)$$

and $J \in Q$ (Theorem 3.10) but $J \notin C$ (Exercise 6.9).

(7.9) Cottle and Dantzig [1970] studied the following generalized LCP. Consider the system

$$w = r + Nz, \qquad w \geq 0, \quad z \geq 0,$$

where N is an $n \times k$ matrix and the variables $w_1, \ldots, w_n$ are partioned into k nonempty sets $S_1, \ldots, S_k$. Find a solution in which exactly one member of $S_l \cup \{z_l\}$ is nonbasic, $l = 1, \ldots, k$. (In (r,M), $k = n$ and $S_l = \{w_l\}$, $l = 1, \ldots, n$.) They show that the problem has a solution when N is positive.

(7.10) Let K be a closed convex cone in R^n, $r \in R^n$, $M \in R^{n \times n}$. A natural geometrical extension of LCP is the following.

Find (or conclude there is no)$z \in K$ such that $w = r + Mz \in K^*$ and $(z,w) = 0$. For examples, see McCallum [1970] and Berman [1974a].

(7.11) Nonlinear complementarity problems NLCP, where $r + Mz$ is replaced by $r + f(z)$, f being a function from R^n to R^n, were studied by, among others, Cottle [1966], Karamardian [1969a,b], Habetler and Price [1971], Moré [1973], and Tamir [1974]. Habetler and Price considered NLCP over convex cones.

REFERENCES

Aharoni, A. [1979]. Combinatorial Problems in Matrix Theory, Ph.D Thesis, Technion, Israel Institute of Technology, Haifa, Israel.

Alefeld, G., and Varga, R. S. [1976]. Zur Konvergenz der symmetrischen relaxationsverfahrens, *Numer. Math.* **25**, 291–295.

Alexandroff, P., and Hopf, H. [1935]. *Topologie*. Springer, New York.

Almon, C., Jr., Buckler, M. B., Horowitz, L. M., and Reimbold, T. C. [1974]. "1985 Interindustry Forecasts of the American Economy." Heath, Lexington, Massachusetts.

Araki, M. [1975]. Applications of M-matrices to the stability problems of composite dynamical systems, *J. Math. Anal. Appl.* **52**, 309–321.

Barker, G. P. [1972]. On matrices having an invariant cone, *Czech. Math. J.* **22**, 49–68.

Barker, G. P. [1973]. The lattice of faces of a finite dimensional cone, *Linear Algebra and Appl.* **7**, 71–82.

Barker, G. P. [1974]. Stochastic matrices over cones, *Linear and Multilinear Algebra* **1**, 279–287.

Barker, G. P., and Carlson, D. [1975]. Cones of diagonally dominant matrices, *Pacific J. Math.* **57**, 15–32.

Barker, G. P., and Foran, J. [1976]. Self dual cones in Euclidean spaces, *Linear Algebra and Appl.* **13**, 147–155.

Barker, G. P., and Loewy, R. [1975]. The structure of cones of matrices, *Linear Algebra and Appl.* **12**, 87–94.

Barker, G. P., and Schneider, H. [1975]. Algebraic Perron-Frobenius Theory, *Linear Algebra and Appl.* **11**, 219–233.

Barker, G. P., and Turner, R. E. L. [1973]. Some observations on the spectra of cone preserving maps, *Linear Algebra and Appl.* **6**, 149–153.

Barker, G. P., Berman, A., and Plemmons, R. J. [1978]. Positive diagonal solutions to the Lyapunov equations, *Linear and Multilinear Algebra* **5**, 249–256.

Beauwens, R. [1976]. Semi-strict diagonal dominance, *SIAM J. Numer. Anal.* **13**, 104–112.

Beckenbach, E. F., and Bellman, R. [1971]. "Inequalities," 3rd ed. Springer-Verlag, Berlin and New York.

Bellman, R. [1960]. "Introduction to Matrix Analysis." McGraw-Hill, New York.

Ben-Israel, A. [1969]. Linear equations and inequalities on finite dimensional, real or complex, vector spaces: A unified theory, *J. Math. Anal. Appl.* **27**, 367–389.

Ben-Israel, A., and Greville, T. N. E. [1973]. "Generalized Matrix Inverses: Theory and Applications." Wiley, New York.

Berge, C. [1976]. "Graphs and Hypergraphs," 2nd ed. North-Holland Publ., Amsterdam.

Berman, A. [1973]. "Cones, Matrices and Mathematical Programming," Lecture Notes in Economics and Mathematical Systems 79. Springer-Verlag, Berlin and New York.

Berman, A. [1974a]. Complementarity problem and duality over convex cones, *Canad. Math. Bull.* **17**, 19–25.

Berman, A. [1974b]. Nonnegative matrices which are equal to their generalized inverse, *Linear Algebra and Appl.* **9**, 261–265.

Berman, A. [1978]. The Spectral radius of a nonnegative matrix, *Canad. Math. Bull.* **21**, 113–114.

Berman, A., and Ben-Israel, A. [1971]. More on linear inequalities with applications to matrix theory, *J. Math. Anal. Appl.* **33**, 482–496.

Berman, A., and Gaiha, P. [1972]. Generalization of irreducible monotonicity, *Linear Algebra and Appl.* **5**, 29–38.

Berman, A., and Neumann, M. [1976a]. Proper splittings of rectangular matrices, *SIAM J. Appl. Math.* **31**, 307–312.

Berman, A., and Neumann, M. [1976b]. Consistency and splittings, *SIAM J. Numer. Anal.* **13**, 877–888.

Berman, A., and Plemmons, R. J. [1972]. Monotonicity and the generalized inverse, *SIAM J. Appl. Math.* **22**, 155–161.

Berman, A., and Plemmons, R. J. [1974a]. Cones and iterative methods for best least squares solutions to linear systems, *SIAM J. Numer. Anal.* **11**, 145–154.

Berman, A., and Plemmons, R. J. [1974b]. Inverses of nonnegative matrices, *Linear and Multilinear Algebra* **2**, 161–172.

Berman, A., and Plemmons, R. J. [1974c]. Matrix group monotonicity, *Proc. Amer. Math. Soc.* **46**, 355–359.

Berman, A., and Plemmons, R. J. [1976]. Eight types of matrix monotonicity, *Linear Algebra and Appl.* **13**, 115–123.

Berman, A., Varga, R. S., and Ward, R. C. [1978]. ALPS: Matrices with nonpositive off-diagonal entries, *Linear Algebra and Appl.* **21**, 233–244.

Birkhoff, G. [1946]. Tres observaciones sobre el algebra lineal, *Univ. Nac. Tucuman Rev. Ser. A***5**, 147–150.

Birkhoff, G. [1967a]. "Lattice Theory," 3rd ed. Amer. Math. Soc. Colloq. Publ. 25, Providence, Rhode Island.

Birkhoff, G. [1967b]. Linear transformations with invariant cones, *Amer. Math. Monthly* **72**, 274–276.

Birkhoff, G., and Varga, R. S. [1958]. Reactor criticality and nonnegative matrices, *J. Soc. Ind. Appl. Math.* **6**, 354–377.

Boedwig, E. [1959]. "Matrix Calculus." Wiley (Interscience), New York.

Borosh, I., Hartfiel, D. J., and Maxson, C. J. [1976]. Answers to questions posed by Richman and Schneider, *Linear and Multilinear Algebra* **4**, 255–258.

Bramble, J. H., and Hubbard, B. E. [1964]. On a finite difference analogue of an elliptic boundary value problem which is neither diagonally dominant nor of nonnegative type, *J. Math. Phys.* **43**, 117–132.

Brauer, A. [1957a]. The theorems of Ledermann and Ostrowski on positive matrices, *Duke Math. J.* **24**, 265–274.

Brauer, A. [1957b]. A new proof of theorems of Perron and Frobenius on nonnegative matrices, I. Positive matrices, *Duke Math. J.* **24**, 367–378.

Brown, D. R. [1964]. On clans of nonnegative matrices, *Proc. Amer. Math. Soc.* **15**, 671–674.

Brualdi, R. A. [1968]. Convex sets of nonnegative matrices, *Canad. J. Math.* **20**, 144–157.

Brualdi, R. A. [1974]. The DAD theorem for arbitrary row sums, *Proc. Amer. Math. Soc.* **45**, 189–194.

Brualdi, R. A. [1976]. Combinatorial properties of symmetric nonnegative matrices, *Proc. Internat. Conf. Combinatorial Theory*, Rome, September, *1973* pp. 99–120. Academia Nazionale dei Lincei, Roma.

Brualdi, R. A., Parter, S. V., and Schneider, H. [1966]. The diagonal equivalence of a nonnegative matrix to a stochastic matrix, *J. Math. Anal. Appl.* **16**, 31–50.

Buzbee, B. L., Golub, G. H., and Nielson, C. W. [1970]. On direct methods for solving Poisson's equations, *SIAM J. Numer. Anal.* **7**, 627–656.

Carlson, D. [1963]. A note on M-matrix equations, *J. Soc. Ind. Appl. Math.* **11**, 1027–1033,

Carlson, D. [1976]. Generalizations of matrix monotonicity, *Linear Algebra and Appl.* **13**, 125–131.

Chakravarti, T. M. [1975]. On a characterization of irreducibility of a nonnegative matrix, *Linear Algebra and Appl.* **10**, 103–109.

Chandrasekaran, R. [1970]. A special case of the complementarity pivot problem, *Opsearch* **7**, 263–268.

Chen, Y. T. [1975]. Iterative methods for linear least squares problems, PhD Thesis. Univ. of Waterloo, Ontario, Canada.

Chung, K. L. [1967]. "Markov Chains with Stationary Transition Probabilities," 2nd ed. Springer, New York.

Ciarlet, P. G. [1968]. Some results in the theory of monnegative matrices, *Linear Algebra and Its Applications* **1**, 139–152.

Cinlar, E. [1975]. "Introduction to Stochastic Processes." Prentice-Hall, Englewood Cliffs, New Jersey.

Clifford, A. H., and Preston, G. B. [1961]. "The Algebraic Theory of Semigroups, Vol. I," Math. Surveys No. 7. American Mathematical Society, Providence, Rhode Island.

Collatz, L. [1942]. Einschliessungenssatz für die charakterischen Zahlen von Matrizen, *Math. Z.* **48**, 221–226.

Collatz, L. [1952]. Aufgaber monotoner Art, *Arch. Math.* **3**, 366–376.

Collatz, L. [1966]. "Functional Analysis and Numerical Mathematics." Academic Press, New York.

Converse, G., and Katz, M. [1975]. Symmetric matrices with given row sums, *J. Combinatorial Theory Ser. A*, **18**, 171–176.

Cooper, D. H. [1973]. On the maximum eigenvalue of a reducible nonnegative real matrix, *Math. Z.* **131**, 213–217.

Cottle, R. W. [1966]. Nonlinear programs with positively bounded Jacobians, *SIAM J. Appl. Math.* **14**, 147–158.

Cottle, R. W. [1972]. Monotone solutions of the parametric linear complementarity problem, *Math. Progr.* **3**, 210–224.

Cottle, R. W. [1975]. On Minkowski matrices and the linear complementarity problem, *In* "Optimization and Optimal Control" (*Proc. Conf. Oberwolfach*) (R. Burlisch, W. Oettli, and J. Stoer, eds.), pp. 18–26. Lecture Notes in Mathematics 477. Springer-Verlag, Berlin and New York.

Cottle, R. W., and Dantzig, G. B. [1968]. Complementary pivot theory of mathematical programming, *Linear Algebra and Appl.* **1**, 103–125.

Cottle, R. W., and Dantzig, G. B. [1970]. A generalization of the linear complementarity problem, *J. Combinatorial Theory* **8**, 79–90.

Cottle, R. W., and Pang, J. S. [1978]. On solving linear complementarity problems as linear programs, *In* "Complementarity and Fixed Points" (M. L. Balinski and R. W. Cottle, eds.), pp. 88–107, Mathematical Programming Study 7. North-Holland Publ., Amsterdam.

Cottle, R. W., and Pang, J. S. [1978]. A least-element theory of solving linear complementarity problems as linear programs, *Math. Prog. Study* **7**, 88–107.

Cottle, R. W., and Sacher, R. S. [1977]. On the solution of large, structured linear complementarity problems: the tridiagonal case, *Appl. Math. Opt.* **4**, 321–340.

Cottle, R. W., and Veinott, A. F. Jr. [1972]. Polyhydral sets having a least element, *Math Progr.* **3**, 238–249.

Cottle, R. W., Golub, G. H., and Sacher, R. S. [1978]. On the solution of large, structured linear complementarity problems: the block tridiagonal case, *Applied Math. Opt.*, to appear.

Crabtree, D. E. [1966a]. Applications of M-matrices to nonnegative matrices, *Duke J. Math.* **33**, 197–208.

Crabtree, D. E. [1966b]. Characteristic roots of M-matrices, *Proc. Amer. Math. Soc.* **17**, 1435–1439.

Cruse, A. B. [1975a]. A note on symmetric doubly stochastic matrices, *Discrete Math.* **13**, 109–119.

Cruse, A. B. [1975b]. A proof of Fulkerson's characterization of permutation matrices, *Linear Algebra and Appl.* **12**, 21–28.

Cryer, C. W. [1971]. The solution of a quadratic programming problem using systematic overrelaxation, *SIAM J. Control* **9**, 385–392.

Csima, J., and Datta, B. N. [1972]. The DAD theorem for symmetric nonnegative matrices, *J. Combinatorial Theory Ser. A* **12**, 147–152.

Dantzig, G. B. [1951]. Application of the simplex method to a transportation problem, *In* "Activity Analysis of Production and Allocation" (T. C. Koopmans, ed.), Cowles Commission Monograph 13, Chapter 23. Wiley, New York.

Dantzig, G. B. [1963]. "Linear Programming and Extensions." Princeton Univ. Press, Princeton, New Jersey.

Dantzig, G. B., and Cottle, R. W. [1967]. Positive (semi) definite programming, *In* "Nonlinear Programming" (J. Abadie, ed.), pp. 55–73. North-Holland Publ., Amsterdam.

Debreau, G., and Herstein, I. N. [1953]. Nonnegative square matrices, *Econometrica* **21**, 597–607.

De Marr, R. [1974]. Nonnegative idempotent matrices, *Proc. Amer. Math. Soc.* **45**, 185–188.

de Oliveira, G. N. [1972]. On the characteristic vectors of a matrix, *Linear Algebra and Appl.* **5**, 189–196.

Djoković, D. Ž. [1970]. Note on nonnegative matrices, *Proc. Amer. Math. Soc.* **25**, 80–82.

Dmitriev, N. A., and Dynkin, E. B. [1945]. On the characteristic numbers of stochastic matrices, *Doklady Akad. Nauk SSSR* **49**, 159–162.

Doob, J. L. [1942]. Topics in the theory of finite Markov chains, *Trans. Amer. Math. Soc.* **52**, 37–64.

Dorfman, R., Samuelson, P. A., and Solow, R. M. [1958]. "Linear Programming and Economic Analysis." McGraw-Hill, New York.

Eaves, B. C. [1971]. The linear complementarity problem, *Management Sci.* **17**, 68–75.

Eberlein, P. [1969]. Remarks on the van der Waerden conjecture II, *Linear Algebra and Appl.* **2**, 311–320.

Eberlein, P. J., and Mudholkar, G. S. [1968]. Some remarks on the van der Waerden conjecture, *J. Combinatorial Theory* **5**, 386–396.

Elsner, L. [1976a]. A note on characterizations of irreducibility of nonnegative matrices, *Linear Algebra and Its Applications* **14**, 187–188.

Elsner, L. [1976b]. Inverse iteration for calculating the spectral radius of a nonnegative irreducible matrix, *Linear Algebra and Appl.* **15**, 235–242.

Engel, G., and Schneider, H. [1977]. The Hadamard-Fischer inequality for a class of matrices defined by eigenvalue monotonicity, *Linear Multilinear Algebra* **4**, 155–176.

Erdös, P., and Minc, H. [1973]. Diagonals of nonnegative matrices, *Linear Multilinear Algebra* **1**, 89–93.

Fadeev, D. K., and Fadeeva, V. N. [1963]. "Computational Methods of Linear Algebra." Freeman, San Francisco, California.

Fan, K. [1958]. Topological proofs for certain theorems on matrices with nonnegative elements, *Monatsh. Math.* **62**, 219–237.

Fan, K. [1964]. Inequalities for M-matrices, *Indag. Math.* **26**, 602–610.

Fan, K. [1969]. "Convex Sets and Their Applications." Lecture Notes, Argonne National Laboratory.

Farahat, H. K. [1966]. The semigroup of doubly stochastic matrices, *Proc. Glasgow Math. Assoc.* **7**, 178–183.

Fiedler, M. [1972]. Bounds for eigenvalues of doubly stochastic matrices, *Linear Algebra and Appl.* **5**, 299–310.

Fiedler, M. [1974a]. Eigenvalues of nonnegative symmetric matrices, *Linear Algebra and Appl.* **9**, 119–142.

Fiedler, M. [1974b]. Additive compound matrices and an inequality for eigenvalues of symmetric stochastic matrices, *Czech. Math. J.* **24(99)**, 392–402.

Fiedler, M. [1975]. A property of eigenvectors of nonnegative symmetric matrices and its application to graph theory, *Czech. Math. J.* **25(100)**, 619–633.

Fiedler, M., and Haynsworth, E. [1973]. Cones which are topheavy with respect to a cone, *Linear and Multilinear Algebra* **1**, 203–211.

Fiedler, M., and Ptak, V. [1962]. On matrices with nonpositive off-diagonal elements and positive principal minors, *Czech. Math. J.* **12**, 382–400.

Fiedler, M., and Ptak, V. [1966]. Some generalizations of positive definiteness and monotonicity, *Numer. Math.* **9**, 163–172.

Fischer, P., and Holbrook, J. A. R. [1977]. Matrices doubly stochastic by blocks, *Canad. J. Math.* **29**, 559–577.

Flor, P. [1969]. On groups of nonnegative matrices, *Compositio Math.* **21**, 376–382.

Forsythe, G. E. [1953]. Tentative classification of methods and bibliography on solving systems of linear equations, *Nat. Bur. Std. Appl. Math. Ser.* **29**, 11–28.

Forsythe, G. E., and Wasow, W. R. [1960]. "Finite Difference Methods for Partial Differential Equations." Wiley, New York.

Frechet, M. [1938]. "Methods des Fonctions Arbitraires. Theórie des Événements en Chaîne dans les cas d'un Nombre Fini d'Etats Possibles." Gauthier Villars, Paris.

Friedland, S. [1974]. Matrices satisfying the van der Waerden Conjecture, *Linear Algebra and Appl.* **8**, 521–528.

Friedland, S. [1977]. Inverse eigenvalue problems, *Linear Algebra and Appl.* **17**, 15–51.

Friedland, S. [1978]. On an inverse problem for nonnegative and eventually nonnegative matrices, *Israel J. Math.* **29**, 43–60.

Friedland, S. [a], The reconstruction of a symmetric matrix from the spectral data, preprint.

Friedland, S., and Karlin, S. [1975], Some inequalities for the spectral radius of nonnegative matrices and applications, *Duke Math. J.* **42**, 459–490.

Friedland, S., and Melkman, A. A. [1979]. On the eigenvalues of nonnegative Jacobi Matrices, *Linear Algebra and Appl.* **25**, 239–254.

Frobenius, G. [1908]. Über Matrizen aus positiven Elementen, *S.-B. Preuss. Akad. Wiss.* (*Berlin*) 471–476.

Frobenius G. [1909]. Über Matrizen aus positiven Elementen, II, *S.-B. Preuss. Akad. Wiss.* (*Berlin*) 514–518.

Frobenius G. [1912]. Über Matrizen aus nicht negativen Elementen, *S.-B. Preuss. Akad. Wiss.* (*Berlin*) 456–477.

Frobenius, G. [1917]. Über zerlegbare Determinanten, *S.-B. Preuss. Akad. Wiss.* (*Berlin*) 274–277.

Fulkerson, D. R. [1956]. Hitchcock transportation problem, *Rand. Corp. Rep. P890.*

Gaddum, J. W. [1952]. A theorem on convex cones with applications to linear inequalities, *Proc. Amer. Math. Soc.* **3**, 957–960.

Gale D. [1960]. "The Theory of Linear Economic Models," McGraw-Hill, New York.

Gale, D., and Nikaido, H. [1965]. The Jacobian matrix and global univalence mappings, *Math. Ann.* **19**, 81–93.

Gantmacher, F. R. [1959]. "The Theory of Matrices, Vols. I and II," Chelsea, New York.

Gantmacher, F. R., and Krein, M. G. [1950]. "Oscillation Matrices and Kernels and Small Vibrations of Mechanical Systems" (English translation, 1961). Office of Technical Services, Dept. of Commerce, Washington D.C.

Gauss, C. F. [1823]. *Brief und Gerling, Werke* **9**, 278–281 (translated by G. E. Forsythe, *Math. Tables Aids Comput.* **5** (1951), 255–258).

Glazman, I. M., and Ljubic, Ju I. [1974]. "Finite Dimensional Linear Analysis." MIT Press, Cambridge, Massachusetts.

Graves, R. L. [1967]. A principal pivoting simplex algorithm for linear and quadratic programming, *Operations Res.* **15**, 482–494.

Grunbaum, B. [1967]. "Convex Polytopes." Wiley, New York.

Gutmanis [1971]. Environmental implications of economic growth in the United States, 1970 to 2000; An input-output analysis, *Proc. IEEE Conf. Decision and Control, New Orleans.*

Habetler, G. J., and Price, A. L. [1971]. Existence theory for generalized nonlinear complementarity problems, *J. Opt. Theory and Appl.* **7**, 223–239.

Hall, M., Jr. [1967]. "Combinatorial Theory." Ginn (Blaisdell), Boston, Massachusetts.

Hall, P. [1935]. On representatives of subsets, *J. London Math. Soc.* **10**, 26–30.

Hansen, B. [1970]. "A Survey of General Equilibrium Systems." McGraw-Hill, New York.

Harary, F. [1969]. "Graph Theory." Addison-Wesley, Reading, Massachusetts.

Harary, F., and Minc, H. [1976]. Which nonnegative matrices are self-inverse, *Math. Mag.* **44**, 91–92.

Hardy, G. H., Littlewood, J. E., and Polya, G. [1929]. Some simple inequalities satisfied by convex functions, *Messenger of Math.* **58**, 145–152.

Hardy, G. H., Littlewood, J. E., and Polya, G. [1952]. "Inequalities," 2nd ed. Cambridge Univ. Press, London and New York.

Hartfiel, D. J. [1974]. Concerning spectral inverses of stochastic matrices, *SIAM J. Appl. Math.* **27**, 281–292.

Hartfiel, D. J. [1975]. Results on measures of irreducibility and full indecomposability, *Trans. Amer. Math. Soc.* **202**, 357–368.

Hartfiel, D. J., Maxson, C. J., and Plemmons, R. J. [1976]. A note on the Green's relations on the semigroup $\mathcal{N}_N$, *Proc. Amer. Math. Soc.* **60**, 11–15.

Hawkins, D., and Simon, H. A. [1949]. Note: Some conditions of macroeconomic stability, *Econometrica* **17**, 245–248.

Haynsworth, E., and Hoffman, A. J. [1969]. Two remarks on copositive matrices, *Linear Algebra and Appl.* **2**, 387–392.

Haynsworth, E., and Wall, J. R. [1979]. Group inverses of certain non-negative matrices, *Linear Algebra and Appl.* **25**, 271–288.

Hensel, K. [1926]. Über Potenzreihen von Matrizen, *J. Reine Angew. Math.* **155**, 107–110.

Hershkowits, D. [1978]. "Existence of Matrices Satisfying Prescribed Conditions," M.S. Thesis, Technion- Israel Inst. of Technology, Haifa, Israel.

Hibbs, N. [1972]. An introduction to NARM, Memorandum, Center for Naval Analysis, Arlington, Virginia.

Hitchcock, F. L. [1941]. Distribution of a product from several sources to numerous localities, *J. Math. Phys.* **20**, 224–230.

Hoffman, A. J. [1967], Three observations on nonnegative matrices, *J. Res. Nat. Bur. Std.* **71B**, 39–41.

Hoffman, A. J. [1972]. On limit points of spectral radii of nonnegative symmetric integral matrices, *Graph. Theory and Appl.* (*Proc. Conf. Western Michigan Univ. Kalamazoo*, 165–172). Lecture Notes in Math. 303, Springer-Verlag, Berlin and New York.

Hofman, K. H., and Mostert, P. S. [1966]. *Elements of Compact Semigroups.* Merril Research and Lecture Series, Columbus, Ohio.

Holladay, J. C., and Varga, R. S. [1958]. On powers of nonnegative matrices, *Proc. Amer. Math. Soc.* **9**, 631–634.

Hoph, E. [1963]. An inequality for positive integral linear operators, *J. Math. and Mech.* **12**, 683–692.

Horn, A. [1954]. Doubly stochastic matrices and the diagonal of a rotation matrix, *Amer. J. Math.* **76**, 620–630.

Householder, A. S. [1958]. The approximate solution of a matrix problem, *J. Assoc. Comput. Mach.* **5**, 204–243.

Householder, A. S. [1964]. "The Theory of Matrices in Numerical Analysis." Ginn (Blaisdell), Waltham, Massachusetts.

Isaacson, D. L., and Madsen, R. W. [1976]. "Markov Chains: Theory and Applications." Wiley, New York.

Jain, S. K., Goel, V. K., and Kwak, E. K. [1979a]. Nonnegative matrices having same nonnegative Moore-Penrose and Drazin inverses, *Linear and Multilinear Algebra*

Jain, S. K., Goel, U. K., and Kwak, E. K. [1979b]. Nonnegative *m*-th roots of nonnegative 0-symmetric idempotent matrices, *Linear Algebra and Its Applic.* **23**, 37–52.

Jain, S. K., Goel, U. K., and Kwak, E. K. [1979c]. Decompositions of nonnegative group-monotone matrices, *Transactions of the A.M.S.*

James, K. K., and Riha, W. [1974]. Convergence criteria for successive overrelaxation, *SIAM J. Numer. Anal.* **12**, 137–143.

Johnson, C. R. [1974a]. A sufficient condition for matrix stability, *J. Res. Nat. Bur. Std.* **73B**, 103–104.

Johnson, C. R. [1974b]. Sufficiency conditions for D-stability, *J. Econ. Theory* **9**, 53–62.

Johnson, C. R. [1977]. A Hadamard product involving M-matrices, *Linear Multilinear Algebra* **4**, 261–264.

Johnson, C. R., Leighotn, F. T., and Robinson, H. A. [1979]. Sign patterns of inverse-positive matrices. *Linear Algebra and Appl.* **24**, 75–84.

Jurkat, W. B., Ryser, H. J. [1966]. Matrix factorizations of determinants and preminants, *J. Algebra* **3**, 1–27.

Kahan, W. [1958]. Gauss-Seidel methods for solving large systems of linear equations, Ph.D. Thesis, Univ. of Toronto, Toronto, Canada.

Kakutani, S. [1941]. A generalization of Brouwer's fixed point theorem, *Duke Math. J.***8**, 457–459.

Kammerer, W. J., and Plemmons, R. J. [1975]. Direct iterative methods for least squares solutions to singular operator equations, *J. Math. Anal. Appl.* **49**, 512–526.

Kancko, K. [1978]. Linear complementarity problems and characterizations of Minkowski matrices, *Linear Algebra and Appl.* **20**, 111–130.

Karamardian, S. [1969a]. The nonlinear complementarity problem with applications, Part I, *J. Optimization Theory Appl.* **4**, 87–98.

Karamardian, S. [1969b]. The nonlinear complementarity problem with applications, Part II, *J. Optimization Theory Appl.* **4**, 167–181.

Karamardian, S. [1972]. The complementarity problem, *Math. Progr.* **2**, 107–129.

Karamata, Y. [1932]. Sur une inégalité relative aux foncitons convexes, *Publ. Math. Univ. Belgrade* **1**, 145–148.

Karlin, S. [1968]. "Total Positivity." Stanford Univ. Press, Stanford, California.

Karpelevich, F. I. [1951]. On the characteristic roots of matrices with nonnegative elements (in Russian), *Isv. Akad. Nauk SSSR Ser. Mat.* **15**, 361–383.

Katz, M. [1970]. On the extreme points of a certain convex polytope, *J. Combinatorial Theory* **8**, 417–423.

Katz, M. [1972]. On the extreme points of a set of substochastic and symmetric matrices, *J. Math. Anal. Appl.* **37**, 576–579.

Keilson, J. H., and Styan, G. P. H. [1973]. Markov chains and M-matrices: inequalities and equalities, *J. Math. Anal. Appl.* **41**, 439–459.

Keller, H. B. [1965]. On the solution of singular and semidefinite linear systems by iteration, *SIAM J. Numer. Anal.* **2**, 281–290.

Kellog, R. B. [1971]. Matrices similar to a positive or essentially positive matrix, *Linear Algebra and Appl.* **4**, 191–204.

Kemeny, J. G., and Snell, J. L. [1960]. "Finite Markov Chains." Van Nostrand-Reinhold, Princeton, New Jersey.

Keynes, J. M. [1936]. "General Theory of Employment, Interest and Money." Macmillan, New York.

Klee, V. L. [1959]. Some characterizations of convex polyhedra, *Acta Math.* **102**, 79–107.

Koopmans, T. C. [1949]. Optimum utilization of the transportation system, *Econometrica* **17**, Supplement.

Krasnoselskii, M. A. [1964]. "Positive Solutions of Operator Equations." Noordhoff, The Netherlands.

Krein, M. G., and Rutman, M. A. [1948]. Linear operators leaving invariant a cone in a Banach space (in Russian), *Usp. Mat. Nauk (N.S.)* **3**, 3–95 (*English transl.: Amer. Math. Soc. Transl. Ser. I* **10** (1962), 199–325.)

Kuo, I-wen [1977]. A note on factorizations of singular M-matrices, *Linear Algebra and Appl.* 217–220.

Kuttler, J. R. [1971]. A fourth order finite-difference approximation for the fixed membrance eigenproblem, *Math. Comp.* **25**, 237–256.

Lawson, L. M. [1975]. Computational methods for generalized inverse matrices arising from proper splittings, *Linear Algebra and Appl.* **12**, 111–126.

Lederman, W. [1950]. Bounds for the greatest latent root of a positive matrix, *J. London Math. Soc.* **25**, 265–268.

Leff, H. S. [1971]. Correlation inequalities for coupled oscillations, *J. Math. Phys.* **12**, 569–578.

Lemke, C. E. [1965]. Bimatrix equilibrium points and mathematical programming, *Management Sci.* **11**, 681–689.

Lemke, C. E. [1968]. On complementary pivot theory, *In* "Mathematics of the Decision Sciences" (G. B. Dantzig and A. F. Veinott, Jr., eds.). American Mathematical Society, New York.

Lemke, C. E., and Howson, J. T., Jr. [1964]. Equilibrium points of bimatrix games, *SIAM J.* **12**, 413–423.

Leontief, W. W. [1936]. Quantitative input and output relations in the economic system of the United States, *Rev. Econ. Statist.* **18**, 100–125.

Leontief, W. W. [1941]. "The Structure of the American Economy." Harvard Univ. Press, Cambridge, Massachusetts.

Leontief, W. W. [1953]. "Studies in the Structure of the American Economy." Oxford Univ. Press, London and New York.

Leontief, W. W. [1966]. "Input-Output Economics." Oxford Univ. Press, London and New York.

Levinger, B. W. [1970]. An inequality for nonnegative matrices, *Notices Amer. Math. Soc.* **17**, 260.

Lewin, M. [1971a]. On exponents of primitive matrices, *Numer. Math.* **18**, 154–161.

Lewin, M. [1971b]. On nonnegative matrices, *Pacific J. Math.* **36**, 753–759.

Lewin, M. [1974]. Bounds for exponents of doubly stochastic primitive matrices, *Math. Z.* **137**, 21–30.

Lewin, M. [1977]. On the extreme points of the polytope of symmetric matrices with given row sums, *J. Combinatorial Theory Ser. A***23**, 223–231.

Loewy, R., and London, D. [1978], A note on an inverse problem for nonnegative matrices, *Linear and Multilinear Algebra*, 6, 83–90.

Loewy, R., and Schneider, H. [1975a]. Indecomposable Cones, *Linear Algebra and Appl.* **11**, 235–245.

Loewy, R., and Schneider, H. [1975b]. Positive operators on the n-dimensional ice cream cone, *J. Math. Anal. Appl.* **49**, 375–392.

London, D. [1964]. Nonnegative matrices with stochastic powers, *Israel J. of Math.* **2**, 237–244.

London, D. [1966a]. Inequalities in quadratic forms, *Duke Math. J.* **33**, 511–522.

London, D. [1966b]. Two inequalities in nonnegative symmetric matrices, *Pacific J. Math.* **16**, 515–536.

London, D. [1971]. On matrices with a doubly stochastic pattern, *J. Math. Anal. Appl.* **34**, 648–652.

Lyapunov, A. [1892]. Problème géneral de la stabilite du mouvement, *Comm. Math. Soc. Kharkov*, reproduced in *Ann. Math. Stud.* **7**, Princeton Univ. Press, Princeton, New Jersey, 1947.

Lynn, M. S. [1964]. On the Schur product of H-matrices and nonnegative matrices and related inequalities, *Proc. Cambridge Philos. Soc.* **60**, 425–431.

Lynn, M. S., and Timlake, W. P. [1969]. Bounds for Perron eigenvectors and subdominant eigenvalues of positive matrices, *Linear Algebra and Appl.* **2**, 143–152.

Maier, G. [1970]. A matrix structural theory of piecewise linear elastoplasticity with interacting yield planes, *Mechanica* **5**, 54–66.

Maier, G. [1972]. Problem 72-2, A parametric linear complementarity problem, *SIAM Rev.* **14**, 364–365.

Mangasarian, O. L. [1968]. Characterizations of real matrices of monotone kind, *SIAM Rev.* **10**, 439–441.

Mangasarian, O. L. [1970]. A convergent splitting for matrices, *Numer. Math.* **15**, 351–353.

Mangasarian, O. L. [1976a]. Linear complementarity problem solvable by a single linear program, *Math. Progr.* **10**, 263–270.

Mangasarian, O. L. [1976b]. Solution of linear complementarity problems by linear programming, In "Numerical Analysis Dundee 1975." (G. A. Watson, ed.), Lecture notes in mathematics 506. Springer-Verlag, Berlin, 166–175.

Mangasarian, O. L. [1978]. Characterizations of linear complementarity problems as linear programs, 74–87. Mathematical Programming Study 7. North-Holland Publ., Amsterdam.

Marchuk, G. I., and Kuznetzov, Yu. A. [1972]. "Stationary Iterative Methods for the Solutions of Systems of Linear Equations With Singular Matrices." Science Press, Novosibirsk.

Marcus, M., and Minc, H. [1964]. "A Survey of Matrix Theory and Matrix Inequalities." Allyn and Bacon, Rockleigh, New Jersey.

Marcus, M., and Newman, M. [1959]. On the minimum of the permanent of a doubly stochastic matrix, *Duke Math. J.* **26**, 61–72.

Marcus, M., and Newman, M. [1962]. The sum of the elements of the powers of a matrix, *Pacific J. Math.* **12**, 627–635.

Marcus, M., and Newman, M. [1965]. Generalized functions of symmetric matrices, *Proc. Amer. Math. Soc.* **16**, 826–830.

Marcus, M., and Ree, R. [1959]. Diagonals of doubly stochastic matrices, *Quart. J. Math. Oxford Ser.* (*2*) **10**, 295–302.

Marcus, M., Minc, H., and Moyls, B. [1961]. Some results on nonnegative matrices, *J. Res. Nat. Bur. Std.* **65B**, 205–209.

Marek, I. [1966]. Spektrale eigenschaften der K-positiven operatoren und einschliessungssätze für der spectralradius, *Czechoslovak. Math. J.* **16**, 493–517.

Marek, I. [1970]. Frobenius theory of positive operators; comparison theorems and applications, *SIAM J. Appl. Math.* **19**, 607–628.

Marek, I. [1971]. A note on K-stochastic operators, *Casopis Pest. Mat.* **96**, 239–244.

Markham, T. L. [1972]. Factorizations of nonnegative matrices, *Proc. Amer. Math. Soc.* **32**, 45–47.

Markov, A. A. [1908]. Limiting distribution theorem for computing the probability of a sum of variables forming a chain (in Russian), *Izbrannye Tr.* (1951), 365–397.

Maxfield, J. E., and Minc, H. [1962]. A doubly stochastic matrix equivalent to a given matrix, *Notices Amer. Math. Soc.* **9**, 309.

Maybee, J. S., and Quirk, J. [1969]. Qualitative problems in matrix theory, *SIAM Rev.* **11**, 30–51.

McCallum, C. J., Jr. [1970]. The linear complementarity problem in complex space, Ph.D. Thesis, Stanford Univ.

Mehmke, R. [1892]. Über das Seidelsche Verfahren, um lineare Gleichungen bei Einer sehr grossen Anzahl der Unbekannten durch sukzessive Annäherungauflisen, *Moskov. Math. Samml.* **16**, 342–345.

Melendez, G. [1977]. "Block Iterative SOR Methods for Large Sparse Least Squares Problems," M.S. Thesis, Univ. of Tennessee.

Menon, M. V. [1968]. Matrix links, an extremization problem, and the reduction of a nonnegative matrix to one with prescribed row and column sums, *Canad. J. Math.* **20**, 225–232.

Menon, M. V., and Schneider, H. [1969]. The spectrum of a nonlinear operator associated with a matrix, *Linear Algebra and Appl.* **2**, 321–344.

Meyer, C. D., Jr. [1975]. The role of the group generalized inverse in the theory of finite Markov chains, *SIAM Rev.* **17**, 443–464.

Meyer, C. D., Jr. [1978]. An alternative expression for the first passage matrix, *Linear Algebra and Its Applic.* **22**, 41–48.

Meyer, C. D., Jr., and Plemmons, R. J. [1977]. Convergent powers of a matrix with applications to iterative methods for singular linear systems, *SIAM J. Numer. Anal.* **14**, 699–705.

Meyer, C. D., Jr., and Stadelmaier, M. W. [1978]. Singular M-matrices and inverse-positivity, *Linear Algebra and Appl.* **22**, 139–156.

Micchelli, C. A. and Willoughby, R. A. [1979]. On functions which preserve the class of Stieltjes matrices. *Linear Algebra and Appl.* **23**, 141–156.

Miernyk, W. H. [1965]. "The Elements of Input–Ouput Analysis." Random House, New York.

Minc, H. [1970]. On the maximal eigenvector of a positive matrix, *SIAM J. Numer. Anal.* **7**, 424–427.

Minc, H. [1974a]. Irreducible matrices, *Linear and Multilinear Algebra* **1**, 337–342.

Minc, H. [1974b]. The structure of irreducible matrices, *Linear and Multilinear Algebra* **2**, 85–90.

Minc, H. [1974c]. Linear transformations of nonnegative matrices, *Linear Algebra and Appl.* **9**, 149–153.

Minc, H. [1978]. "Permanents" (Encyclopedia of Mathematics and its Applications.) Addison-Wesley, Reading, Massachusetts.

Minkowski, H. [1900]. Zur Theorie der Einkerten in den algebraischen Zahlkörper, *Nachr. K. Ges. Wiss. Gött, Math.-Phys. Klasse* 90–93.

Minkowski, H. [1907]. "Diophantische Approximationen." Teubner, Leipzig.

Mirsky, L. [1963]. Results and problems in the theory of doubly-stochastic matrices, *Z. Wahrscheinlichkeitstheorie und Verw. Gebiete* **1**, 319–334.

Mirsky, L. [1964]. Inequalities and existence theorems in the theory of matrices, *J. Math. Anal. Appl.* **9**, 99–118.

Mohan, S. R. [1976]. On the simplex method and a class of linear complementarity problems, *Linear Algebra and Appl.* **14**, 1–9.

Montague, J. S., and Plemmons, R. J. [1973]. Doubly stochastic matrix equations, *Israel J. Math.* **15**, 216–229.

Moré, J. R. [1973]. Classes of functions and feasibility conditions in nonlinear complementarity problems, Tech. Rep. No. 73-174, Dept. of Computer Sciences, Cornell Univ.

Moreau, J. J. [1962]. Decomposition orthogonale d'un espace hilbertien selon deux cones mutuellement polaires, *C. R. Acad. Sci. Paris* **255**, 233–240.

Moylan, P. J. [1977]. Matrices with positive principal minors, *Linear Algebra and Appl.* **17**, 53–58.

Mulholland, H. P., and Smith, C. A. B. [1959]. An inequality arising in genetical theory, *Amer. Math. Mo.* **66**, 673–683.

Murty, K. G. [1972]. On the number of solutions of the complementarity problems and spanning properties of complementarity cones, *Linear Algebra and Appl.* **5**, 65–108.

Nekrasov, P. A. [1885]. Die Bestimmung der Unbekannten nach der Methode der Kleinstan Quadrate bei einer sehr grossen Anzahl der Unbekannten, *Mat. Sb.* **12**, 189–204.

Neumann, M. [1979]. A note on generalizations of strict diagonal dominance for real matrices, *Linear Algebra and Appl.*

Neumann, M., and Plemmons, R. J. [1978]. Convergent nonnegative matrices and iterative methods for consistent linear systems, *Numer. Math.* **31**, 265–279.

Neumann, M., and Plemmons, R. J. [1979]. Generalized inverse-positivity and splittings of M-matrices, *Linear Algebra and Appl.* **23**, 21–26.

Neumann, M., and Plemmons, R. J. [1979]. M-matrix characterizations II: general M-matrices, preprint.

Nikaido, H. [1968]. "Convex Structures and Economic Theory." Academic Press, New York.

Oldenburger, R. [1940]. Infinite powers of matrices and characteristic roots, *Duke Math. J.* **6**, 357–361.

Ortega, J. M. [1972]. "Numerical Analysis: A Second Course." Academic Press, New York.

Ortega, J. M., and Plemmons, R. J. [1979]. Extensions of the Ostrowski-Reich theorem for SOR iterations, *Linear Algebra and Appl.*

Ortega, J. M., and Rheinboldt, W. [1967]. Monotone iterations for nonlinear equations with applications to Gauss-Seidel methods, *SIAM J. Numer. Anal.* **4**, 171–190.

Ostrowski, A. M. [1937]. Über die Determinanten mit überwiegender Hauptdiagonale, *Comment. Math. Helv.* **10**, 69–96.

Ostrowski, A. M. [1952]. Bounds for the greatest latent root of a positive matrix, *J. London Math. Soc.* **27**, 253–256.

Ostrowski, A. M. [1954]. On the linear iteration procedures for symmetric matrices, *Rend. Mat. e Appl.* **14**, 146–163.

Ostrowski, A. M. [1956]. Determinanten mit überwiegender Hauptdiagonale und die absolut Konvergenz von linearen Iterationsprozessen, *Comment. Math. Helv.* **30**, 175–210.

Ostrowski, A. M. [1960/61]. On the eigenvector belonging to the maximal root of a nonnegative matrix, *Proc. Edinburgh Math. Soc.* **12**, 107–112.

Ostrowski, A. M. [1963]. On positive matrices, *Math. Annalen* **150**, 276–284.

Ostrowski, A. M., and Schneider, H. [1960]. Bounds for the maximal characteristic root of a nonnegative irreducible matrix, *Duke Math. J.* **27**, 547–553.

Pang, J. S. [1976]. "Least Element Complementarity Theory," Ph.D. thesis, Stanford Univ.

Pang, J. S., Kaneko, K., and Hallman, W. P. [1977]. On the solution of some (parametric) linear complementarity problems with application to portfolio analysis, structural engi-

neering and graduation, Working paper 77-27, Dept. of Industrial Engineering, Univ. of Wisconsin, Madison, Wisconsin.
Pearl, M. [1973]. "Matrix Theory and Finite Mathematics." McGraw-Hill, New York.
Perfect, H. [1953]. Methods of constructing certain stochastic matrices I, *Duke Math. J.* **20**, 395–404.
Perron, O. [1907]. Zur Theorie der Über Matrizen, *Math. Ann.* **64**, 248–263.
Plemmons, R. J. [1973]. Regular nonnegative matrices, *Proc. Amer. Math. Soc.* **39**, 26–32.
Plemmons, R. J. [1974]. Direct iterative methods for linear systems using weak splittings, *Acta Univ. Car.* **1**–8.
Plemmons, R. J. [1976a]. Regular splittings and the discrete Neumann problem, *Numer. Math.* **25**, 153–161.
Plemmons, R. J. [1976b]. M-matrices leading to semiconvergent splittings, *Linear Algebra and Appl.* **15**, 243–252.
Plemmons, R. J. [1977]. M-matrices characterizations I: Nonsingular M-matrices, *Linear Algebra and Appl.* **18**, 175–188.
Plemmons, R. J. [1979]. Adjustment by least squares in Geodesy using block iterative methods for sparse matrices, *Proc. of the U.S. Army Conf. on Numerical Anal. and Computers*, El Paso, Texas.
Plemmons, R. J., and Cline, R. E. [1972]. The generalized inverse of a nonnegative matrix, *Proc. Amer. Math. Soc.* **31**, 46–50.
Poole, G., and Boullion, T. [1974]. A survey on M-matrices, *SIAM Rev.* **16**, 419–427.
Price, H. S. [1968]. Monotone and oscillation matrices applied to finite difference approximation, *Math. Comp.* **22**, 484–516.
Pullman, N. J. [1971]. A geometric approach to the theory of nonnegative matrices, *Linear Algebra and Appl.* **4**, 297–312.
Pullman, N. J. [1974]. A note on a theorem of Minc on irreducible nonnegative matrices, *Linear and Multilinear Algebra* **2**, 335–336.
Rao, P. S. [1973]. On generalized inverses of doubly stochastic matrices, *Sankhya Ser. A***35**, 103–105.
Reich, E. [1949]. On the convergence of the classical iterative method of solving linear simultaneous equations, *Ann. Math. Statist.* **20**, 448–451.
Rheinboldt, W. C., and Vandergraft, J. S. [1973]. A simple approach to the Perron-Frobenius theory for positive operators on general partially-ordered finite-dimensional linear spaces, *Math. Comp.* **27**, 139–145.
Richman, D. J., and Schneider, H. [1974]. Primes in the semigroup of nonnegative matrices, *Linear Multilinear Algebra* **2**, 135–140.
Richman, D. J., and Schneider, H. [1978]. On the singular graph and the Weyr characteristic of an M-matrix, *Aequationes Math.* **17**, 208–234.
Robert, F. [1966]. Recherche d'une M-matrice, parmi les minorantes d'un operateur lineaire, *Numer. Math.* **9**, 189–199.
Robert, F. [1973]. "Matrices Non-negatives et Normes Vectorielles." Lecture Notes, Scientific and Medical Univ. at Grenoble.
Rockafellar, R. T. [1970]. "Convex Analysis." Princeton Univ. Press. Princeton, New Jersey.
Romanovsky, V. [1936]. Recherches sur les chains de Markoff, *Acta Math.* **66**, 147–251.
Rothblum, U. G. [1975]. Algebraic eigenspaces of nonnegative matrices, *Linear Algebra and Appl.* **12**, 281–292.
Rothblum, U. G. [1979]. An index classification of M-matrices, *Linear Algebra and Appl.* **23**, 1–12.

Saigal, R. [1971]. Lemke's algorithm and a special linear complementarity problem, *Opsearch* **8**, 201–208.

Salzmann, F. L. [1972]. A note on eigenvalues of nonnegative matrices, *Linear Algebra and Appl.* **5**, 329–338.

Samelson, H., Thrall, R. M., and Wesler, O. [1958]. A partitioning theorem for Euclidean *n*-space, *Proc. Amer. Math. Soc.* **9**, 805–807.

Sandberg, I. W. [1974a]. A global non-linear extension of the Le Chatelier-Samuelson principle for linear Leontief models, *J. of Economic Th.* **7**, 40–52.

Sandberg, I. W. [1974b], Some comparative-status results for nonlinear input–output models in a multisectored economy, and related results, *J. Economic Theory* **8**, 248–258.

Sarma, K. [1977]. An input–output economic model, *IBM Systems J.* **16**, 398–420.

Sawashima, I. [1964]. On spectral properties of some positive operators, *Nat. Sci. Rep. (Ochanormizu Univ.)* **15**, 53–64.

Schaefer, H. H. [1960]. Some spectral properties of positive linear operators, *Pacific J. Math.* **10**, 1009–1019.

Schaefer, H. H. [1971]. "Topological Vector Spaces," 3rd printing. Springer, New York.

Schaefer, H. H. [1974]. "Banach Lattices and Positive Operators." Springer, New York.

Schneider, H. [1953]. An inequality for latent roots applied to determinants with dominant principal diagonal, *J. London Math. Soc.* **28**, 8–20.

Schneider, H. [1956]. The elementary divisors, associated with 0, of a singular M-matrix, *Proc. Edinburgh Math. Soc.* **10**, 108–122.

Schneider, H. [1965]. Positive operators and an inertia theorem, *Numer. Math.* **7**, 11–17.

Schneider, H. [1977]. The concepts of irreducibility and full indecomposability of a matrix in the works of Frobenius, König and Markov, *Linear Algebra and Appl.* **18**, 139–162.

Schneider, H., and Turner, R. E. L. [1972]. Positive eigenvectors of order-preserving maps, *J. Math. Anal. Appl.* **7**, 508–519.

Schneider, H., and Vidyasagar, M. [1970]. Cross-positive matrices, *SIAM J. Numer. Anal.* **7**, 508–519.

Schröder, J. [1961]. Lineare Operatoren mit positive Inversen *Arch. Rational Mech. Anal.* **8**, 408–434.

Schröder, J. [1970]. Proving inverse-positivity of linear operators by reduction, *Numer. Math.* **15**, 100–108.

Schröder, J. [1972]. Duality in linear range-domain implications, *In* "Inequalities III" (O Shisha, ed.), pp. 321–332. Academic Press, New York.

Schröder, J. [1978]. M-matrices and generalizations, *SIAM Rev.* **20**, 213–244.

Schumann, J. [1968]. *Input–Output Analyse*. Springer-Verlag, Berlin and New York.

Schur, I. [1923]. Über eine Klasse von Mittelbildungen mit Anwendungen auf die Determinantentheorie, *Sitzber. Berl. Math. Ges.* **22**, 9–20.

Schwarz, S. [1967]. A note on the semigroup of doubly stochastic matrices, *Mat. Casopis Sloven. Akad. Vied.* **17**, 308–316.

Seneta, E. [1973]. "Non-Negative Matrices." Wiley, New York.

Sierksma, G. [1979]. Nonnegative matrices; the open Leontief model, *Linear Algebra and Appl.*

Sinkhorn, R. [1964]. A relationship between arbitrary positive matrices and doubly stochastic matrices, *Ann. Math. Statist.* **35**, 876–879.

Sinkhorn, R. [1974]. Diagonal equivalence to matrices with prescribed row and column sums II, *Proc. Amer. Math. Soc.* **45**, 195–198.

Sinkhorn, R., and Knopp, P. [1967]. Concerning nonnegative matrices and doubly stochastic matrices, *Pacific J. Math.* **21**, 343–348.

Smith, J. H. [1974]. A geometric treatment of nonnegative generalized inverses, *Linear and Multilinear Algebra* **2**, 179–184.

Southwell, R. V. [1946]. "Relaxation Methods in Theoretical Physics." Oxford Univ. Press (Clarendon), London and New York.

Stein, P. [1952]. Some general theorems on iterants, *J. Res. Nat. Bur. Std.* **48**, 82–83.

Stein, P., and Rosenberg, R. L. [1948]. On the solution of linear simultaneous equations by iteration, *J. London Math. Soc.* **23**, 111–118.

Stoer, J., and Witzgall, C. [1970]. "Convexity and Optimization in Finite Dimensions I." Springer, New York.

Stone, D. [1970]. "An Economic Approach to Planning the Conglomerate of the 70's. Auerbach Publ. New York.

Stolper, W., and Samuelson, P. A. [1941]. Protection and real wages, *Rev. Econ. Studies* **9**, 58–73.

Strang, G. [1976]. "Linear Algebra and its Applications." Academic Press, New York.

Suleimanova, H. R. [1949]. Stochastic matrices with real eigenvalues, *Soviet Math. Dokl.* **66**, 343–345.

Sylvester, J. J. [1883]. On the equation to the secular inequalities in the planetary theory, *Philos. Mag.* **16**, (5) 267–269.

Tam, B. S. [1977]. "Some Aspects of Finite Dimensional Cones," Ph.D. Thesis, Univ. of Hong Kong.

Tamir, A. [1974]. Minimality and complementarity properties associated with Z-functions and M-functions, *Math. Progr.* **7**, 17–31.

Tartar, L. [1971]. Une nouvelle characterization des M-matrices, *Rev. Francaise Informat. Recherche Operationnelle* **5**, 127–128.

Taskier, C. E. [1961]. "Input–Output Bibliography." United Nations Publ., New York.

Taussky, O. [1958]. Research problem, *Bull. Amer. Math. Soc.* **64**, 124.

Thomas, B. [1974]. Rank factorization of nonnegative matrices, *SIAM Rev.* **16**, 393.

Tucker, A. W. [1963]. Principal pivotal transforms of square matrices, *SIAM Rev.* **5**, 305.

Ullman, J. L. [1952]. On a theorem of Frobenius, *Michigan Math. J.* **1**, 189–193.

Uekawa, Y., Kemp, M. C., and Wegge, L. L. [1972]. P and PN-matrices, Minkowski and Metzler Matrices, and generalizations of the Stoepler-Samuelson-Rybezynski Theorems, *J. Internat. Econ.* **3**, 53–76.

Vandergraft, J. S. [1968]. Spectral properties of matrices which have invariant cones, *SIAM J. Appl. Math.* **16**, 1208–1222.

Vandergraft, J. S. [1972]. Applications of partial orderings to the study of positive definiteness, monotonicity, and convergence, *SIAM J. Numer. Anal.* **9**, 97–104.

van der Waerden, B. L. [1926]. Aufgabe 45, *Jber. Deutsch. Math. Verein.* **35**, 117.

Varga, R. S. [1959]. p-cyclic matrices: a generalization of the Young–Frankel successive over-relaxation scheme, *Pacific J. Math.* **9**, 617–628.

Varga, R. S. [1960]. Factorization and normalized iterative methods, *in Boundary Problems in Differential Equations* (R. E. Langer, ed.), pp. 121–142. Univ. of Wisconsin Press, Madison, Wisconsin.

Varga, R. S. [1962]. "Matrix Iterative Analysis." Prentice-Hall, Englewood Cliffs, New Jersey.

Varga, R. S. [1968]. Nonnegatively posed problems and completely monotonic functions, *Linear Algebra and Appl.* **1**, 329–347.

Varga, R. S. [1976]. On recurring theorems on diagonal dominance, *Linear Algebra and Appl.* **13**, 1–9.

Varga, R. S. [a], Revision of Varga [1962], preprint.

Vitek, Y. [1975]. A bound connected with primitive matrices, *Numer. Math.* **23**, 255–260.

Vitek, Y. [1977]. Exponents of primitive matrices and a Diophantine problem of Frobenius, Ph.D. Thesis, Technion-Israel Institute of Technology, Haifa, Israel.

von Mises, R., and Pollaczek-Geiringer, H. [1929]. Praktische Verfahren der Gleichungsauflösing, *Z. Angew. Math. Mech.* **9**, 58–77.

von Neumann, J. [1945/46]. A model of a general economic equilibrium, *Rev. Econ. Stud.* **13**, 10–18.

Wall, J. R. [1975]. Generalized inverses of stochastic matrices, *Linear Algebra and Appl.* **10**, 147–154.

Wallace, V. L. [1974]. Algebraic techniques for numerical solutions of queueing networks, *Proc. Conf. Math. Methods in Queueing Theory*, Lecture Notes in Economics and Math. Systems 98. Springer-Verlag, Berlin and New York.

Wielandt, H. [1950]. Unzerlegbare, nicht negative Matrizen, *Math. Z.* **52**, 642–648.

Willoughby, R. A. [1977]. The inverse M-matrix problem, *Linear Algebra and Appl.* **18**, 75–94.

Willson, A. N. [1971]. A useful generalization of the P_0 matrix concept, *Numer. Math.* **17**, 62–70.

Yan, C. [1969]. "Introduction to Input–Output Economics." Holt, New York.

Young, D. M. [1950]. "Iterative Methods for Solving Partial Difference Equations of Elliptic Type," Ph.D. Thesis, Harvard Univ., Cambridge, Massachusetts.

Young, D. M. [1972]. "Iterative Solution of Large Linear Systems." Academic Press, New York.

INDEX